# THE AUTHORS

GARETH JONES B.Sc., Ph.D., is a Member of the Institute of Environmental Science and is presently Senior Lecturer in the Department of Geography at the University of Strathclyde, having previously held posts in Belfast, New Zealand and South Africa. His previous publications include *Vegetation Productivity* and *Conservation of Ecosystems and Species*, and his current research interests include conservation and geographic information systems.

ALAN ROBERTSON B.Sc., M.Phil., M.R.T.P.I. has worked for several local authorities in Scotland in the field of town planning and is presently working with an Edinburgh-based economics, planning and transportation consultancy.

JEAN FORBES B.A., M.Sc., Dip.T.P., M.R.T.P.I., M.Ed. was a founder member and sometime Vice Chairman of the Scottish Environmental Education Council and from 1974–84 was a member of the Local Government Boundary Commission for Scotland. She is currently Senior Lecturer in the Centre for Planning at the University of Strathclyde and is the author of several publications on the role of environmental education in promoting better public participation in town planning.

GRAHAM HOLLIER B.A., M.A., Ph.D is an economic geographer with research interests in African rural development, agricultural and marketing systems, and the wider problems of Third World development, on which he has written in professional journals and contributed textbooks. He is currently a lecturer in the Department of Geography at the University of Strathclyde, having previously held posts at the universities of Sussex and Leeds.

# COLLINS

## DICTIONARY OF
# ENVIRONMENTAL SCIENCE

### Gareth Jones, Alan Robertson, Jean Forbes & Graham Hollier

Adviser: Professor Joy Tivy, Senior Research Fellow,
Dept. of Geography and Topographic Science,
University of Glasgow

HarperCollins*Publishers*

First published 1990

© HarperCollins*Publishers* 1990

Diagrams by Karen Glover

ISBN 0 00 434348 4

Reprint 10 9 8 7 6 5 4 3 2

**British Library Cataloguing in Publication Data**
Dictionary of environmental science.
   1. Environmental studies
   I. Jones, Gareth
   333.7

Typeset by Burgess & Son (Abingdon) Ltd in Bembo 10/11pt

Printed and bound in Great Britain by
HarperCollins*Manufacturing*, Glasgow

# PREFACE

At no other time in our history has the need to understand the environment been more urgent. The sophistication, magnitude and diversity of the ways in which we alter the biosphere continues to grow, often in advance of our ability to deal with any resulting side-effects; the bio-engineering of new strains of plants and animals, the removal of vast tracts of the rain forests, the extensive use of synthetic agricultural chemicals, and the effects of the many types of pollution are just some of the ways by which the character of life on this planet is being altered.

As the pace of this change increases, so too does the study of it. The past two decades have witnessed a growing awareness of the effects of human activity upon our planet's resources and during this period environmental science has evolved as a multidisciplinary field of study to examine the interaction of people and their environment. In the UK, as elsewhere, at the secondary and particularly the tertiary level, courses are being offered in a wide range of subjects which are all covered by the blanket term of environmental science, including environmental planning, heritage management, ecology, conservation policy, and countryside management. However, concern for, and interest in, the world in which we live is not the prerogative of the academic; as ecological issues increasingly attract media attention, so terms which were previously used only by environmental scientists have become articles of everyday speech, for example, CFCs, the greenhouse effect and reactor core melt-down.

The Dictionary is thus intended to be of use to a variety of readers. Whilst primarily aimed at those studying environmental science or any of its component topics either in the final year at school or at university, it is hoped that it will also prove to be an invaluable source book for the lay reader as the public debate over environmental issues becomes increasingly technical. The Dictionary covers four major areas of study: the physical world, the biological world, the built environment, and the agro-economic infrastructure. In dealing with each of these, the underlying objectives have been to emphasize those topics which demonstrate the interaction of people and their environment and where possible to place each term within an international context. In compiling the Dictionary the authors took particular care to minimize problems likely to be caused by international variations in the interpretation of widely used terms and, as far as possible, to include the newest environmental terminology. Also, to help the reader set any particular headword within the broader context of its field, the authors have made extensive use of cross-referencing: text which appears in SMALL CAPITALS indicates a cross-referred headword in which the reader will find supplementary information while *italicized* text indicates subsidiary terms which are defined within the entry in which they appear.

As senior author I would like to express my thanks to the other contributors for their thoroughness and dedication in the preparation of their portions of the text. I would also like to thank Professor Joy Tivy of the University of Glasgow, and Dr Peter Batey of the University of Liverpool for their constructive criticisms and suggestions for improving the manuscript. The burden of word processing our text was chiefly borne by Lorraine Nelson of the Department of Geography to whom we are extremely grateful. We were well served by Jim Carney of Collins who provided constant help and encouragement throughout the long preparation of the Dictionary. Finally, the authors wish to thank sincerely their families and colleagues for their tolerance and forbearance during the many hours we have given to this work.

Gareth Jones
Glasgow, 1989

# A

**abiota,** see ECOLOGY.

**abiotic,** *adj.* (of an ecosystem) without life. The abiotic elements of an ecosystem comprise its climatic, geological and pedologic components. They are characterized by low energy, high ENTROPY materials and form the essential base from which energy and matter is transferred to the BIOTIC components of an ecosystem (see Fig. 30).

**abrasion,** see FLUVIAL EROSION, WIND EROSION.

**absolute drought,** see DROUGHT.

**absolute humidity,** *n.* the quantity of WATER VAPOUR present in a given volume of air, measured in grams per cubic metre. The absolute humidity of any parcel of air at a given temperature and

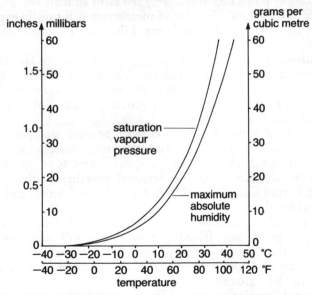

Fig. 1. **Absolute humidity.** Maximum vapour pressure and absolute humidity.

pressure is fixed and varies directly with temperature, that is, the warmer the air, the higher its absolute humidity (see Fig. 1). Compare RELATIVE HUMIDITY. See HUMIDITY.

**absorbed dose,** see RADIATION DOSE.

**abstraction,** see RIVER CAPTURE.

**abyssal benthic zone,** see BENTHIC ZONE.

**abyssal deposit,** see DEEP-SEA DEPOSIT.

**abyssal environment,** *n.* the environment at the bottom of deep oceans, that is, greater than 2000 m. It represents one of the most constant marine BIOMES: temperature is stable between 0 and 2°C while SALINITY is also relatively constant at about 35%. See DEEP-SEA PLAIN.

**abyssal plain,** see DEEP-SEA PLAIN.

**abyssal rock,** see INTRUSIVE ROCK.

**acceptable daily intake (ADI),** *n.* the daily intake of a substance consumed over the entire life span of an organism which will not produce an adverse effect on the health of that organism. Acceptable daily intake values are most usually calculated for humans in relation to nutrient or vitamin requirements although they are also applied to toxic material intake such as HEAVY METALS or air-borne pollutants. Intensively fed farm animals will also have specific ADI values. The unit of measurement is usually expressed as milligrams per chemical per kilogram of organism body weight.

**accessibility,** *n.* the degree to which services and/or facilities can be reached from a specific location. The ease with which specific services (such as shopping, education, public transport, banks) can be reached influences the development potential of a location. Accessibility is a function of:

(a) *physical accessibility*, the existence of such transport means, routes and provision as will permit the user to travel to the service point within tolerable time and cost; the degree to which these are tolerable will vary with the personal mobility of the traveller which in turn is conditioned by personal fitness, income and degree of need for the service.

(b) *psychological accessibility*, the degree to which the service is provided in a 'user friendly' way. This will vary with the confidence of the user and his or her degree of capability to transact the business in hand, both of which are conditioned by education and social background.

**accidental species,** *n.* any plant species which is present in less than 25% of a STAND.

**acid deposition** or **acid rain,** *n.* the precipitation of dilute solutions of strong mineral acids from the atmosphere. These acid solutions are formed by the mixing in the atmosphere of various industrial pollutants (sulphur dioxide, nitrogen oxides, hydrogen chloride and other minor compounds) with naturally occurring oxygen and water vapour. These acid solutions then precipitate as rain, snow or fog. The acidity of 'pure' rain water is about pH 6.5 whereas acid rain usually has a pH of below 4.5 (with an extreme low of pH 2.5).

Acidification of the atmosphere is a major threat to the environment of the northern hemisphere. The major urban and industrialized areas produce an estimated 90 million tonnes of sulphur dioxide ($SO_2$) pollution p.a., the majority of which is released directly into the atmosphere. Power stations and industries dependent upon FOSSIL FUELS are major producers of $SO_2$. Vehicle exhausts provide the main source of nitrogen oxides ($NO_x$ group). Highly polluted exhaust gases are liberated from tall chimneys at high temperatures (greater than 50°C) and high efflux velocity (greater than 5 m/sec). These gases are transported by prevailing winds for many hundreds of kilometres until the pollutants are washed out of the atmosphere by precipitation. Southern Sweden, Norway, parts of central Europe and the eastern seaboard of North America have shown the highest levels of acid rain. The acid fallout is believed to be directly responsible for the death of fresh water fish and the progressive decline, and ultimately, the death of trees.

Highly acidic deposition can cause:

(a) the direct leaching of potassium, calcium and magnesium from the leaves of trees;

(b) plant roots to become unhealthy and unable to extract nutrients from the soil. This appears to be due to the increased acidity mobilizing aluminium in the soil. The high levels of aluminium (and not the increased acidity level *per se*) are responsible for killing the plant rootlets. In times of high rainfall, large amounts of aluminium (and other metals) can be flushed from the soil into water courses where aquatic animals can be killed from metal poisoning and suffocation.

Considerable disagreement exists amongst scientists as to the real danger of acid deposition upon the environment. This has been partly due to conflicting field and laboratory evidence. It would appear that the impact of acid deposition is very site-specific and is highly dependent upon the 'buffering capacity' of the soil. Some soils, notably limestone or sandstone, appear to be able to neutralize the effect of acid deposition whereas thin glacial soils or thick granite bedrock (as in Scandinavia and Canada) fail to provide an adequate buffer against the enhanced acidity levels.

Action to reduce the amount of acid deposition shows great variation and depends upon the extent of damage already caused by this hazard. In West Germany, for example, financial losses to the timber industry from acid deposition have been calculated at $800 million per year, while agricultural losses from acidification of the soil amount to $600 million. It has been estimated that 50% of the acid deposition in West Germany originates from outside that country.

Estimates suggest that damage from acid precipitation could become ten times more severe by the year 2000 unless stringent action is taken to reduce pollution output. Most European governments (except that of the UK) have joined the *Thirty Percent Club*, whose members have pledged that sulpher emissions will be reduced by at least 30% by 1989 and up to 60% by the end of the century.

**acid mine drainage,** *n.* the seepage of sulphuric acid solutions (pH 2.0-4.5) from mines and their removed wastes dumped at the surface. These solutions result from the interaction of GROUNDWATER and percolating precipitation with sulphide minerals exposed by mining.

**acid rain,** see ACID DEPOSITION.

**acid rock,** *n.* any IGNEOUS ROCK with a high silica content, for example, granite.

**action area,** *n.* any area within a town or city that is selected for priority treatment within a larger plan which has been prepared for the town or city. The area may require total renewal or rehabilitation by reason of its physical condition, or may have been awarded special priority because of the concentration of an environmentally related social problem, such as CROWDING. See LOCAL PLAN, STRUCTURE PLAN.

**activated sludge process,** *n.* a biological treatment of SEWAGE waters in which microorganisms are encouraged to grow under favourable conditions of oxygenation and nutrients, and in so doing, cleanse the waters. Dissolved or suspended organic matter in the polluted water acts as a nutrient base for the microorganisms (chiefly BACTERIA) which feed on the organic pollutants and secrete enzymes to digest and oxidize the absorbed material and so purify the wastes.

The actual removal of the pollutants from the liquid and the absorption by the activated sludge occurs within a few minutes of contact between sewage and the activated sludge while the oxidation of the absorbed material takes much longer. Normally the process of absorbtion and oxidation take place within the same sewage tank, although numerous variations can be incorporated into the process. See TRICKLE FILTER.

**active layer,** *n.* the upper layer of soil in PERMAFROST regions which thaws during the summer. The depth of the active layer is dependant on such features as the duration of the summer, the air temperature, the nature and extent of surface vegetation, the soil moisture content, and the mineralogical and organic composition of the soil itself. Active layers rarely exceed depths of 5 m and are often poorly drained due to the presence of permanently frozen underlying ground.

Human activities in permafrost regions, especially during the summer, can damage the delicate ecological balance of the active layer, turning it into a quagmire and causing the movement of the permafrost table through THERMAL EROSION.

**activity node,** *n.* any place at which people congregate to pursue a particular activity, such as sport or shopping.

**adaptive radiation,** see RADIATION.

**additive,** *n.* any substance which is added to a separate product to bring about a change in the chemical and/or physical state of that product. Additives are commonly found in foodstuffs to enhance the visual appearance or to extend the life of the product. Other additives are introduced to petroleum to improve the combustion process in modern high compression engines. See ANTIKNOCK ADDITIVE, GRAS.

**ADI,** see ACCEPTABLE DAILY INTAKE.

**adiabatic process,** *n.* a thermodynamic process in which a change of temperature occurs in a mass of gas without a transfer of heat between the gas and its surrounding environment. The rate at which this change of temperature takes place is termed the LAPSE RATE. The adiabatic expansion and compression of air is responsible for the cooling of rising air and the warming of descending air respectively. See ADIABATIC WIND, DRY ADIABATIC LAPSE RATE, SATURATED ADIABATIC LAPSE RATE.

**adiabatic wind,** *n.* the movement of air which results either from the expansion of rising and cooling air (a process called *adiabatic cooling*) or from the compression of descending and warming air (a process called *adiabatic heating*). The characteristics of some local

# ADVANCED GAS-COOLED REACTOR (AGR)

winds found in mountainous terrain such as the Fohn, the Chinook and the Southerner are associated with this phenomenon. See ADIABATIC PROCESS.

**advanced gas-cooled reactor (AGR),** *n.* a type of NUCLEAR REACTOR in which the fuel source comprises enriched uranium dioxide and the coolant comprises gaseous carbon dioxide. Graphite CONTROL RODS are used to limit the rate of the nuclear reaction. Operating conditions in AGR differ from the earlier MAGNOX REACTOR design in that reactor temperatures are substantially higher at 675°C.

Unlike most other designs of nuclear reactors, fuel can be loaded while the reactors continue to provide power for electricity generation. Each reactor in British AGRs contains 332 vertical fuel channels and on average five channels will be replaced each month. See NUCLEAR WASTE.

The first AGR in Britain to produce electrical power to the NATIONAL GRID was Hinckley Point B in Somerset (1976). A series of similar stations now exists throughout Britain, each twin reactor station using between 30-40 tonnes of fuel per year (see Fig. 44).

**advance factory,** *n.* any factory built to a standard design for rental to an incoming business user as part of the larger REDEVELOPMENT of an area. Advance factories are not built with any particular industrial process in mind but are designed to accommodate a wide range of both type and size of enterprise. The existence of advance factories, ready for immediate occupation, has been a valuable asset in the efforts to regenerate the local economy of older industrial areas throughout the UK since the 1950s. Those in the NEW TOWNS especially have attracted a wide range of entrepreneurs, often from overseas.

**advection fog,** *n.* a type of FOG formed when a warm, moist air mass passes over a colder sea or land surface causing water vapour to condense. Sea fog around the Grand Banks off the Newfoundland coast is formed in this way when warm moist air associated with the northward moving NORTH ATLANTIC DRIFT converges with the air above the cold southward moving Labrador current. Compare RADIATION FOG, STEAM FOG.

**advertisement control,** *n.* the exercise of statutory authority to limit or prevent the display of advertising material in public places where advertising might be deemed to give offence to the population at large or where the siting of the advert is out of character with its surroundings. All advertising items are required to seek planning consent before they may be displayed. Local planning authorities can keep an area almost free of advertisements by declaring it an *Area of Special Control*.

Attitudes to the control of advertisement vary greatly from country to country. Controls in the UK are much stricter than in North America, especially in relation to advertising along highways, where, in the UK, public safety is felt to be endangered.

**aeolian process** or **eolian process,** *n.* an erosional process involving the action of wind. It is normally confined to arid and semi-arid locations but is also a problem in areas of INTENSIVE AGRICULTURE where cultivation has left the soil unprotected against wind in early spring.

**aerobe,** *n.* any organism (typically a microorganism) that requires free oxygen or air for respiration. Compare ANAEROBE.

**aerolite,** see METEORITE.

**aerosol,** *n.* any substance such as a paint, pesticide or cleaning fluid dispensed from a canister as a spray mist by a propellant under pressure. These propellents or *carrier solvents* are usually members of the CHLOROFLUOROCARBON (CFC) group which are odourless, have low toxicity for mammals and are inflammable. However, there is a growing concern about, and awareness of, the environmental problems stemming from the continuing use of products dependent upon CFCs which accumulate in the UPPER ATMOSPHERE and in the process, alter the OZONE content of the ionosphere with consequent adverse changes in the heat balance of the BIOSPHERE. Alternative carrier solvents are now being used in many aerosols.

**afforestation,** *n.* the planting of trees on land which was formerly used for land uses other than forestry; this contrasts with REAFFORESTATION which is the restocking of existing woodlands which have been depleted.

In the developed world apart from the USA, afforestation and reafforestation in recent decades has exceeded the rate of forest removal. This has been particularly so in the urbanized countries of the northern hemisphere, where timber has been in short supply for centuries. Fig. 2 shows the percentage growth in forest cover for 25 European countries. Only Spain has recorded a negative value while Turkey has planted 150 000 ha p.a. during the 1970s alone. In the UK, state and private forestry has accounted for 30-40 000 ha of new forest per annum during the decade 1972-81. In 1984 the EC GATTO REPORT on forestry recommended a major afforestation programme with emphasis on re-establishing forests in Mediterranean countries. Recent trends (1985-86) in overproduction of foodstuffs within the European Community have encouraged more land to be transferred from agriculture to forestry use.

In other areas of the world afforestation has not exceeded the rate of timber removal. In the USA woodland area declined from 307 to 290 million ha between 1963 and 1980. In tropical regions, DEFORESTATION rates have exceeded reafforestation by 10-20 times in recent years. See TROPICAL RAIN FOREST.

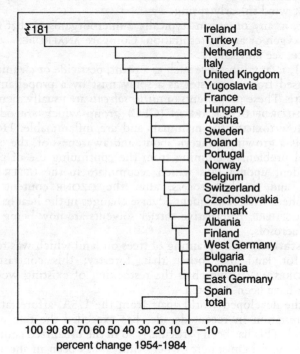

Fig. 2. **Afforestation.** Percentage growth in forest cover for 25 European countries.

**afro-alpine,** see PARAMO.

**afterburner,** *n.* a device fitted to the exhaust flues of furnaces and to the exhaust systems of motor vehicles in order to remove noxious gases through incineration. See CATALYTIC CONVERTER, AUTOMOBILE EMISSIONS, PHOTOCHEMICAL SMOG.

**Agent Orange,** *n.* a compound HERBICIDE comprising 2,4-D and 2,4,5-T widely used by US forces during the Vietnam War (1962-73) for defoliation and crop destruction. Defoliation of rainforest cover was so great that the ALBEDO rate was altered thus causing disruption of a major natural BIOME. Application rates of Agent Orange in Vietnam comprised 33.7 litres per hectare, a figure approximately ten times as high as recommended rates for use in the USA. When agricultural crops were to be destroyed a sodium salt of cacodylic acid (an organic arsenical containing about 47% arsenic) was added. This additive was not BIODEGRADABLE and the arsenic remained in the FOOD CHAIN. Agent Orange is also blamed for illnesses among veterans of the war from both sides, together with deformities in their offspring.

Agent Orange was one of three widely used herbicides in Vietnam; the others were Agent White (2,4-D and picloram) used for forest defoliation, and Agent Blue (cacodylic acid and its sodium salt) used for destruction of rice and other crops. Their names originate from the colour of the coded stripe around the canister containing the herbicide.

| Year | Forest defoliation (ha) | Crop destruction (ha) |
|---|---|---|
| 1962 | 2000 | 300 |
| 1963 | 9995 | 100 |
| 1964 | 33785 | 4198 |
| 1965 | 62972 | 26688 |
| 1966 | 299967 | 42081 |
| 1967 | 601534 | 89560 |
| 1968 | 617676 | 68795 |
| 1969 | 568305 | 46632 |
| Total | 2196234 | 278354 |

Fig. 3. **Agent Orange.** Results in areas treated with herbicides in Vietnam.

**age structure,** see ZERO POPULATION GROWTH.

**aggregate,** *n.* **1.** (Pedology) a mass of soil particles which cluster together such that they behave mechanically as a single body. See also PED, FLOCCULATION.

**2.** sand, gravel or other inert material mixed with cement and water to form concrete. The aggregate imparts stability, volume and anti-erosion qualities to the finished product.

**AGR,** see ADVANCED GAS-COOLED REACTOR.

**agrarian reform,** *n.* any of a wide range of measures designed to bring about major changes in the agricultural sector of an area or country, including both STRUCTURAL REFORM and the reform or development of institutions complementary to and necessary for rural progress. These include the provision of AGRICULTURAL EXTENSION SERVICES, credit, AGRICULTURAL COOPERATIVES, labour legislation, improved marketing systems, mechanisms of price support, and public services such as health care and education, without which LAND REFORM may fail either to effect a significant redistribution of income and power, or to maintain and increase levels of production.

The term agrarian reform is often used interchangeably with land reform, especially in Spanish-speaking areas of Latin America where *reforma agraria* has been a prominent policy issue throughout the present century.

**agribusiness,** *n.* a farming enterprise in which the ethics, principles and highly profit-oriented goals of modern industrial business management and accounting are applied to agricultural production. Agribusinesses are generally characterized by a high degree of centralized control in which key DECISION MAKING regarding the strategic direction of the business, forward planning and investment is divorced from the day-to-day routine of on-farm management.

A feature of modern agricultural development has been the increasing capital intensity of operations stemming from technological and production innovations in IRRIGATION, planting techniques, SELECTIVE BREEDING and GENETIC ENGINEERING, the increased application of AGROCHEMICALS, and the development of FEEDLOT and FACTORY FARMING systems. These advances have resulted from the increasing inroads being made into the agricultural production process by institutional investors, and by AGRO-INDUSTRIES anxious to gain control over the land on which their products ultimately depend. See AGRICULTURAL INDUSTRIALIZATION.

**agricultural cooperative,** *n.* an association of individual farmers acting in combination for mutual benefit with regard to some or all aspects of production, storage, marketing, services and the supply of farmers' requirements, including credit, market information, and access to specialist knowledge and innovations.

The main benefits of cooperation for members lie in greater security, improved returns and new market outlets together with the saving of time and stress. By pooling resources, a cooperative also allows small farmers to enjoy the benefits of bulk buying and selling normally associated with larger landowners. The disadvantages include some loss of independence and flexibility, especially in the opportunity to deal on the open market and to excerise personal judgement in any transaction.

In many developing countries, cooperatives have been established with government backing to promote agricultural MECHANIZATION, to facilitate credit transactions, and to stimulate CASH CROP and craft production. In parts of Nigeria and Tanzania, for example, farm and village settlement schemes have adopted cooperative structures based on the Israeli *moshav* model in which SMALLHOLDERS retain their independence and the right to their own profits but are supported by a strong multi-purpose cooperative responsible for marketing produce, purchasing farm and domestic requisites, and providing credit and other services.

**agricultural cycle,** *n.* the sequence and timing of farming activities undertaken during the course of the year, particularly in the production of CROPS. The cycle of predominant activities on any farm enterprise is regulated by the number and diversity of crops grown, and the specific cultivation requirements of each. Four main stages in the agricultural cycle may be recognized: soil preparation, including land clearance and seedbed preparation (see TILLAGE); planting; weeding; and harvesting. In modern farming systems in DEVELOPED COUNTRIES, the application of AGROCHEMICALS will also have a place in the agricultural round, generally in place of weeding. The agricultural cycle is characterized by seasonal peaks in labour demands and capital inputs. In many forms of tropical agriculture, for example, hand weeding assumes a critical and labour-intensive role in the sequence of activities leading to a successful harvest.

**agricultural extension services,** *n.* a range of services supplied to individuals or groups of farmers that are designed to educate, advise on agricultural practices and disseminate innovations in order to facilitate agricultural improvements and promote RURAL DEVELOPMENT. Agricultural extension services includes farm visits by agricultural officers, village meetings, courses at local training centres, and media presentations. Services may be organized by government departments, or extended by commercial enterprises interested in promoting products such as AGROCHEMICALS, or CONTRACT FARMING. Agricultural extension services in many DEVELOPING COUNTRIES have been criticized for concentrating effort and resources on larger and more progressive farmers at the expense of smaller or geographically less accessible farmers.

**agricultural hearth,** *n.* any of 12 areas in which the earliest agricultural developments and innovations occurred. The majority were in Africa, Asia and the Mediterranean Basin, although there were at least three centres of origin in the New World. Most were situated in mountainous regions between 10°S and 40°N. Major advances in crop PRODUCTIVITY were subsequently achieved in more fertile lowland areas through careful management of water resources. The main cereal-growing areas were in south-west Asia where wheat and barley came into cultivation some 10 000 years ago, northern China (millet), south-east Asia (rice), and southern Mexico (maize), with a minor agricultural hearth in north-east Africa (finger millet and sorghum). The development of *vegeculture*, that is, plant reproduction by cultivation of cuttings as opposed to the planting of seeds, occurred in the tropics along the forest-grassland fringes of West Africa (yams), south-east Asia (taro, banana) and the Americas (potato, cassava, sweet potato). However, the pre-eminent source area, from which there was the widest diffusion of crops and livestock, was a region of considerable geographical diversity in south-west Asia, based on the FERTILE CRESCENT but extending into the Nile valley of Egypt and eastwards to the Indus valley of modern Pakistan.

There is much debate over the independence of agricultural hearths. It is not yet clear, for example, whether maize cultivation emerged independently in Peru or as a result of diffusion from meso-America, or if there was independent plant domestication in sub-Saharan Africa as opposed to diffusion from Egypt and the Maghreb. It is, however, now generally accepted that northern China and south-east Asia were agricultural hearths in their own right and not dependent on south-west Asia for the diffusion of seed agriculture. See DOMESTICATION OF PLANTS AND ANIMALS.

**agricultural industrialization,** *n.* the application of biotechnical innovations, increased mechanization and industrial management practices to traditional, small-scale farming systems. Since the mid-19th century, many traditional FAMILY FARMS, have been transformed into large-scale, highly capitalized and specialized production units as a result of CONTRACT FARMING, FACTORY FARMING, the operation of FEEDLOTS, and AGRIBUSINESS. Furthermore, the natural constraints on agricultural production have been overcome by the use of GENETIC ENGINEERING, AGROCHEMICALS and by the increasing automation of agricultural activities. It can be argued that agricultural industrialization constitutes a significant AGRICULTURAL REVOLUTION, and has been the major change affecting modern agriculture in most DEVELOPED COUNTRIES.

**agricultural region,** *n.* any geographical area in which there is a clearly defined and predominant agricultural structure producing broadly similar types of farming, such as the Mediterranean region, the prairie provinces of Canada and areas of wet-rice cultivation in South-East Asia. The delimitation of agricultural regions is closely linked to the CLASSIFICATION OF FARMING SYSTEMS.

**agricultural revolution,** *n.* any fundamental transformation or innovation in agricultural practice or technology. The term is commonly applied to the fundamental changes in agriculture that occurred in 18th century Europe but is also applicable to several other periods. At the global level, four agricultural revolutions have been identified:

(a) the beginnings of agriculture from the time of the first DOMESTICATION OF PLANTS AND ANIMALS some 10 000 years ago, and its spread across the globe from several AGRICULTURAL HEARTHS to replace hunting and gathering as the basis of societal existence. See HUNTER-GATHERER.

(b) the European medieval revolution, placed variously between the 6th and 9th centuries AD and the late 8th and 12th centuries, in which significant changes were made in FIELD SYSTEMS and plough technology, including the replacement of the ox by the horse.

(c) the shift from essentially SUBSISTENCE FARMING to a more COMMERCIAL FARMING system after the mid-17th century, embracing the ENCLOSURE movement, advances in CROP ROTATION and agricultural technology, the development of MIXED FARMING, and the increased use of MANURE.

(d) the process of AGRICULTURAL INDUSTRIALIZATION in which farming became increasingly dominated by the industrial model of organization (see AGRO-INDUSTRY). Some researchers date this revolution from the late 1920s while others see the process beginning a century earlier with the growing use of industrially processed farm inputs and the adoption of labour-saving machinery (see MECHANIZATION) and extending to the present day.

Historians argue that none of these changes represent revolution so much as evolution. Much depends on how the time-scale over which change was effected is interpreted, and the extent of the time-lag between innovation and widespread adoption.

**agricultural typology,** *n.* any scheme for the identification, classification, and regionalization of types of farming. The main aim of an agricultural typology is to be comprehensive enough to encapsulate all the major variations in farming systems deemed to be appropriate at a given geographical scale of inquiry. Early typologies drawn up to delimit AGRICULTURAL REGIONS were essentially descriptive, and tended to be both selective and qualitative, while other approaches focused on differences in cultivation systems. More ambitiously, the Commission on Agricultural Typology established by the International Geographical Union in 1964 has worked towards a standardized classification of world farming systems. This identifies over 100 different types of agriculture, organized in a hierarchy of three orders based on social, operational, production and structural attributes. See also CROP COMBINATION ANALYSIS.

**agriculture,** *n.* the occupation or science of cultivating land and rearing CROPS and livestock. Agricultural systems are complex and diverse, ranging from SHIFTING CULTIVATION and nomadic PASTO-RALISM to the technologically sophisticated and increasingly industrialized farming systems of parts of Europe and North America. These wide variations in type are determined by such factors as LAND TENURE and FARM SIZE; the operational methods of farm holdings, and the intensity of inputs, whether of labour or capital; PRODUCTIVITY and the degree of commercialization; the structural characteristics regarding farm layout and the proportion of land devoted to particular crops or activities.

Agriculture is the most important of the world's economic activities. Some 35% of the world's land area (excluding Antarctica) is devoted to agricultural activities in the form of ARABLE FARMING, land under permanent crops, and PASTURE (but excluding wooded areas).

Approximately 45% of the economically active population of the world were engaged in agriculture in 1980, varying from 7% in the industrialized market economies of the DEVELOPED COUNTRIES to 62% in DEVELOPING COUNTRIES, rising to 75% in Sub-Saharan Africa. In the EC in 1986, 8.3% of the labour force were employed in agriculture, ranging from 2.6% in the UK to 29% in Greece. Agriculture continues to make a significant contribution to the gross domestic product (GDP) of Sub-Saharan Africa (36%) and developing countries overall (19%) but is responsible for less than

3% of GDP in the industrialized market economies. See also INTENSIVE AGRICULTURE, EXTENSIVE AGRICULTURE, COMMERCIAL FARMING, ANIMAL HUSBANDRY, GLASSHOUSE CULTIVATION, AGRONOMY, FERTILIZER, IRRIGATION, TILLAGE.

**agrochemical,** *n.* any inorganic, artificial or manufactured chemical substance used in agricultural production processes. The most important agrochemicals include FERTILIZERS, HERBICIDES, PESTICIDES and INSECTICIDES, and are the product or by-product of the chemical and petrochemical industries.

**agroforestry, farm forestry** or **social forestry,** *n.* any land management practice in which farmers are encouraged to incorporate the cultivation of trees and shrubs along with CROP production and ANIMAL HUSBANDRY. Although not confined to DEVELOPING COUNTRIES, agroforestry methods have been most widely applied in tropical regions where scientific management techniques have been integrated with traditional land use practices. At one level, SHIFTING CULTIVATION can be described as a form of agroforestry, as can the tree crop, grazing and arable cropping regime of Mediterranean intercultures. Increasingly, agroforestry involves the deliberate cultivation of woody plants with particular qualities such as fast growth, economic value, fuelwood yield, or NITROGEN-FIXATION.

Agroforestry includes the planting and maintenance of trees and shrubs near dwellings, along the boundaries of farmland as SHELTERBELTS, in cultivated fields as a form of MIXED CROPPING to assist in SOIL CONSERVATION and water retention, and in FALLOW plots to aid the process of soil regeneration. *Alley cropping*, in which rows of crops are alternated with rows of fast-growing trees and shrubs, permits the land to be utilized continuously without recourse to fallow, while ensuring a supply of fuelwood and generally increased levels of total land productivity.

**agro-industry,** *n.* any sector of industry supplying farm inputs or engaged in the processing and distribution of agricultural products. Agro-industries are frequently large-scale corporate enterprises, with multinational interests, which control not only the upstream supply of farm inputs and the downstream processing of farm produce, but also transportation, marketing and distribution, as well as the wholesale and retail outlets for their produce. This level of vertical integration is more prevalent in the USA and the Third World than in Europe. Almost one-third of US farm output is accounted for by direct AGRIBUSINESS production or forward CONTRACT FARMING, and almost all the production of vegetables

AGRONOMY

for processing, potatoes, citrus fruits and broiler chickens is sold to closed markets.

Modern agro-industrial systems provoke mixed public reactions ranging from acceptance of their inevitability or necessity in highly urbanized societies to concern over the environmental impact of the increased use of AGROCHEMICALS, MECHANIZATION, and the trend towards specialized MONOCULTURES which they promote.

**agronomy,** *n.* the science of land cultivation, crop production and soil management. Its major concerns include crop heredity, physiology and biology and the characteristics, properties, classifications, uses and conservation of soils.

An agronomic classification of plants specifies how a crop will be utilized, for example, GRAIN CROPS, FODDER CROPS, BREAK CROPS, CATCH CROPS, etc., as distinct from their LINNEAN CLASSIFICATION.

**agrotown,** *n.* any settlement chiefly populated by the agricultural workers from a surrounding rural area. Agrotowns are commonly found throughout Mediterranean Europe and arose originally from political insecurity or land-holding patterns based on large estates with tenanted labour. Agrotowns may support populations of up to 20 000, and act as self-sufficient service centres largely independent of other settlements.

**A-horizon** or **E-horizon,** *n.* the uppermost layer in a SOIL (see Fig. 55). Mineral and organic material (including soil fauna and the roots of vegetation) accumulate in the A-horizon, while soluble salts and clays are removed as a result of ELUVIATION (hence the alternative term of E-horizon). It is the horizon of greatest significance for agricultural land use. Compare B-HORIZON, C-HORIZON. See also H-HORIZON.

**aid,** *n.* the flow of resources from governments and public institutions of DEVELOPED COUNTRIES and from international agencies to governments of DEVELOPING COUNTRIES with the object of promoting economic development and welfare.

Formerly, the term included private transactions with the THIRD WORLD, such as investments, export credits and bank lending by private companies based in the 'donor' country. Today, the Development Assistance Committee (DAC) of the Organization for Economic Cooperation and Development (OECD) classifies aid or *overseas development assistance* (ODA) as only those disbursements of loans and grants made on concessional terms by official agencies of its members, together with technical cooperation and assistance. It thus includes soft or low-interest, long-term loans from the International Development Association of the WORLD BANK, but not lending on near-commercial terms from the World Bank itself and other international agencies which are designated 'non-concessional flows'.

The amount of official aid is relatively small. ODA from DAC members rose from $7 billion in 1970 in nominal terms, that is, before allowing for inflation, to $27.3 billion in 1980, and $36.7 billion in 1986. At constant 1980 prices, the value of aid rose from $20.7 billion to $28.8 billion between 1965 and 1985. As a percentage of gross national product (GNP), ODA fell from 0.48% in 1965 to 0.35% in 1986. Only the Scandinavian countries and the Netherlands exceeded 0.8% while the USA's contribution, the largest monetary amount at $9.6 billion in 1986, was only 0.24% of its GNP. Members of the ORGANIZATION OF PETROLEUM EXPORTING COUNTRIES (OPEC) have contributed significantly to aid flows in recent years, exceeding $9.6 billion in 1980 with Saudi Arabia contributing some 5% of its GNP as development assistance. Since 1981 the value of OPEC aid has fallen sharply, to $4.6 billion in 1986.

Approximately one quarter of ODA is channeled through multilateral agencies such as the World Bank, the European Development Fund, several regional banks, and the specialized agencies of the United Nations such as the FAO and the WORLD FOOD PROGRAMME. The bulk of aid is bilateral, that is, directly from one government to another. More than half of this is project aid with funds made available to finance or part-finance specific development projects. Typically, grants or low-interest, long-term loans are negotiated for the initial or construction phase of the project, but not its subsequent operation and maintenance. The remainder of bilateral aid is in the form of sector aid (finance for local development banks or for small projects within a particular region or sector), FOOD AID, and general non-project or programme aid.

More than half of all ODA, and some three-quarters of UK and US aid is 'tied' in the sense that it must be spent on importing goods and sevices from the donor country. There is, thus, a considerable return flow of capital back to the donor. At least half of all French aid, for example, is thought to be repatriated in the form of salaries paid to French technical assistants. Far from being an entirely charitable gesture, tied aid may prove more costly than borrowing on the open market for it tends to be focused on high-profile projects based on imported skills and materials rather than local enterprise and goods. It may incorporate non-competitive tendering for contracts, and ensure a long-term dependence on the donor to maintain and replace equipment. Aid dependence hardens existing political and economic relations between donors and recipients, while political selectivity leaves left-wing governments critically short of development funds. The bulk of bilateral aid is directed towards a few middle-income developing countries, as defined by the WORLD BANK, while net bilateral flows to low-income countries from the OECD has failed to reach even 0.1% of donor GNP during the 1980s.

**air corridor,** *n.* a prescribed routeway along which commercial and military aircraft are guided by an AIR TRAFFIC CONTROL SYSTEM.

**air frost,** see FROST.

**air mass,** *n.* a large homogenous body of air which has almost

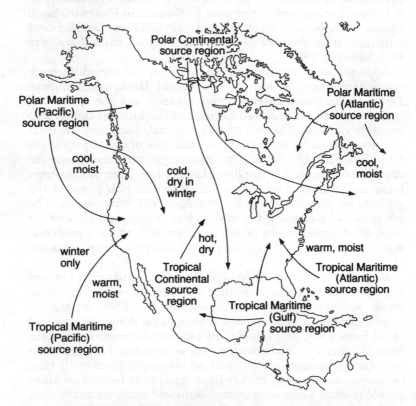

Fig. 4. **Air mass.** The air mass source regions of North America.

uniform conditions of temperature and humidity. Air masses are contained by FRONTS and may cover tens of thousands of square kilometres.

Air masses can be classified according to their characteristics which reflect the underlying land and sea source regions (see Fig. 4). These characteristics and the associated weather become modified as the air mass is transported as part of the ATMOSPHERIC CIRCULATION system. Four main types of air mass are recognized:

(a) *tropical continental air masses* originate in the HORSE LATITUDES in areas such as the Sahara and arid regions of the USA. These air masses are hot and dry, and are unstable particularly in summer, although the low humidity prevents the large scale development of cloud with associated precipitation.

(b) *tropical maritime air masses* have source regions in the subtropical, high-pressure systems over the oceans, for example the Pacific and Atlantic Oceans. High humidity, warmth and INSTABILITY, particularly in summer, are characteristic features of these air masses. Both types of tropical air mass move polewards.

(c) *polar continental air masses* develop over the Antarctic, the Arctic, Siberia and Northern Canada during the winter. Air masses from these areas are usually cold, very dry and stable. During the summer, the lower layers of these masses may become warmed, allowing some cloud to form.

(d) *polar maritime air masses* originate over the Pacific and Atlantic oceans in high latitudes. Cool, moist and unstable in their lower levels, these air masses usually cause rainy and cloudy weather.

See also STABILITY.

**air pollution,** *n.* any toxic or radioactive gases and PARTICULATE MATTER introduced into the atmosphere, principally as a result of human activity. Air pollution is usually associated with the emissions from car exhausts, power stations, factories, incineration plants and the domestic burning of FOSSIL FUELS in urban and industrial areas. Pollutants may also originate in the countryside from pesticide sprays and the dust generated by mining and agricultural practices. However, air pollution can also originate from natural sources, for example dusts generated by strong winds over deserts, grit, ash and dust from volcanic explosions, and sea-salt nuclei blown inland by strong winds. Natural gaseous pollution originates from volcanic explosions, FUMAROLES, bogs, marshes and decomposing matter.

Pollutants are transported by winds and rising air currents with the larger particles rapidly returning to the earth's surface under gravity (FALLOUT), whilst the smaller particles are removed from the atmosphere by rain (*washout*).

Air pollution levels are normally minimized due to dispersal of the pollutants by winds. However, concentrations can form in the calm weather conditions caused by stationary ANTICYCLONES and can be exacerbated by the presence of low-level INVERSION layers which trap the pollutants in the atmosphere beneath them.

Air pollution can have a serious detrimental effect on the environment. In urban areas the effects of atmospheric pollutants can be seen in the deterioration of building materials and the tarnishing of metals due to the acidity of pollution, while in rural areas, ACID DEPOSITION can cause considerable damage to crops and vegetation. Many health problems have been closely linked with air pollution ranging from simple eye irritations to complex pulmonary diseases. High atmospheric lead concentrations in cities, caused by AUTOMOBILE EMISSIONS, may result in brain damage and lowered intelligence of children.

On both a national and international basis, efforts are being made to reduce the major environmental problems of air pollution. Many countries in the developed world are enacting legislation to limit air pollution: factories, power stations and incinerator plants are required to fit cleansing devices to allow the almost complete sanitation of waste products; clean-air legislation encourages the use of smokeless domestic fuels and cars are increasingly being run on UNLEADED PETROL and fitted with CATALYTIC CONVERTERS to limit the release of pollutants in exhaust fumes. However, in the developing world, similar measures have yet to be implemented on a such a scale.

After a period of scientific uncertainty the global effects of air pollution have now become clear. Increasing levels of carbon dioxide and other gaseous pollutants are generating a GREENHOUSE EFFECT which is partly offset by a natural and gradual atmospheric cooling. This cooling is due to a reduction in INSOLATION caused by an increase in the reflection of solar radiation by clouds, and its diffusion as a result of an accumulation of atmospheric particulate matter. Furthermore, the rapid build-up of atmospheric CHLORO-FLUOROCARBONS has resulted in international action to restrict the sources of this form of atmospheric pollution. See also SMOKE CONTROL ZONE, CLEAN AIR ACTS, BACKGROUND CONCENTRATION.

**airport,** *n.* any large, specially equipped area where commercial aircraft can land and take off, and where facilities exist to handle the onward traffic of passengers and freight. Take-off and landing runways are constructed to align with the most frequently occuring winds in the locality, and are supplemented by a network of taxiing tracks along which aircraft can be guided safely to their appropriate departure runway or to the terminal building. All traffic movement on the runway and taxi tracks is under the airport's AIR TRAFFIC CONTROL SYSTEM. Airports which handle international traffic must have two runways, one not less than 3750 m for take-off and a second, not less than 2500 m for landing. Airports designed solely for short take-off and landing aircraft, such as the City Airport in Docklands, London, may have a runway as short as 1220 m.

**air pressure** or **atmospheric pressure,** *n.* the pressure exerted in all directions by the weight of the ATMOSPHERE. Air pressure decreases with altitude (see Fig. 5) and is measured in millibars (mb). Standard sea-level pressure is 1013.2 mb; at an altitude of 5639 m the air pressure is only half that at sea level. See also PRESSURE GRADIENT, BAROMETER.

| Altitude range (metres) | Rate of pressure decrease (per 300 m) |
|---|---|
| 0 – 600 | 4% |
| 600 – 1500 | 3% |
| 1500 – 3000 | 2.5% |

Fig. 5. **Air pressure.** Table showing the variation of air pressure with altitude.

**Air Quality Act (1967),** *n.* US legislation which established air quality control regions, gave recommendations for AIR QUALITY STANDARDS and set time schedules by which these should be met. The act, in its original form, was deficient in many ways and was amended in 1970 to include national emission standards for hazardous air pollutants from existing as well as new plants, automobile and aircraft emission standards, and a clause which allowed the private citizen to file a law suit against any person who violated an emission standard.

**air quality standard,** *n.* a prescribed level of an atmospheric pollutant permitted during a specified time in a specific geographical area. Air quality standards are usually set by the regional public health department and enforced by officers who regularly monitor the gases which are released from chimneys. In some instances, national government may legislate for air quality standards (as in the UK) although enforcement of the regulations is devolved to local government. The limits for air quality standards vary from area to area and are notoriously difficult to enforce due to the frequent invisibility of gaseous pollutants, their ephemeral release patterns, the variable dilution rates which are dependent upon climatic conditions, and the level of tolerance of a polluted environment by the general public. Standards have usually been set as a result of specific air pollution episodes; for example, the London smog of 1952 resulted in the UK CLEAN AIR ACT (1956) while the PHOTOCHEMICAL SMOG problem of California stimulated public and political concern which led to the passing of the AIR QUALITY ACT (1967) in the USA. See AIR POLLUTION. See also SMOKE CONTROL ZONE, BEST PRACTICAL MEANS.

**air traffic control system,** *n.* any system by which the movement of aircraft through a specified airspace, or on the ground at airports, is organized and controlled. There are several levels of control exercised over the movement of aircraft. Every airfield has its own control tower which regulates the altitude, speed and direction of aircraft, both on the ground and in the air when they are within its 'circuit'. Larger areas of control cover several AIRPORTS and circuits; the Scottish Control Area, for example, covers the UK north of the latitude 55°N. All aircraft flying across the North Atlantic are controlled from the UK westward to the mid-way point by international control units at Prestwick (Scotland) and London, until they pass into the control of similar units at Gander (Newfoundland) and New York.

**albedo,** *n.* the proportion of solar radiation which reaches the earth's surface and is immediately reflected back into the atmosphere. The average albedo for the earth's surface is 40% but this figure varies greatly according to the nature of the surface. For example, the albedo for new snow is around 80% whilst the figure for forests and wet soil ranges between 5% and 10%. See also INSOLATION.

**aldrin,** *n.* 1,2,3,4,10,10-hexachloro-1,4,4a,5,8,8a-hexahydro-*exo*-1,4-*endo*-5,8-dimethanonapthalene, a chlorinated hydrocarbon used as potent contact INSECTICIDE against soil insects. It rapidly disappears from the ground to which it has been applied, a feature which contributed to it being described as a 'safe' insecticide. However, aldrin is rapidly converted to *dieldrin*, which is extremely persistent and it was the accumulation of dieldrin in the bodies of animals located at the top of complex FOOD CHAINS, particularly birds of prey, which caused a massive reduction in the their numbers.

Stringent restrictions on the use of aldrin were introduced in many countries during the mid-1960s, for example, in the UK the Pesticides Safety Protection Scheme banned its use apart from in very special circumstances. Since that time most afflicted species have shown a gradual recovery in numbers.

**algae,** *n.* simple plants that lack true stems, roots and leaves but possess chlorophyll and other pigments and so are capable of PHOTOSYNTHESIS. They show diverse life forms ranging from simple unicellular forms to complex colonial and filamentous forms. Algae can be found throughout the BIOSPHERE although usually they have a close association with water.

Algae are particularly numerous in environments which higher plants have failed to colonize. They can form symbiotic relationships with lichen and in this form are able to colonize bare soil and even rock. Algae form the first step in the colonization of land by plants, adding organic material and vital trace elements necessary for the growth of higher plants.

The photosynthesized products from algae form low-grade sources of energy and, for the most part, are unsuited for human consumption. Indirectly, however, algae are of very considerable importance to humans as they fix large quantities of atmospheric nitrogen in soils, thus improving and maintaining soil fertility for the higher plants, including agricultural crops.

**algal bloom,** *n.* the temporary, rapid growth of ALGAE in fresh water. During the early part of the growing season an abundance of nutrients in the surface waters combined with an increase in water temperature allows algae (and other aquatic plants) to rapidly multiply until one nutrient becomes scarce or limiting, usually nitrogen or phosphorus. The phase of rapid growth is replaced by a massive and sudden mortality of algae which rise to the surface of the lake forming a green scum. The decomposing algae consume large quantities of oxygen dissolved in the water causing the lake waters to become anaerobic, killing most other aquatic life (primarily fish). The waters gradually undergo a self-cleansing process as oxygen levels are re-established. It is possible for a second bloom to occur under favourable autumn conditions.

The occurrence of algal blooms has increased in recent years due to enhanced levels of nitrogen and phosphatic materials leaching from agricultural fields into water courses. Reservoirs and shallow lakes located in intensive agricultural areas are susceptible to algal blooms. Reservoirs must sometimes be drained and pipework and filters carefully cleaned to remove the green slime produced by the algal bloom.

**algorithm,** *n.* a procedure or series of stages, usually computational steps within a computer program which, if correctly followed, will provide the optimal solution to a problem. In ecology, algorithms are commonly used in SIMULATION STUDIES or in techniques involving linear programming wherein the provision of initial values are used to construct a MODEL of behaviour for plants or animals.

**Alkali Inspectorate,** the official body in the UK charged with monitoring industrial AIR POLLUTION and enforcing clean-air legislation. The kinds of pollution dealt with by the inspectorate are found particularly in connection with the petrochemical industry and aluminium smelting. Processes which produce PARTICULATE MATTER such as mineral grit and dust are also subject to the scrutiny of the inspectors.

The Alkali Inspectorate was initiated through the Alkali Acts of 1863 and 1906 in which it was recognized that it is not always possible to eliminate all noxious emissions but that as far as possible gases should be rendered harmless and inoffensive before discharge into the atmosphere. See BEST PRACTICAL MEANS.

**alkalization,** *n.* the accumulation of sodium salts in a soil resulting in a soil pH greater than 7.0. Alkalization usually occurs in arid and semi-arid regions, especially in coastal areas and where the underlying PARENT MATERIAL has a high sodium content. The limited effects of LEACHING in these environments results in the selective removal of the more soluble potassium and magnesium salts from the soil through ELUVIATION and a gradual concentration of the less soluble sodium salts. Alkaliztion is greatly accelerated by the upward movement of saline groundwater through CAPILLARITY. High air and ground temperatures cause evaporation and the deposition of sodium salts from the groundwater. The process of alkalization causes the formation of SOLONETZ soils. See SALINIZATION. See also CALCIFICATION.

**alley cropping,** see AGROFORESTRY.

**allotment,** *n.* a small portion of ground rented annually by an individual for the purpose of growing flowers or vegetables for personal use or recreation. Allotments are most commonly found in the UK and are worked by amateur gardeners who do not have land attached to their own house. The Agriculture Act (1947) set the maximum size of allotments at 0.1 ha.

**alluvial fan,** *n.* a fan-shaped mass of rock debris formed when a fast-flowing mountain stream leaves its constricted course and enters a broad valley or gently sloping plain. The sudden reduction in the speed of flow causes large scale deposition of debris which in turn causes the stream to divide and change course and which results in the formation of a conical fan. Well-developed alluvial fans are most commonly found in arid or semi-arid climates where rapid evaporation and PERCOLATION are additional factors favouring the formation of these features. Where closely spaced streams discharge from a mountainous region onto a PIEDMONT, it is possible for deposits to converge to create a piedmont *alluvial plain* or *bahada*, such as the Indo-Gangetic Plain. Soils on alluvial fans can be of great agricultural value although in semi-arid areas, the emergence of flash floods on fans can cause rapid and severe SOIL EROSION. These floods can also pose a serious threat to settlements and lines of communication built within the vicinity of EPHEMERAL STREAMS.

**alluvial plain,** see ALLUVIAL FAN.

**alluvial soil, fluvent** or **fluvisol,** *n.* a soil type of highly variable nature, found in river channels or FLOODPLAINS on recently deposited ALLUVIUM. Many alluvial soils are highly productive, with a nutrient content maintained at very high levels due to the regular flushing in of nutrients by flood waters. The lower courses of large rivers frequently contain alluvial soils.

**alluvium,** *n.* the fragmented and unconsolidated material transported and then deposited by a river. It consists mainly of clay, silt, sand and gravel and, once deposited, can be reworked by the river to form features such as DELTAS, ALLUVIAL FANS and FLOODPLAINS. Some of the world's most fertile soils are derived from alluvium. See FLUVIAL TRANSPORTATION, FLUVIAL DEPOSITION, ALLUVIAL SOIL.

**alp,** *n.* gently sloping land above the steep sides of a U-SHAPED VALLEY; the term is sometimes restricted to such land in the European Alps. Alps are glacial in origin and as such, are often covered with glacial till. During the summer, alps are used for livestock grazing as their elevation permits long hours of sunshine while being free from the insect pests of the valley bottoms. In winter, alps are snow-bound and improved access, often by cable car or chair lift, has made them ideal sites for winter sports.

**alpha emitters,** see ALPHA PARTICLE.

**alpha particle,** a nucleus of helium-4, consisting of two protons and two neutrons emitted by certain radioactive substances called *alpha emitters*. Alpha radiation can penetrate a few centimetres of air and can be stopped by a single sheet of paper. Thus, alpha radiation is normally not hazardous to humans unless ingested directly into the body.

**alpine,** *adj.* (of an ecosystem) characterized by an extremely short (10-12 week) GROWING SEASON. Alpine conditions are generally found on mountains at mid- and high latitudes; they can also be found at very high altitudes (above 4000 m) in tropical latitudes, for example, on the Pic Bolivar mountain range in Venezuela. See ALPINE COMMUNITY.

**alpine community,** *n.* any plant and animal groupings to be found immediately below the permanent SNOWFIELDS of mountainous areas. The height at which alpine communities can be found rises gradually with decreasing latitude. In southern Alaska (lat. 60° N) the lower limit of the alpine zone is 1200 m above sea level, in the Sierra Nevada (lat. 40° N) the lower limit is 3000 m, while in Colorado and new Mexico (lat. 35° N) the limit occurs at 3650 m.

Alpine vegetation shows marked zonation patterns which are dependent upon the rapidly changing environmental conditions. Variation is due to the frequent changes in aspect, angle of slope and degree of exposure/shelter afforded by the topography. These physical characteristics will determine the speed with which the snow will melt in late spring and re-form in early autumn. The snow-free period varies between 16 and 24 weeks. Average daily temperatures in summer rarely exceed 10°C, and even in mid-summer, night-time frosts are common. Total amounts of INSOLATION in the long summer days can be very high and leaf surface temperatures can attain 45°C. PERMAFROST conditions are not usually present (see TUNDRA) although freeze-thaw action can create major soil disturbance (see PHYSICAL WEATHERING).

Alpine PRIMARY PRODUCTIVITY is low, with net primary productivity values of only 10 and 400 g/m² per year being commonplace.

Vegetation communities are dominated by herbaceous forms many of which are XEROPHYTES due to the high rate of insolation. Trees are notably absent apart from creeping forms of willow, birch and rhododendron. Alpine flora is remarkably colourful and varied. In the short summer the alpine meadows can be transformed into patches of vivid colour. Insects and small mammals abound while birds of prey (eagles, buzzards and hawks) find abundant supplies of food. Herds of goats and deer graze the summer pasture while the more accessible grasslands are grazed by cattle and some sheep driven up from the valleys (see TRANSHU-MANCE).

Alpine communities occur in regions of great scenic and natural beauty, as a result of which they attract large numbers of walkers, climbers and skiers. However, alpine communities are very FRAGILE ECOSYSTEMS and hence, are vulnerable to human pressure. The thin soils and sparse vegetation found in these regions take many years to recover from damage caused by indiscriminate recreational pursuits. Consequently, many alpine areas have been made into NATIONAL PARKS or CONSERVATION AREAS.

**alpine farming,** *n.* a system of EXTENSIVE AGRICULTURE based on the raising of livestock on high mountain PASTURES, often above the treeline. The shortened GROWING SEASON brought about by the climatic effects of altitude means that adequate pasture may be available for only three months a year, leading to patterns of TRANSHUMANCE. In the EC, such areas are eligible for economic assistance. See HILL FARMING.

**alternative energy,** *n.* energy obtained from sources other than NUCLEAR POWER or traditional FOSSIL FUELS such as coal, oil and gas. Unlike most energy resources currently in use, alternative energy resources are usually renewable and non-polluting. The main alternative energy sources are tidal and WAVE POWER, SOLAR ENERGY, GEOTHERMAL POWER, and BIOMASS fuels (such as METHANE and ETHANOL). To this list can be added fuelwood and hydro power although both have been extensively used for many thousands of years.

The search for alternative energy supplies has been due to:

(a) massive increases in costs of traditional fuels;

(b) the increase in the level of awareness of the POLLUTION caused by fossil and nuclear fuels;

(c) shortages of energy due to exhaustion of traditional resources and artificial scarcity due to political factors;

(d) an EXPONENTIAL GROWTH in energy consumption by all nations apart from the poorest.

At present, it is in DEVELOPING COUNTRIES that alternative energy supplies are most important. Nearly half the world's population rely on BIOMASS energy (mostly from firewood) for heating (see Fig. 6). In treeless areas, as for example in Bangladesh and rural China, crop residues and animal dung provide as much as 90% of household energy (see DUNG FUEL).

In DEVELOPED COUNTRIES alternative energy provision involves the use of high-technology engineering features such as solar panels, heat pumps and architectural designs which re-use waste heat from lighting and ventilation motors for space heating. On average, industry and commerce now use between 10 and 15% less energy compared with the early 1970s as a result of energy management. Energy management has proved to be far more cost effective than investment in alternative energy sources. The latter provide barely 3% of the energy demands of western nations.

| Country | Year | % biomass energy |
|---|---|---|
| Burkina Faso | 1980 | 94 |
| Malawi | 1980 | 94 |
| Mozambique | 1980 | 89 |
| Sri Lanka | 1981 | 75 |
| Kenya | 1981 | 68 |
| India | 1979 | 42 |
| Phillipines | 1981 | 38 |
| Columbia | 1978 | 24 |
| Portugal | 1981 | 7 |

Fig. 6. **Alternative energy.** The percentage of biomass in the total energy consumption for selected countries.

**altocumulus,** *n.* a medium-altitude CLOUD consisting of white/ grey, fleecy, globular masses. Altocumulus is associated with fair weather.

**altostratus,** *n.* a dark grey, continuous layer of medium-altitude CLOUD associated with the approach of the WARM FRONT of a DEPRESSION.

**amenity** or **environmental quality,** *n.* **1.** the fact or condition of being pleasant or agreeable. An important purpose of town planning is to maintain existing amenity and ensure that new developments are of sufficiently good design and construction to enhance local amenity. DEVELOPMENT CONTROL is the means of protecting existing amenity through the refusal to permit damaging developments; positive promotion of better amenity is exercised through DEVELOPMENT PLANS. The term takes in many subjective values relating to aesthetic judgements, prevailing fashion in design, expected minimum levels of local quality of life, and public security.

**2.** any edifice or service which enhances the quality of life. A cathedral is an amenity which is both a source of aesthetic pleasure and a place of spiritual comfort. An efficient railway service is a social amenity, as is a landscaped park or a well-preserved historic building. Public standards of judgement about amenity are rising and this is expressed in the increasing numbers of local 'amenity societies' which act as pressure groups in defence or in promotion of an aspect of local amenity. In the UK, the formation of these amenity societies has been fostered by a national organization, the Civic Trust, founded in 1957. The Trust provides advice and information for local societies, as well as a model constitution to aid administration. One consequence of the growth in such societies is that public support is now more readily obtained for the spending of money on the enhancement of amenity.

See also AREA OF OUSTANDING NATURAL BEAUTY, CONSERVATION AREA, CIVIC AMENITIES ACT.

**amphibian,** *n.* any cold-blooded, four-limbed vertebrate of the class Amphibia, typically living a part of their life history on dry land but breeding in water. The class is subdivided into three orders: Gymnophiona (limbless amphibians, 160 species), Anura (frogs and toads, 2500 species) and Urodela (newts and salamanders, 300 species). The most common characteristic of the class is the exposed, water-permeable skin, rich in glands that secrete mucus, which is sometimes poisonous. The amphibian lifestyle has produced a breathing system designed for the uptake of oxygen via external or internal gills, simple lungs, a moist skin and a mucus membrane lining the mouth. Most amphibians lay small, round eggs protected by a gelatinous mass and rely on external fertilization. Some species are *oviviparous*, that is, the larvae hatch and develop within the female body. Adult amphibians are usually CARNIVORES feeding upon insects and beetles. In turn, they provide food for large REPTILES and some birds. Many members of this class are highly poisonous and indicate this condition by vivid colouration.

**anabatic wind** or **valley wind,** *n.* a localized wind which flows up valley slopes, usually in mountainous regions, on summer afternoons. Anabatic winds are caused by the convectional heating of air above the valley slopes which subsequently rises to be replaced by cooler air from the valley floor. Anabatic winds are often only poorly developed compared to KATABATIC WINDS.

**anaerobe,** *n.* any organism that does not require free oxygen or air for respiration, obtaining energy from the degradation of glucose in ANAEROBIC RESPIRATION. Compare AEROBE.

**anaerobic respiration,** *n.* a type of cellular respiration that takes place in ANAEROBES whereby glucose in food reserves is chemically broken down and energy for other cellular processes is released in the absence of oxygen. This absence may vary from short-lasting (several hours) episodes to the permanent absence of oxygen.

**ancient monument,** *n.* any officially protected site or edifice, usually more than 1000 years old, regarded as part of the cultural inheritance of a nation. Ancient monuments are generally preserved and protected by the same legislation as HISTORIC BUILDINGS. They have no functional use and are maintained primarily for scientific study and public education.

**angiosperm,** *n.* any plant of the division Angiospermae, the most highly evolved group of vascular plants, characterized by the production of true flowers which, after pollination, form a well protected seed. Angiosperms, like the simpler plants, are agglomerations of cells but the angiosperms differ in that the cells have undergone complex specialization to form distinct tissue types. The mature plant can be divided into four parts: stem, root, leaf and flower.

Two groups of angiosperms can be distinguished: the MONOCOTYLEDONS, and the highly varied DICOTYLEDONS. The distinction is based mainly on the differences in floral structure. The monocotyledons include sedges, rushes and grasses, and they have proved to be of great significance to humans as they have been bred to produce all of our cereal crops. The dicotyledons include all the colourful flowering herbs and the HARDWOOD trees such as oak, beech, teak and rosewood.

**anhydrite,** *n.* a SEDIMENTARY ROCK composed of a calcium sulphate mineral precipitated by evaporating seawater. Anhydrite is used in the manufacture of fertilizers and cement.

**animal husbandry,** *n.* the rearing and management of livestock to facilitate the optimum conversion of PASTURE and fodder into products of use to man. The method of animal husbandry depends on the reasons for keeping livestock, the nature of the physical environment, the cultural milieu, and the personal preferences and expertise of the farmer in GRAZING MANAGEMENT.

Animal husbandry embraces a range of practices from the keeping of small numbers of a few species primarily for domestic use, and MIXED FARMING or integrated animal husbandry, to more

COMMERCIAL FARMING and FACTORY FARMING. On more marginal lands *stock farming* is still mostly the business of various types of NOMADISM and RANCHING, requiring skills of extensive RANGE MANAGEMENT.

**animal realm** or **faunal region,** *n.* any one of eight biological divisions of the earth, each containing faunal species (particularly vertebrates and insects) endemic to it (see Fig. 7 overleaf). These indigenous species are prevented from migrating to other regions by various natural barriers such as deserts, mountains, seas or oceans. Species common to all, or many regions, are few in number and usually comprise small protozoan forms able to overcome the physical barriers to migration by being easily spread by wind or water or through association with human activity, frequently as parasites on domestic species. It is now almost certain that the animal realms owe much of their origin to the original geographical arrangement of the land masses prior to the start of CONTINENTAL DRIFT some 1700 million years ago. See also WALLACE LINE.

**anion,** see ION.

**annual,** *n.* a plant which completes its life cycle from seed to seed in one year. Many weed species of agricultural fields are annuals. The spread of annual weeds can be conveniently prevented through the use of selective HERBICIDES applied before pollination has occured. Compare BIENNIAL, PERENNIAL.

**annual work units,** see MAN-DAY.

**Antarctic Treaty,** an agreement signed in 1959 by New Zealand, Norway, South Africa, the UK, the USA, the USSR, Argentina, Australia, Belgium, Chile, France, and Japan, whereby the Antarctic continent was designated an area devoted to peaceful purposes. The Antarctic (approximately 10% of the earth's land surface) is demilitarized, nuclear-free and devoted to research. The original group of twelve nations was later joined by four others (Poland, 1977; West Germany, 1979; Brazil and India, 1983).

The Antarctic Treaty represents a successful experiment in international cooperation between world superpowers and major industrialized nations. A further 13 countries have 'acceding nation status' to the Treaty, but may not participate in decision making.

The Treaty is under continual stress due to the desire by some nations to undertake limited exploitation of the extensive mineral (oil, coal, iron ore) and biological resources (especially KRILL) which occur in the Antarctic region.

**antecedent drainage,** *n.* an established river drainage system able to maintain its original course by downcutting at the same rate as

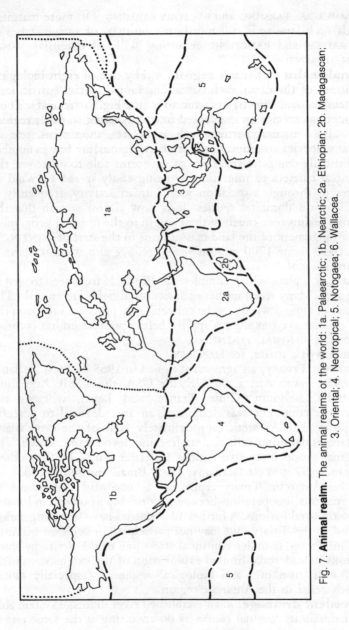

Fig. 7. **Animal realm.** The animal realms of the world: 1a. Palaearctic; 1b. Nearctic; 2a. Ethiopian; 2b. Madagascan;
3. Oriental; 4. Neotropical; 5. Notogaea; 6. Wallacea.

the surrounding land surface is uplifted and folded by earth movements. Examples of antecedent drainage include the Indus and Brahmaputra in the Himalayas and the Colorado in the USA. See also SUPERIMPOSED DRAINAGE, DRAINAGE PATTERN.

**anthracite,** *n.* a high-grade, hard, shiny black coal containing up to 98% carbon. Of all types of coal anthracite gives off the most heat and least smoke. The high carbon content of anthracite is a result of METAMORPHISM during the process of COALIFICATION. Anthracite is particularly favoured for domestic heating. The mining of anthracite is usually expensive due to the highly faulted nature of the seams. Compare LIGNITE, BITUMINOUS COAL.

**anthropogenic climax,** see CLIMAX COMMUNITY.

**anthropogenic factor,** *n.* the influence exerted upon a HABITAT or environment by humans. The anthropogenic factor covers all processes by which the natural or semi-natural vegetation of an area is converted to a managed landscape, and includes the use of FIRE, DEFORESTATION, pasturing, mowing, draining, IRRIGATION, and grazing.

The anthropogenic factor has become the most important or *master factor* in the creation of an organized landscape, and in many areas now supercedes the climatic, topographic, geological and biological factors that were formally responsible for defining the plant and animal characteristics of a region. Compare BIOTIC FACTOR.

**anticline,** *n.* an archlike, upwardly convex part of a FOLD in rock strata, often the result of great horizontal pressure exerted by tectonic movements in the earth's crust. Compare SYNCLINE.

**anticyclone** or **high,** *n.* a body of moving air of higher pressure than the surrounding air, in which the pressure decreases away from the centre. Anticyclones are formed by the convergence of air in the upper layers of the atmosphere and the subsequent subsidence and gradual divergence of this air at lower levels. Anticyclones are characterized by conditions of STABILITY: they are usually slow-moving or stationary systems associated with calm and settled weather. In the northern hemisphere, winds circulate in a clockwise direction around the centre of an anticyclone whilst in the southern hemisphere, the movement is anticlockwise. During the summer in temperate latitudes, periods of warm, dry weather are associated with the formation of temporary anticyclones whilst in winter, the weather is cold and clear or occasionally foggy. Precipitation is usually limited when highs develop due to the relatively low humidity of the air subsiding from the upper atmosphere.

Semi-permanent anticyclones exist over the oceans in the HORSE LATITUDES around 30°N and a seasonal anticyclone exists over Siberia during the winter.

On weather maps, anticyclones are represented by closed, widely spaced circular or oval ISOBARS.

**antiknock additive,** *n.* substances added to petroleum (gas) to minimize pre-ignition (*knocking* or *pinking*) in internal combustion engines. Tetraethyl lead has been the most widely used additive. Research has shown that most of the antiknock additives are released, unchanged, in AUTOMOBILE EMISSIONS. The level of lead in air, water and soil has been shown to be highly damaging to all life forms. In humans, especially the young, the old and the chronically sick, enhanced lead levels in the blood can produce brain damage. International action against antiknock additives has resulted in government legislation directed towards the ultimate eradication of lead in petrol. See also UNLEADED PETROL.

**antinuclear,** *adj.* denoting any person, organization, argument, etc. opposed to the use of NUCLEAR POWER both for civil and military purposes. In Europe, the antinuclear argument is most forceably made by CND. See also GREEN POLITICS.

**anvil cloud,** see CUMULONIMBUS.

**AONB,** see AREA OF OUTSTANDING NATURAL BEAUTY.

**appropriate technology,** *n.* any technology that maximizes the use of the factors of production (natural resources, capital and labour) that are locally plentiful and minimizes the use of those that are locally scarce. Appropriate technology should meet the test of fitness for the purpose by enabling projects to be managed on a limited and realistic availability of funds and technological expertise, and with control and self-reliance.

Appropriate technology for DEVELOPING COUNTRIES may be *intermediate technology* in being more expensive and efficient than many indigenous and traditional methods, yet smaller in scale, more labour-intensive, less dependent on imported materials and less sophisticated than technology transferred from DEVELOPED COUNTRIES in the course of normal commercial trading relations. However, not all intermediate technology is necessarily appropriate. To be competitive, many industrial processes, for example steel-making and sugar processing, depend on sophisticated machinery and capital intensive techniques. In developing countries the demand for appropriate technologies has come from the desire to make the most efficient and socially acceptable use of their resources, particularly labour, thereby increasing employment and production.

**aquaculture,** *n.* the management and use of water environments for the raising and harvesting of plant and animal food products. Whereas FISHING and WHALING remain essentially hunting activities, aquaculture represents a more intensive utilization of water resources more akin to RANCHING and AGRICULTURE in which fish and shellfish are reared in enclosed ponds, tanks and cages, or on protected beds.

Aquaculture is almost exclusively confined to inland waters and estuarine or other near-shore coastal waters. Developments further out on the CONTINENTAL SHELF remain at the experimental stage. The use of ponds to breed and raise fish such as carp dates back several thousand years in parts of China and Southeast Asia, but the extension of aquaculture to marine areas is a recent development. The raising of freshwater or marine fish for commercial purposes is often separately classed as FISH FARMING. Marine aquaculture enterprises have concentrated on raising shellfish, especially molluscs such as oysters, mussels and clams which are relatively immobile and command high market prices. With some crustacea, for example shrimps, and in salmon fish farming it is necessary to catch wild stock to raise to commercial standards in pens.

**aquiclude,** *n.* a layer of porous rock, such as shale, that absorbs water slowly but that will not allow its free passage. The term is most commonly used in North America and is sometimes incorrectly used as a synonym for AQUIFUGE. Compare AQUIFER.

**aquifer,** *n.* a layer of permeable rock, sand or gravel that absorbs water and allows it free passage through the interstices of the rock. When the underlying rock is impermeable, an aquifer acts as a GROUNDWATER reservoir which may be tapped by wells for domestic, agricultural or industrial use. Aquifer contamination is a serious environmental problem in many countries and can result from the seepage of sewage or toxic material from waste dumps. The overutilization of groundwater in coastal areas can allow incursion of salt water into the aquifer. See ARTESIAN WELL. Compare AQUICLUDE, AQUIFUGE.

**aquifuge,** *n.* a layer of impermeable rock that will neither absorb water nor allow its free passage. The term is most commonly used in North America. Compare AQUICLUDE and AQUIFER.

**arable farming,** *n.* the practice of cultivating land to produce CROPS. Arable land is usually cultivated annually or is at least fit for ploughing, and may be distinguished from PASTURES and land devoted to PERMANENT CROPS of trees and bushes. In MIXED CROPPING regimes, however, arable farming of annual crops may be undertaken between the rows of tree crops (see also AGROFORESTRY).

The geographic boundaries of arable farming are determined by locational deficiencies, specifically in terms of climatic conditions, soil types and relief. The margins of cultivation are reached at about 70°N where the cropping of spring barley and potatoes becomes limited by the shortness of the GROWING SEASON. In parts of the Andes, the potato is grown at heights of 4300 m, while in the SAHEL arable farming of millet survives on barely 250 mm of annual rainfall. In any farming system dependent at least partly on CASH CROPS a further limiting factor on the distribution of arable farming is the distance of the farm from market. See also TILLAGE, CROPLAND.

**arcade,** *n.* an arched or vaulted passageway within a city centre block, lined by small shops usually of a specialized nature and dealing in high value goods. The Burlington Arcade in London is a famous example.

**Archaeozoic era,** see PRECAMBRIAN ERA.

**archibenthic zone,** see BENTHIC ZONE.

**arcuate delta,** see DELTA.

**Area of Outstanding Natural Beauty (AONB),** *n.* any area of considerable scenic beauty and scientific importance selected for protection and enhancement by local planning authorities in the UK. AONB are so designated by the Countryside Commission under the NATIONAL PARKS AND ACCESS TO THE COUNTRYSIDE ACT (1949) and represent the second tier of conservation sites in the UK after NATIONAL PARKS. Originally not intended for use as public recreation areas their often considerable scenic value combined with their coastal and/or rural location has subjected them to intense visitor usage. Thirty-two AONB have been designated in England and Wales; the smallest is Dedham Vale (57 km²), and the largest, North Wessex Downs (1738 km²). The latter is larger than most British national parks. AONB have no strict legal status and are not protected by such comprehensive regulatory powers as the national parks; however, high standards of DEVELOPMENT CONTROL are applied so as to preserve and enhance the natural beauty of the landscape.

**Area of Special Control,** see ADVERTISEMENT CONTROL.

**arenaceous rock,** see SEDIMENTARY ROCK.

**arête, coombe ridge** or **grat,** *n.* a narrow, steep-sided and sharp-edged ridge lying between adjacent CIRQUES. Arêtes are formed by the headward extension of cirques through GLACIAL EROSION, rock falls and freeze-thaw action (see PHYSICAL WEATHERING). See Fig. 66.

**argillaceous rock,** see SEDIMENTARY ROCK.

**arid climate,** *n.* any climate which has a severe moisture deficiency due to:

(a) a rate of potential evaporation in excess of actual precipitation;

(b) very low precipitation levels brought about by a RAIN SHADOW or by being a great distance from an ocean.

Arid areas usually have only sparse vegetation and the potential for agriculture is limited unless IRRIGATION schemes can be developed. In the USA, an arid climate is defined as one which receives less than 250 mm of precipitation per annum. Examples of arid climates are found in Death Valley in California, USA and in many parts of North Africa and the interior of Australia. Depending upon the strict definition of arid climate, estimates of the area of the arid climatic zone vary from about 15% to 30% of the world's land surface. See SEMI-ARID CLIMATE, MEDITERRANEAN, SAVANNA. See also DUST BOWL.

**arid region,** *n.* see DESERT.

**artesian basin,** *n.* a basin-shaped area of land underlaid by strata of impermeable rock containing an AQUIFER. On the margins of the artesian basin, precipitation infiltrates the exposed aquifer bed and descends under gravity to form a GROUNDWATER reservoir. When wells are drilled, hydrostatic pressure forces the groundwater to the surface without the aid of pumping (see ARTESIAN WELL).

Water from artesian basins may be used for agricultural and industrial purposes, and is of great importance in arid and semi-arid areas. Such water frequently requires extensive treatment to remove chemical residues leached from the aquifer rocks before it is suitable for domestic consumption.

Extensive artesian basins are found in North America, the Sahara and under many areas of Australia.

**artesian well,** *n.* a perpendicular boring sunk into oblique strata to tap water-bearing rock (or AQUIFER) lying beneath a relatively impermeable stratum. If the WATER TABLE at the margins of the aquifer is higher than the outlet of the well, water will flow out of the boring under pressure. It may be necessary to cap the well in order to regulate supply for IRRIGATION purposes. Where the wellhead is above the water level in the aquifer the water has to be pumped to the surface. This is referred to as a *sub-artesian well* to distinguish it from the naturally flowing artesian well.

**arthropod,** *n.* any invertebrate of the ancient animal phylum, Artheopoda. The bodies of arthropods are segmented and strengthened with chitin, each segment bearing a pair of appendages, (antennae, feet or mandibles). The phylum includes the CRUSTACEANS, insects, arachnids, and centipedes. Arthropods comprise 75% of the earth's animal species and their immense evolutionary complexity and infinite diversity of form has allowed them to colonize every habitat on the planet.

The arthropod phylum contains species which are both of valued importance to humans (especially flies and bees which perform the essential task of cross-pollinating plants) and others which have brought extreme misery through disease and famine, such as the migratory locust which has caused incalculable damage to agricultural crops since ancient times.

**artificial fertilizer,** see FERTILIZER.

**asbestos,** *n.* a substance formerly in wide use in the building industry for thermal insulation, as sound deading material, for electrical insulation and prevention of electrical fires, and as roofing material. There are three main types:

(a) blue asbestos or *crocidolite* which is extremely toxic and is banned in many countries;

(b) white asbestos which can be spun or woven into thread or tape and which can be safely used;

(c) brown asbestos which has little usefulness as a resource.

Asbestos comprises fibrous magnesium silicate minerals, minute particles of which can be inhaled by persons who come in contact with the material. However, many years often elapse between exposure to asbestos fibres and evidence of respiratory system damage. Its toxicity was recognized as early as 1931, although it was not until the 1970s that it was positively identified as a CARCINOGEN. After many years of neglect asbestos legislation is now rigorously enforced. Most problems with asbestos now occur when old buildings are demolished or when ships, submarines or railway carriages are broken up.

**association,** *n.* a group of similar plants that grow in a uniform environment and contain one or more dominant species. A plant association usually comprises a stable COMMUNITY of definite floristic composition. The term is applied to the largest recognizable natural vegetation groups. See FORMATION.

| Constituent gas | Percentage of atmosphere by volume |
|---|---|
| Nitrogen | 78.08 |
| Oxygen | 20.95 |
| Argon | 0.93 |
| Carbon dioxide | 0.03 |
| Trace gases (including helium, hydrogen, xenon and methane) | 0.01 |

Fig. 8. **Atmosphere.** The composition of dry air.

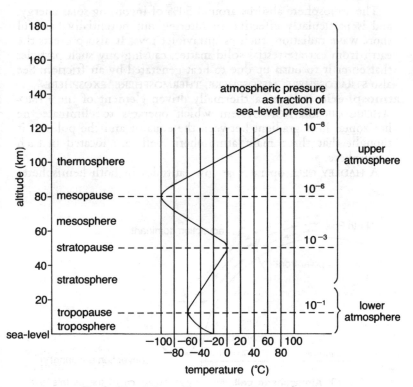

Fig. 9. **Atmosphere.** The zonation of the atmosphere.

# ASTHENOSPHERE

**asthenosphere,** *n.* the partially molten zone within the MANTLE, found at depths between 100 km and 250 km, and which permits the slow movement of LITHOSPHERIC PLATES. See PLATE TECTONICS.

**atmosphere,** *n.* a 500 km thick composite layer of colourless, odourless gases, known as air, which surrounds the earth and which is kept in place by gravitational forces. The proportion of most gases in the LOWER ATMOSPHERE is relatively stable (see Fig. 8 on p. 39) although the level of water vapour (which is not a gas but is usually grouped with them) can vary greatly.

The atmosphere shows distinct vertical zonation, each of which has a different temperature range (see Fig. 9 on p. 39). Most of the meteorological phenomena of concern to humans occur contained within the lowest zone, the TROPOSPHERE, which extends to an altitude of some 16 km at the equator and thins to 8 km towards the poles.

The atmosphere absorbs around 50% of incoming solar energy, and is particularly effective in filtering out potentially harmful short wave radiation, such as ultraviolet rays. It also protects the earth from extraterrestial solid matter, causing any such particles that enter it to burn up due to heat generated by air friction. See also STRATOSPHERE, MESOSPHERE, THERMOSPHERE, EXOSPHERE

**atmospheric cell,** *n.* a thermally driven element of the ATMO-SPHERIC CIRCULATION system which operates to eliminate the horizontal heat gradient between the equator and the poles. It is thought that three main atmospheric cells are located in each hemisphere.

A HADLEY CELL operates at low latitudes in both hemispheres

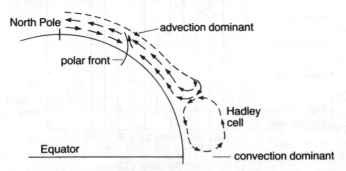

Fig. 10. **Atmospheric cell.** An idealized diagram of the earth's atmospheric cells.

whilst a high-latitude cell functions between the polar ANTI-
CYCLONE and the POLAR FRONT (see Fig. 10). There is an
equatorward low-level air flow within the high-latitude cell, with
a compensating high-level poleward flow. The mid-latitude cell is
less strongly developed and is probably largely maintained by the
Hadley and high-latitude cells. This mid-latitude cell is located
between the subtropical anticyclone belt and the polar front. Air
flow at low levels is poleward in the mid-latitude cell, with a
compensating high-level return flow. The presence of JET STREAMS
and the passage of a DEPRESSION or a migratory anticyclone can
modify the operation of atmospheric cells.

**atmospheric circulation,** *n.* the general planetary movement and
circulation of air through the atmosphere. Atmospheric circulation
is driven by the horizontal heat gradient between the equator and
the poles, and operates to correct the imbalance by the poleward
movement of heat. The circulation system is largely manifested in
several ATMOSPHERIC CELLS and various WIND BELTS.

**atmospheric pressure,** see AIR PRESSURE.

**atoll,** see CORAL REEF.

**atom,** *n.* an extremely small particle of MATTER that with other
atoms form the basic building blocks of all material. The atom is
the smallest part of a chemical element that can take part in a
chemical reaction. All atoms comprise a positively charged nucleus
surrounded by negatively charged electrons which circulate in a
prescribed orbit around the nucleus.

**atomic power,** see NUCLEAR POWER.

**ATP,** *abbrev. for* adenosine triphosphate, the energy storage and
transferring compound of most biological systems. ATP is a
phosphorylating agent capable of transfering a phosphate group to
another molecule, releasing energy that is utilized by the living cell
in which the transfer occurs. During this transfer, adenosine
diphosphate (ADP) forms and when a phosphate group is added to
ADP, ATP is regenerated. The importance of this chemical cycle
lies in the role that it plays in an organism's *metabolism*, that is, the
conversion of energy derived from sunlight and fuels (such as
carbohydrates) into energy utilized by the cell to undertake
chemical, osmotic and mechanical functions.

**attrition,** see FLUVIAL EROSION.

**aurora australis** or **southern lights,** *n.* a spectacular display of
coloured light in the skies of the southern hemisphere, most
commonly observed in latitudes higher than 70°S. The aurora
australis results from the ionization of air molecules in the

THERMOSPHERE by solar X-rays and ultraviolet radiation. Compare AURORA BOREALIS.

**aurora borealis** or **northern lights,** *n.* a spectacular display of coloured light most frequently observed in latitudes higher than 70°N during summer months. The phenomenon is due to the ionization of air molecules in the THERMOSPHERE by solar X-rays and ultraviolet radiation. Compare AURORA AUSTRALIS.

**autecology,** *n.* the study of the ECOLOGY of individual species. Compare SYNECOLOGY.

**autobahn,** see MOTORWAY.

**automobile emissions,** *n.* the collection of car exhaust gases comprising the $NO_X$ group (nitrogen oxides), carbon monoxide, lead, unburnt hydrocarbons, water vapour, carbon dioxide and aldehydes. The first three pollutants in this list can now be found in major concentration in most urban areas throughout the world. Traffic congestion can produce atmospheric carbon monoxide levels of up to 100 parts per million (ppm) compared to clean air concentrations of 0.1 ppm.

In many of the wealthiest world cities, such as Los Angeles, automobile emissions, rubber from car tyres and tarmacadam particles from the road surfaces combine in the presence of high sunshine levels to produce a highly toxic form of pollutant mist called PHOTOCHEMICAL SMOG.

The problem of automobile exhaust emissions has been exacerbated by the trend towards the highly stressed, high compression petrol engine (10:1 compression ratio) much favoured by European motor manufactures in their search for improved performance. Petroleum additives such as ANTIKNOCK ADDITIVES necessary for the efficent operation of the petrol engine are especially hazardous when released to the environment.

Most industrial nations have now applied legislation to reduce the level of automobile emissions. UNLEADED PETROL, exhaust gas cleaners (after-burners or CATALYTIC CONVERTORS) and LEAN-BURN ENGINES are the main methods so far used to combat high pollution emission levels. California has some of the most stringent legislation against exhaust emissions, and carbon monoxide has been reduced by 87%, hydrocarbons by 95% and $NO_X$ by 75% since the introduction the AIR QUALITY ACT (1967). Sweden, West Germany and Canada have also strictly enforced automobile emission standards.

**autostrada,** see MOTORWAY.

**autotroph,** *n.* any self-feeding, organic green cell or plant capable of combining energy from the sun with simple inorganic salts from the soil to form complex, manufactured sugars via the process of PHOTOSYNTHESIS. Autotrophs form the first TROPHIC LEVEL in a FOOD CHAIN, and as such are of fundamental importance to all HETEROTROPHS.

**avalanche,** *n.* the rapid downslope movement of snow and ice in steep mountain areas. The influence of gravity on the accumulated weight of newly fallen uncompacted snow or on thawing older snow leads to avalanches which may be triggered in a variety of ways including earthquakes, gun shots and the movements of animals or skiers. Avalanches are most common during winter or spring but glacier movements may cause ice avalanches during summer. Avalanches may cause considerable loss of life and can destroy settlements, roads, railways and forests. In many areas, regular avalanche tracks can be identified and precautions taken to minimize damage, such as the prevention of development in these areas, the construction of avalanche sheds over existing roads and railways and the use of tunnels for new road and rail links. Avalanches present the greatest threat in areas where their path cannot be predicted and may cause a major hazard for skiers and mountaineers. The term is sometimes used for the movement of rock debris on steep mountain slopes.

**azonal soil,** *n.* any IMMATURE SOIL displaying little, if any, differentiation between SOIL HORIZONS. This results from the processes of soil formation having insufficient time to develop the characteristics found in ZONAL SOILS. Typically, azonal soils lack a defined B-HORIZON with the A-HORIZON lying directly above the parent material of the C-HORIZON. See also INTRAZONAL SOIL.

# B

**background concentration,** *n.* **1.** the general level of AIR POLLUTION in a region with all local sources of pollution ignored. For example, the background pollution levels in an area in which one coal-fired thermal power station existed would comprise the total pollution levels for the area minus the pollution produced by the power station.
**2.** the general level of background radioactivity in an area. This value usually comprises the radioactive emissions from COUNTRY ROCK plus short-wave solar radiation.

**backwash,** *n.* the mass of water receding seaward under the influence of gravity after a WAVE has broken on a beach. The backwash is usually weaker than the SWASH, as much of the water infiltrates permeable beach sand, but can still sweep sediment seaward.

**bacteria,** *n.* a large group of diverse, unicellular microorganisms which exist singly, in chains or in clusters. They form the smallest of the living organisms (1-10 microns in size) and together with FUNGI, form the DECOMPOSER group of organisms. They can be subdivided into the non-photosynthetic filamentous gliding forms, such as Myxobacteria, and the true bacteria, Aubacteria. The former group are common in decaying plant materials and are the active decomposers. The Aubacteria contain bacteriochlorophylls and carry out photosynthesis anaerobically.

Bacteria occur in soil, water and air, as symbiants, parasites or pathogens of humans, animals, plants and other microorganisms. Saprophytic species are important in the major biospheric movements of matter (for example, the nitrogen and sulphur cycles). Certain species form symbiotic relationships with higher plants (see ROOT NODULES). While some bacteria are vital to the continuation of human life, other forms are responsible for highly dangerous human diseases such as anthrax, tetanus and tuberculosis.

**badland,** *n.* **1.** any area of rugged topography dissected by large numbers of gullies and ravines. Badlands result from severe SOIL EROSION caused by the action of heavy rainstorms on sparsely vegetated slopes underlaid by impermeable rock. The removal of surface vegetation through overgrazing may sometimes lead to the development of badlands. Usually found in areas of low, or unreliable rainfall, badlands are of low agricultural value and are best used for extensive rangeland grazing.

**2. Badlands.** a deeply eroded barren region covering parts of South Dakota and Nebraska in the USA

**bahada,** see ALLUVIAL FAN.

**bar-built estuary,** see ESTUARY.

**barchan,** see DUNE.

**barometer,** *n.* an instrument which measures AIR PRESSURE. The principle of operation is best illustrated by a mercury barometer. At sea level, a standard air pressure of 1 atmosphere (101 kPa) is assumed to exist; 1 atmosphere can support a column of mercury 760 mm high, which is equivalent to standard sea level air pressure of 1013.2 millibars. Variation in air pressure is reflected in a change in height of the column of mercury.

Aneroid barometers have now largely replaced mercury barometers. In this type, a vacuum is enclosed within a small metal box to which is attached a spring that contracts and expands according to changes in atmospheric pressure and drives a pointer around a scale marked in millibars.

**barometric gradient,** see PRESSURE GRADIENT.

**barrier beach,** see OFFSHORE BAR.

**barrier island,** see OFFSHORE BAR.

**barrier reef,** see CORAL REEF.

**basalt,** *n.* a dark-coloured, fine-grained basic EXTRUSIVE ROCK formed by the solidification of LAVA. Basaltic lavas are usually extruded from fissures and occasionally from volcanoes. Basalt is the commonest volcanic rock and is often characterized by hexagonal jointing such as at Fingal's Cave on the island of Staffa in the Scottish Hebrides and the Giant's Causeway in Northern Ireland. Extensive lava flows may innundate surrounding land-scapes and result in the formation of *flood basalts*, such as the Deccan Plateau in India and the Columbia-Snake River Plateau in the USA. Basalt is often highly resistant to erosion and hence, is frequently used for construction purposes.

**base level,** *n.* the lowest theoretical level to which a river can erode its bed, which is usually considered to be SEA LEVEL. Local, temporary base levels may occur along a stream's course as it encounters LAKES, WATERFALLS and KNICKPOINTS.

**base status,** *n.* the quantity of positively charged ions (*cations*) available in a soil, the most important of which are the cations of calcium, magnesium,potassium and sodium. The bases are held in the soil by means of a weak electrical bond. The bases are positively charged and the *clay humus complex* is negatively charged. The strength by which these cations are held by the colloidal surface varies and a descending order of tenacity is as follows:

hydrogen > calcium > magnesium > potassium >
nitrogen > sodium

**basic rock,** *n.* any IGNEOUS ROCK of relatively low silica content compared to iron and magnesium minerals such as BASALT. See also ULTRABASIC ROCK. Compare ACID ROCK.

**basin bog,** see BOG.

**batholith,** *n.* a large, dome-shaped mass of INTRUSIVE ROCK (particularly granite) formed by either EMPLACEMENT or GRANITIZATION at great depth within the earth's crust. Batholiths may be hundreds of thousands of square kilometres in extent and may be found at depths of up to 30 km. Always associated with OROGENESIS, batholiths often form the cores of major mountain ranges, as for example the Alps. The formation of a batholith results in the thermal METAMORPHISM of surrounding COUNTRY ROCKS. After long periods of erosion, batholiths may be exposed to the surface to form extensive plateaux such as Bodmin Moor in England and the Idaho batholith in the USA.

**bauxite,** *n.* a clay mineral consisting largely of aluminium oxides and hydroxides. It is usually produced by the CHEMICAL WEATHERING of igneous rocks in tropical areas. Bauxite is the chief ore of aluminium and forms an important export in the economies of tropical countries. Australia is the largest producer of bauxite (24 500 tonnes in 1983), followed by Guinea (11 080 tonnes in 1983).

**beach,** *n.* an area of gently sloping land located between the low tide watermark and the highest level reached by storm waves. Beaches are an accumulation of various types and sizes of material including shingle, sand, mud and fragmented shell. This material is deposited by wave and current action and originates from several sources, such as the erosion of cliffs, sediment from rivers entering the sea and debris recycled from other beaches. A beach profile is generally concave with the coarsest material at the top of the beach. The balance of erosion and deposition on beaches is constantly changing.

Beaches, especially sandy ones, are of considerable human importance. The exploitation of sands and gravels in coastal areas provides the construction industry with valuable raw materials. Beaches are also a major tourist and recreational resource. Tourist developments with a strong emphasis on sunbathing and water-sports are found throughout the world. The pollution of beaches, from the discharge of raw domestic sewage, oil pollution and industrial effluent not only represents an ecological threat but can also threaten the economies of areas orientated towards tourism (see MARINE POLLUTION).

**Beaufort scale,** *n.* an internationally agreed scale of wind force which has 13 standardized categories and associated descriptions (see Fig. 11 overleaf). The scale was originally devised for use at sea by Admiral Sir Francis Beaufort in 1805 but has subsequently been modified for use over land. The Beaufort Scale terminology is commonly used to indicate wind strength in weather forecasting for shipping.

**becquerel,** *n.* the derived SI unit of radioactivity equal to the number of atoms of a radioactive substance that disintegrate per second. Symbol: Bq. The becquerel is much smaller than the previous standard unit, the CURIE and so can be used to measure much smaller doses of radioactivity (1 Bq = $3.7027 \times 10^{-10}$ Ci).

**bedding plane,** *n.* the dividing layer demarcating the boundaries of each STRATUM in a SEDIMENTARY ROCK.

**bedload, bottom load** or **traction load** *n.* solid particles carried by a stream or river along or close to its bed. These particles (mostly gravel, pebbles and boulders) are too coarse to be carried in suspension and therefore are moved along by pushing rolling or bouncing. See FLUVIAL TRANSPORTATION. See also TURBIDITY CURRENT.

**bedrock,** *n.* the unweathered rock underlying the REGOLITH. The depth of bedrock is largely dependent upon the composition of the overlying regolith and the nature and scale of WEATHERING and EROSION.

**before present (BP),** *n.* a timescale devised by geologists, archeologists and palaeoecologists to allow unambiguous dating of events from the PLEISTOCENE EPOCH to the present. The year chosen to represent the beginning of the present has been set at 1950. The term is most frequently used in palaeo-environmental reconstruction studies, for example, POLLEN ANALYSIS, glacial geomorphology and palaeosoil studies.

**beheading,** see RIVER CAPTURE.

# BELT TRANSECT

| Beaufort Force | Name | Average wind speed (km/hour) | Observed effects over land |
|---|---|---|---|
| 0 | Calm | less than 1 | Smoke rises vertically |
| 1 | Light Air | 1–5 | Wind direction shown by smoke but not by wind vanes |
| 2 | Light Breeze | 6–11 | Wind felt on face; leaves rustle; wind vane moves |
| 3 | Gentle Breeze | 12–19 | Leaves and small twigs in constant motion; a flag is extended |
| 4 | Moderate Breeze | 20–29 | Dust raised; small branches moved |
| 5 | Fresh Breeze | 30–39 | Small trees sway; crested waves on lakes |
| 6 | Strong Breeze | 40–50 | Large branches in motion; wind whistles in telephone wires |
| 7 | Moderate Gale | 51–61 | Whole trees in motion |
| 8 | Fresh Gale | 62–74 | Twigs break off trees |
| 9 | Strong Gale | 75–87 | Slight structural damage to buildings; chimney pots and slates removed |
| 10 | Whole Gale | 88–101 | Trees uprooted; considerable structural damage to buildings |
| 11 | Storm | 102–117 | Widespread damage |
| 12 | Hurricane | Above 119 | Devastation |

Fig. 11. **Beaufort scale.** The Beaufort scale modified for use on land.

**belt transect,** see TRANSECT.
**benefit-cost analysis,** see COST-BENEFIT ANALYSIS.
**benthic zone,** *n.* the sea bottom. The benthic zone can be divided into three areas:

(a) the LITTORAL zone, which comprises the shore area between the limits of high- and low-tide.

(b) the *sublittoral zone*, which extends from the low-tide level to the edge of the CONTINENTAL SHELF.

(c) the *deep-sea zone*, which is divisible into the *archibenthic zone*

48

extending from the lower edge of the *continental shelf* to a depth of about 1000 m and the *abyssal benthic zone*, extending beyond 1000 m.

Because of the great underwater pressure and the salinity of the oceans, humans have so far made little use of the benthic zone apart from deep-sea fishing (see BENTHOS). Up to 75% of all light is absorbed in the top 10 m of the ocean, leaving a vast proportion of the benthic zone dark and cold (see EUPHOTIC ZONE). Even at the equator the temperature of the archibenthic zone does not exceed 4°C.

The benthic zone has been used as a *sink area* or dumping ground for some of the most dangerous pollutants, for example, radioactive materials and SEWAGE with a high HEAVY METAL content. Pipelines and telephone cables are also routed along the benthic zone.

**benthos,** *n.* the plant and animal community of the sea bottom. The benthos is a highly varied collection of plant and animal forms, most of which are found in relatively shallow seas less than 200 m deep. The type of life found in the benthos is dependent upon the nature of the BENTHIC ZONE, for example, the proportion of sand and rock, the amount of light, and the temperature of the water. With the aid of extreme specialization, organisms have been able to colonize even the most inhospitable regions of the benthic zone. The absence of light and the low water temperatures are compensated for by the total uniformity of the environmental conditions in this zone. In shallow water can be found the seaweeds (colonial ALGAE), along with molluscs (oysters and mussels), CRUSTACEANS (crabs and lobsters), worms, starfish and bottom-dwelling fish. The latter species, the DEMERSAL FISH such as plaice and cod, form important human food resources. However, apart from fishing, humans have made little attempt to harness the productivity of the benthos.

**best practical means,** *n.* the compromise situation whereby industrial premises are allowed to emit higher than normally acceptable pollution levels due to exceptional local circumstances. The circumstances may include equipment which can no longer meet new pollution standards but which in itself is not life-expired, or a factory which is the sole source of employment in an area of already high unemployment and which would be forced to close or make expensive alterations if it was forced to meet new anti-pollution measures.

The term appears to have been first used by the ALKALI INSPECTORATE in 1906 and has been the cause of much confusion ever since. There is no absolute definition for 'best practical means'; it relies instead on local conditions and the interpretation of the term by the local health inspectorate. As a result, emission standards for similar industries in different regions of the same country can show a wide range of permitted values. Improved methods of monitoring pollution emissions combined with the more stringent application of pollution legislation have enabled a reduction in the occasions when 'best practical means' is allowed to operate.

**beta emitters,** see BETA PARTICLE.

**beta particle,** *n.* a high-energy electron that originates as a result of the decay process in radioactive ISOTOPES known as *beta emitters*, for example, cobalt 60. Beta radiation can penetrate 3-4 cm of air and to a depth of 1 cm in biological tissue causing the malfunction of cells through the formation of tumours or from the failure of DNA to replicate itself. As such it is highly hazardous to humans.

**B-horizon,** *n.* the layer located between the A- and C-HORIZONS of a well developed mineral soil (see Fig. 55). The B-horizon is the zone of ILLUVIATION where minerals and clay particles washed down from the A-HORIZON accumulate. It contains less organic matter than the horizon above and is also characterized by less biological activity. In agriculture, the B-horizon is sometimes referred to as the *subsoil*.

**biennial,** *n.* a plant which extends its life cycle over two years. See CROP.

**biochore,** *n.* a climatic boundary marked by a change in vegetation type. The term was first used in 1916 to define the climatological boundaries of major plant groupings. Four major series were defined:

(a) the *phanerophyte climate* in the tropical zone where precipitation is not deficient,

(b) the *therophyte climate* in regions of the sub-tropics and characterized by winter rains,

(c) the *hemicryptophyte climate* which extends over the cold temperate zone.

(d) the *chamaeophyte climate* of the cold zone.

Confusion over the meaning of the term has arisen through its (incorrect) application by some researchers to describe four major vegetation FORMATIONS, namely forest, savanna, grassland and desert.

**bioconversion,** *n.* the conversion of organic waste products into an energy resource through the action of microorganisms, usually

BACTERIA, for example, the decomposition of organic plant residues to produce METHANE gas. Chemically, this is the reduction of complex organic compounds into simpler, more stable forms. See BIOGAS, BIOGAS DIGESTER, BIODEGRADABLE.

**biodegradable,** *adj.* (of waste materials) capable of being decomposed by BACTERIA or other biological means. Use of the term commonly implies that the residues of decomposition will be non-toxic and will not accumulate in FOOD CHAINS. Most organic wastes (paper, woollen garments, leather, wood) are biodegradable whereas most plastics are, at present, non-biodegradable.

**bioecology,** *n.* the study of the relationships between plants and animals and the environments in which they occur. Bioecology differs from ECOLOGY in that greater attention is given to the impact of humans upon ECOSYSTEMS and species. See also ANTHROPOGENIC FACTOR

**biofertilizers,** *n.* any naturally occurring organic substance applied to the soil for the purpose of maintaining or improving fertility. They include animal MANURES, NIGHT SOIL, leguminous crops and compost. See FERTILIZER.

**biogas,** *n.* a METHANE-rich gas produced from the fermentation of animal dung, human excreta or crop residues in an air-tight container. The gas can be used to heat stoves, light lamps, run small machines and generate electricity. Biogas fuels are usually pollution-free and are derived from RENEWABLE RESOURCES. As such, their future potential utility to mankind may far exceed their present day value.

The greatest application of biogas energy has been made by developing nations. For example, in China up to 40% of a commune's electricity can be produced from a methane gas

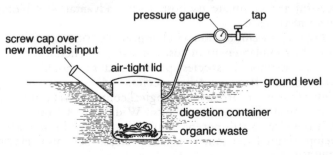

Fig. 12 **Biogas digester.**

converter. The aim is to increase this figure to 80% by the deliberate cultivation of fast-growing plants such as water hyacinth and Napier grass. Other plants suitable for biogas production include eucalyptus trees, sugar cane, sorghum, manioc and radiata pine. The residue left after biogas production can be used as a low-grade organic fertilizer.

**biogas digester,** *n.* an airtight vessel in which ANAEROBIC DIGESTION of organic waste material takes place and from which METHANE may be drawn off (see Fig. 12 on p. 51). See BIOGAS.

**biogeochemical cycle,** see NUTRIENT CYCLE.

**biogeography,** *n.* that part of geography concerned with the study of the inhabited areas of the BIOSPHERE. It includes the study of plants, animals, soils, water and climate along with the inter-relationships of these components with mankind. Biogeographical studies embrace both geographical and biological sciences, the former because of its concern with distribution patterns, and the latter through its involvement with living organisms. See ZOOGEO-GRAPHY.

**biological benchmark,** *n.* a plant or animal species used to indicate either:

(a) the general level of pollution in an area, or

(b) the presence of a specific type of pollutant in the environment.

LICHENS and mosses have been widely used to indicate the presence of $SO_2$ while Western Red Cedar (*Thuja plicata*), a North American conifer tree, has proved an accurate indicator of general AIR POLLUTION levels.

**biological control,** *n.* the control of pests and weeds by natural predators, parasites, disease-carrying bacteria or viruses. Unlike chemical control, once established, biological control should become self-perpetuating providing that it does not become too successful and eliminate its host. Other advantages of biological control methods are:

(a) the non-accumulation of chemicals in FOOD CHAINS,

(b) a low implementation cost compared to chemical control,

(c) the method is species-specific, and so should not create unwanted side effects on other species.

One of the most successful biological controls has been the use of the myxomatosis virus on rabbits. Widely used from the early 1960s the virus is still potent despite the ability of some rabbits to develop limited resistance to the virus. See also PESTICIDE, HERBICIDE.

**biological monitoring,** *n.* the comparison of species number, reproductive capacity and BIOMASS following an environmental change usually brought about as a result of human activity. Ideally, biological monitoring should take place both prior to and following any environmental change so that the full magnitude of the changed circumstances can be identified. Biological monitoring is a much used technique in ENVIRONMENTAL IMPACT ASSESSMENT.

**biological oxygen demand (BOD),** *n.* the amount of dissolved oxygen consumed by micoorganisms as they decompose organic material in polluted water. Measurement of the rate of oxygen take-up is used as a standard test to detect the polluting capacity of effluent; the greater the BOD value (and hence the greater presence of oxygen-consuming microorganisms) the greater the volume of pollutant present. The test measures the mass (in grams) of dissolved oxygen consumed per litre of water when a sample is incubated in a dark chamber at 25°C for five days. Examples of BOD values expressed in $g/m^3$ are as follows:

| | |
|---|---|
| domestic sewage | 350 |
| brewing | 550 |
| distilling | 7000 |
| paper pulp mill | 25000 |

**biological weathering** or **organic weathering,** *n.* the breakdown of rocks and their constituent minerals by the action of plants and animals. The physical element of biological weathering involves the wedging apart of rocks by the growth of tree roots and the activity of burrowing animals such as rabbits, worms and ants, and in bringing soil and rock to the surface where it can be weathered by other processes. The chemical element entails the production of HUMIC ACIDS from decaying organic matter which assists the CHEMICAL WEATHERING of rock, and the combination of carbon dioxide from plant respiration and soil water to cause CARBONATION. Biological weathering also incorporates the disintegration of rock due to human activities such as the construction of roads and railways, mining and quarrying.

**biomass,** *n.* the quantity of living plant and animal material in a given area. Biomass values are usually quoted as oven dry weight in tonnes per hectare or kilograms per square metre, although it can also be calculated in terms of the amount of carbon or fresh (green) weight, or as units of heat. Biomass values represent the amount of organic material which has accumulated within an ecosystem. In general terms, higher biomass values are associated with more favourable environmental conditions; for example, warm, damp environments have higher biomass values than cool and/or dry environments which result in a period of inactivity (see Fig. 13).

| Ecosystem | Biomass value (dry weight in kg/m$^2$) |
|---|---|
| Extreme desert, rock and ice | 0.02 |
| Tundra and alpine | 0.6 |
| Desert scrub | 0.7 |
| Algae and estuaries | 1.0 |
| Agricultural land | 1.0 |
| Temperate grassland | 1.5 |
| Tropical savanna | 4.0 |
| Woodland and shrub | 6.0 |
| Boreal forest | 20.0 |
| Temperate forest | 30.0 |
| Tropical forest | 45.0 |

Fig. 13. **Biomass**. Examples of plant biomass for the major world ecosystems.

Great care should be taken when comparing biomass values obtained from different sources as methodological variability can cause considerable variation. The biomass value should ideally include both plant and animal material for both above and below ground locations. However, because of the difficulty associated with catching animals and the problems inherent in collecting the underground biomass, most published biomass values refer only to above ground plant biomass.

The calculation of biomass values demands destructive sampling of the sample. As such, it is usual to select small sample areas for assessment of biomass and to apply a 'scaling-up factor' to these figures. For this reason, biomass values are only approximations, and it is customary for a low and a high biomass estimate to be given for each ecosystem.

**biome,** *n.* any major ecological community of organisms, both plant and animal, usually characterized by the dominant vegetation type, for example TUNDRA biomes, TROPICAL RAIN FOREST biomes, etc. Biomes are defined in terms of the entire community of living oganisms and of their inter-relationships with their immediate environment (and not only with the botanical habitat). Biomes extend over large areas and broadly correspond with climatic regions; characteristic biomes have been identified for all the major climatic regions, emphasising the ability of living organisms to adapt to a wide variety of environments.

**biosphere,** *n.* that part of the earth's surface and its immediate atmosphere that is inhabited by living organisms. The biosphere fulfills three primary functions for plants and animals:

(a) it provides a safe HABITAT within which an individual organism can complete its life cycle;

(b) it provides a stable habitat within which the evolution of species can occur;

(c) it forms a self-regenerating system in which energy is provided by the sun and the materials essential for life are recycled from within the system.

The biosphere represents a complex series of inter-relationships between the soil, rock, water and air and the living organisms contained therein. Within the biosphere can be found myriads of different ECOSYSTEMS. Each ecosystem inter-relates with its neighbour; a change in one ecosystem creates a ripple effect bringing change to adjacent systems. See also GAIA CONCEPT.

**biosphere reserve,** *n.* any terrestrial or coastal environment that has been internationally recognized as an area for CONSERVATION, study and sustained development (as distinct from exploitative development). Biosphere reserves form an international network of protected areas approved by the International Coordinating Council of UNESCO's MAN AND THE BIOSPHERE PROGRAMME. In conjunction with the CONVENTION ON INTERNATIONAL TRADE IN ENDANGERED SPECIES, a total of 194 different biogeographical provinces have been identified and for each of which at least one biosphere reserve is considered necessary. By December 1985, 243 biosphere reserves had been established in 65 different countries and covering 100 of the biogeographical provinces. Marine provinces, however, are poorly represented.

Each reserve must contain an ECOSYSTEM that is typical of a biogeographical realm in terms of its naturalness, diversity and effectiveness as a conservation unit. Each reserve must exhibit minimal disturbance. Within each reserve at least one *core area* must exist within which no interference with the natural ecosystems is permitted. Surrounding the core area there is a *transition zone* within which experimental research is permitted and beyond this lies a *buffer zone* which protects the whole biosphere reserve from agricultural, industrial, and urban land use pressures. Some NATIONAL PARKS have been designed in accordance with biosphere reserve plans, for example, le Parc National des Ecrins in southeast France.

**biota,** see ECOLOGY.

**biotechnology,** *n.* **1.** the application of scientific and technical knowledge to the provision of long term, sustainable solutions to BIOSPHERE problems. For example, the oil crises of 1973 and 1979 have stimulated much research into alternative, sustainable energy sources such as methane gas convertors and the alkaline battery. Advances in biotechnology have relied upon the growing understanding of natural ecosystems. For example, river basin planning has benefited as a result of research into the effects of deforestation and the reafforestation of river catchment areas; flood control has similarly benefited. Other areas in which biotechnical solutions have been used include the bacterial clean-up of oil spills in transport depots and from a bacteria which 'eats' rust deposits on metals.

**2.** the industrial use of living microorganisms such as bacteria and other biological agents to perform chemical processing, as of waste and water, or to produce other materials such as animal feedstuffs. See also GENETIC ENGINEERING.

**biotic,** *adj.* (of an ecosystem) of or relating to living organisms. The biotic elements of an ecosystem consist of its plants and animals (see Fig. 30). Compare ABIOTIC.

**biotic climax,** see PLAGIOCLIMAX.

**biotic factor,** *n.* the influence exerted upon a HABITAT by the plant and animal organisms which inhabit an area. Biotic influences include grazing, trampling, manuring, predation, parasitism, migration and the territorial behaviour patterns of animals.

The role of humans as a biotic factor may vary. Originally, human influence was confined to the hunting and gathering of foodstuffs. As such, the human impact on habitats was nontechnological and could be included in the total collection of biotic influences. Gradually, as humans discovered FIRE, DEFORESTATION, ploughing, the DOMESTICATION OF PLANTS AND ANIMALS, and building skills, they became a unique biotic factor known as the ANTHROPOGENIC FACTOR.

**biotype,** *n.* any plant which has an identical genetic structure to its parent stock. Such plants can only be obtained by means of vegetative reproduction. In natural vegetation, this is achieved through the spread of bulbs, corms, tubers, rhizomes, suckers, runners and adventitious roots. Plant populations of the same biotype can be particularly successful in stable environments such as aquatic environments. Many weeds found on agricultural land resort to vegetative reproduction, for example, couch grass (*Agropyron repens*) and bracken (*Pteridium aquilinum*).

Vegetative reproduction produces a population with a constant phenotype. In a changing environment, plants which show this characteristic are disadvantaged in that they lack the capacity for genetic variability to allow for adaptation to constantly changing conditions. See CLONE.

**bird's foot delta,** see DELTA.

**birth rate,** *n.* the ratio of live births in a specified area, group, etc. to the population of that area, etc. It is customary to express the *crude birth rate* as the number of births per 1000 population per year. The crude rate does not take into account the age and sex distribution of the population. It is the easiest birth rate statistic to calculate but usually the crude birth rate is in excess of the statistically more accurate *standardized birth rate* or *fertility rate*, the number of births per 1000 females of reproductive age. For humans, this range is taken to be 15-44 years of age. The standardized birth rate must be distinguished from *fecundity* which is the potential level of births over time for a population. Thus, while the fertility rate for a human population may be one birth per seven years per female of child-bearing age, the fecundity rate for humans is one birth per 9-11 months per female of child-bearing age. It is permissible to make comparisons between standardized birth rates (for the same species) whereas comparison of crude birth rates is often meaningless due to the variation between population groups. Compare DEATH RATE.

**bituminous coal,** *n.* a medium-grade, soft, shiny black COAL containing around 80% carbon and 20% oxygen. Bituminous coal gives off more heat and less smoke than LIGNITE. Bituminous coal is the commonest form of coal and is used for a variety of purposes including the generation of electricity, the production of coal and for domestic heating. It is sometimes called *steam coal* due to predominate use for raising steam in boilers. Compare ANTHRACITE.

**black earth,** see CHERNOZEM.

**black ice,** see GLAZED FROST.

**blanket bog,** see BOG.

**blizzard,** *n.* a severe snow storm characterized by poor visibility. Blizzards usually occur in high-latitude and mountainous regions but occasionally develop in mid-latitude areas.

**Blueprint for Survival, A** *n.* a radical review of human environmental problems published in the *Ecologist* magazine in January, 1972. The *Blueprint* examined the changes that would be necessary to create a society based primarily upon stability and re-use of materials instead of growth and development. The document was supported by 34 distinguished scientists who charged governments throughout the developed world with refusing to face the relevant facts about environmental deterioration and with briefing their scientists in such a way that the seriousness of the situation could be played down.

The starkness of the arguments contained in the *Blueprint* stimulated major debate amongst scientists and politicians. Although there were many critics of the *Blueprint*, it undoubtedly raised the awareness of politicians, economists and industrialists to the need for greater ENVIRONMENTAL MANAGEMENT. In retrospect, *A Blueprint for Survival* can be criticized for its unsubstantiated arguments and its sometimes naive statements. However, the document also contributed to a major change in social attitudes towards the bioshere, the environment and conservation. As such, it deserves a prominent place in the history of environmental science.

**BOD,** see BIOLOGICAL OXYGEN DEMAND.

**bog,** *n.* an area of undrained land which supports wet, spongy vegetation consisting mainly of mosses, sedges, rushes and some grasses. Bogs have many specialized names, for example, *peat bog*, *muskeg*, *mire*, FEN, or SWAMP. Bogs usually result from the occurrence of special physical conditions. For example, *basin bogs* can form in surface depressions in which water accumulates, while *blanket bogs* can form in areas of very high rainfall and low evaporation. All bogs undergo definite stages of evolution. In the early stages extensive open pools alternate with infrequent 'islands' of vegetation. In later stages the vegetation extends to form an extensive *hummock surface* separated by small pools. The final stage is attained when the poorly decomposed vegetation accumulates to a considerable thickness (often in excess of 10 m) and the surface is raised above the WATER TABLE level. The bog surface becomes drier and new species, such as heather (*Calluna vulgaris*) and Erica species replace the sphagnum mosses which predominate in the earlier phases. Trees, such as birch, willow and some pine may also appear.

Bogs are most extensive in maritime regions in latitudes beyond 50°N and S of the equator where high rainfall and winter cold prevents the decomposition of organic material. High-latitude and high-elevation bogs are usually acidic (pH 3.5-5.0) and OLIGO-TROPHIC (nutrient-poor). Lowland, estuarine bogs can occur at both high and low latitudes; they are often alkaline (pH 7.0) and are EUTROPHIC (nutrient-rich). MANGROVE swamps are a special form of low-latitude bogs.

In northern Europe there has been a long history of bogs being drained and brought into agricultural use. In Ireland and parts of Scotland peat bogs have been used as important sources of domestic fuel. In Ireland, peat bogs provide the fuel source for electricity generating stations, and unsuccessful attempts were made in the 1950s to develop peat-driven railway locomotives. Furthermore, bogs have long been unrecognized and underused WILDERNESS AREAS.

**bog-burst,** *n.* the rupture of a BOG and subsequent release of water and black organic matter, often over a wide area. Bogs become swollen because of increased water retention brought about by fallen vegetation; heavy rainfall exacerbates this situation until over-saturation causes the margins of the bog to collapse and spill its contents. See MUDFLOW.

**boiling water reactor,** see WATER-COOLED REACTOR.

**boreal forest** or **taiga,** *n.* a major vegetation FORMATION which extends in a continuous belt between latitudes 50°-70° in the northern hemisphere. The most extensive area of taiga extends from Scandinavia across the northern extent of the USSR. The boreal forest of North America is less extensive yet contains a wider species variety than its Soviet counterpart.

In both continents, the forest is dominated by the severity of winter. For at least six months of the year, mean temperatures fall below 0°C. Snowfall is heavy and long-lasting while the soils are severely affected by PERMAFROST. The growing season is short (50-100 days p.a.) although conditions in the long daylight hours of summer can be favourable for plant growth.

The characteristic vegetation form is the coniferous tree. Four species dominate the flora: spruce (*Picea* species); pine (*Pinus* species); larch (*Larix* species) and fir (*Abies* species). In addition, small numbers of HARDWOODS, mainly alder (*Alnus* species), birch (*Betula* species) and willow (*Salix* species) can be found growing in small numbers alongside the conifers. The hardwoods occur most frequently wherever the boreal forest has been disturbed by burning or more commonly, by felling.

The boreal forest also forms an important SOFTWOOD timber resource (see Fig. 14).

Apart from its timber, boreal forests are important resources for hunting and trapping (beaver, sable, mink, fox), for industrial resources (coal, natural gas, oil, metaliferrous and non-metaliferrous ores) and provide suitable locations for major hydro-electric power stations. Because of the extent of boreal forests they also play an important role in stabilizing the $O/CO_2$ and water vapour balance of the atmosphere.

Because of their immense economic value, boreal forests have become highly managed vegetation units. Major replanting schemes are rigorously followed in order that this natural resource will be able to yield timber well into the next century.

| Country | Extent of Forest and Woodland 1980 ('000 ha) | | Reforestation 1980s ('000 ha p.a.) | Managed Closed Forest ('000 ha) | Protected Forest Areas 1980 ('000 ha) |
|---|---|---|---|---|---|
| | Open | Closed | | | |
| USSR | 137000 | 791600 | 4540 | 791600 | 20000 |
| Finland | 3340 | 19885 | 158 | 10578 | 294 |
| Sweden | 3442 | 24400 | 207 | 14301 | 230 |
| Norway | 1066 | 7635 | 79 | 1130 | 60 |
| Canada | 172300 | 264100 | 720 | n.a. | 4870 |
| USA* | 102820 | 195356 | 1775 | 102362 | 31198 |

* includes data for all forest types.

Fig. 14 **Boreal forest.** Boreal forest resources.

**bottom load,** see BEDLOAD.

**bottom-up approach,** *n.* a strategy of development based on the notion of greater territorial integration, that is, the use of an area's resources by its residents to meet their own needs, in contrast to the more functional TOP-DOWN APPROACH in which an area's potential is exploited only because of the role it plays in the larger national or international economy.

Bottom-up approaches, or development 'from below', stress greater regional or areal self-reliance to break the centre's dominance. Development programmes are oriented towards problems of poverty and should be motivated and initially controlled from below. Projects are basic-needs-oriented, labour-intensive, small-scale using APPROPRIATE TECHNOLOGY, regional-resource-based and often rural-centred.

Bottom-up approaches emerged, mostly after the mid-1970s, out of disillusion with the progress of the top-down approaches that have dominated national and regional development in most DEVELOPING COUNTRIES in recent decades, and which have done little to meet the needs and aspirations of the mass of the population. As yet, few countries have been willing to devolve power and to implement development from below.

**boulder clay,** see GLACIAL DEPOSITION.

**BP,** see BEFORE PRESENT.

**brackish water,** *n.* water that contains too much salt to be drinkable but not enough to be classified as seawater. Its average salt content ranges between about 0.5 and 1.7%.

**Brandt Commission,** see NEW INTERNATIONAL ECONOMIC ORDER.

**break crop,** *n.* any LEAFY CROP grown primarily to break the likely build up of disease in continuous arable cultivation or PERMANENT CROPPING of cereals.

Root crops such as potato and sugar beet often have this role in temperate zone CROP ROTATIONS, although any grasses sown in this way would provide time for the host material in the soil to decompose before the planting of the next cereal crop. In most cases, a one-year break is sufficient, but LEY FARMING is sometimes a more effective measure against arable weeds.

Arable break crops, as compared to ley-rotations, can have significant adverse effects on both the structural condition and chemical fertility of soils. The harvesting of potatoes and sugar beet, for example, can increase impaction, and neither crop provides much residue on harvest, thus reducing organic matter and facilitating nutrient loss.

**breakwater,** *n.* an artificial barrier extended into the sea, usually made of large boulders or concrete and constructed to protect harbours and coastlines by dissipating the destructive energy of breaking waves. Breakwaters may be built parallel or at an angle to the coast and can sometimes be used to limit the effects of LONGSHORE DRIFT. See COASTAL PROTECTION.

**breccia,** *n.* a SEDIMENTARY ROCK consisting of angular fragments of rock debris within a matrix of finer material. Several types of breccia are recognized according to their origin, including fault breccias, volcanic breccias and glacial breccias.

**breeder reactor,** *n.* a type of NUCLEAR REACTOR that produces more nuclear fuel than it consumes by converting non-fissionable uranium-238 into fissionable plutonium-239. The breeding gain per unit of fuel is usually between 1.2 and 1.4. Prototype fast breeder reactors are at work in France, the UK and the USSR. Breeder reactors are needed because the supply of uranium-238 is finite and the earlier types of reactor quickly consume their uranium fuel. See also WATER-COOLED REACTOR, MAGNOX REACTOR, ADVANCED GAS-COOLED REACTOR.

**brown earth, brown forest soil, sol brun, cambisol** or **inceptisol,** *n.* a medium-textured, brown soil with high organic content commonly found beneath DECIDUOUS FORESTS in humid temperate zones. Characteristic features of this soil type include a B-HORIZON rich in iron compounds brought about by LEACHING from the free-draining A-HORIZON which has a high organic content. However, zonation may be difficult to detect from simple visual inspection.

Because of the rich HUMUS content of their upper horizons, brown earth soils are very fertile and have been in continuous agricultural use in Europe since the Middle Ages. They have proved to be of immense value, providing a base for intensive arable agriculture with few management problems.

**brown forest soil,** see BROWN EARTH.

**brownfield site,** *n.* any area of land, usually within a city, which has been cleared of the former buildings or waste materials of its previous use and is available for REDEVELOPMENT. In recent years such sites have been increasingly used for private sector housing. The attraction of these sites is that they are already accessible to the social and physical INFRASTRUCTURE of the city. However, they are seldom large sites and tend to occur in older and more run-down parts of the city. They are thus slightly more difficult to deal with, technically and financially, than GREENFIELD SITES.

**bryophyte,** *n.* any plant of the division Bryophyta, a group of primitive, simple plants including mosses and liverworts. Along with the ALGAE and FUNGI, they comprise the non-vascular plants which do not have a specialized internal physiology to transport materials from one part of the plant to another. Bryophytes are generally confined to damp locations such as BOGS. The life history of bryophytes consists of a complex two-stage cycle, the thallus stage being totally dependent on water for completion of sexual reproduction, and the non-sexual stage comprising a hardy vegetative growth phase.

Bryophytes form an important component of pioneer vegetation communities, and form a major source of summer grazing for caribou and reindeer in latitudes north of the Arctic Circle. Like LICHENS they are very vulnerable to damage from AIR POLLUTION.

**buffer zone,** see BIOSPHERE RESERVE.

**built environment,** *n.* that part of the physical surroundings created and organized as a result of human activity. The term is mostly used in relation to buildings within urban settlements, and the other structures which are functionally connected with them such as roads, bridges, docks or sports grounds. It is debatable whether or not those features of the rural landscape created by humans should also be considered as part of the built environment, for example, farmsteads. See also CITY FORM, INFRASTRUCTURE, TRANSPORT NETWORK, TOWNSCAPE.

**bush,** *n.* any uncultivated or sparsely settled land, especially forested land in varying conditions of naturalness, as found in Africa, Australia, and New Zealand. Bush may vary from open shrubby country to dense rainforest.

**bus lane,** *n.* one lane of a multilane road, normally that next to the footpath, reserved for the exclusive use of bus traffic during the peak travel times of the morning and evening commuter rush. Although bus lanes seldom extend continuously for more than a few blocks, this is sufficient to allow buses to by-pass lanes of stopped traffic, thereby minimizing disruption to other traffic. See TRAFFIC LANE, TRAFFIC MANAGEMENT.

**butte,** *n.* an isolated, steep-sided, flat-topped hill that is a remnant of a PLATEAU which has undergone prolonged FLUVIAL EROSION and SLOPE RETREAT. Buttes are more eroded and hence less extensive than MESAS.

# C

**caatinga,** *n.* a type of thorn woodland or thorn scrub found in northeast Brazil. It is related to a similar woodland type which extends widely over Venezuela and Colombia. Caatinga is *microphyllous*, that is, dominated by low, bushy trees and tall shrubs, many of which have small evergreen leaves. These are probably an adaptation to the extensive dry season. Scattered throughout the caatinga can be found *barrigudos* or Brazilian bottle trees (*Cavanillesia arborea*) whose great trunks, often swollen to a diameter of 5 m, act as water reservoirs. The soils beneath the caatinga are usually very sandy and hence pervious to water. Caatinga has often been cleared to form dry range land for beef cattle.

**CAI,** see CURRENT ANNUAL INCREMENT.

**Cainozoic era,** see CENOZOIC ERA

**calcareous,** *adj.* composed of or containing calcium compounds, particularly calcium carbonate.

**calcareous rego black soil,** *n.* a RENDZINA-like soil as listed in the Canadian soil classification system.

**calcimorphic soil,** *n.* a type of soil found above calcium-rich PARENT MATERIAL. Typical examples include RENDZINA and TERRA ROSA soils.

**calcification,** *n.* the accumulation of calcium carbonate in the B-HORIZON of a soil. Calcification usually occurs in the low-rainfall continental interiors such as the New South Wales grasslands, the North American PRAIRIES, and the Soviet STEPPES. Limited LEACHING results in the downward movement of calcium carbonate but only as far as the B-horizon, where a DURICRUST may form; this calcium carbonate duricrust is known as *calcrete* or *caliche*. Calcification is greatly accelerated by the upward movement of calcium-rich GROUNDWATER through CAPILLARITY; high air and ground temperatures cause evaporation and the deposition of calcium salts from the groundwater. See also ALKALIZATION, SALINIZATION.

**calcrete,** see CALCIFICATION.

**caliche,** see CALCIFICATION.

**cambisol,** *n.* the term used in the FAO soil classification system for a BROWN EARTH soil.

**Cambrian period,** *n.* the earliest geological PERIOD in the PALAEOZOIC ERA. It commenced about 600 million years ago and ended about 500 million years ago when the ORDOVICIAN PERIOD began. Invertebrate marine species were abundant during the Cambrian period and their fossils can be used to date and correlate rocks. This period obtains its name from the Cambrian Mountains in Wales, the area in which Cambrian lithology was first studied. For many years Cambrian slate was extensively used as roofing tiles whilst the QUARTZITE found in Cambrian rocks is used for road metal. See EARTH HISTORY.

**canyon,** see GORGE.

**CAP,** see COMMON AGRICULTURAL POLICY.

**capillarity,** *n.* the ability of a soil to retain a film of water around individual soil particles and in pores through the action of surface tension. This is achieved despite the action of gravity which serves to move water molecules downwards through the SOIL PROFILE. In some soils in semi-arid regions, capillary water is drawn upwards and dissolved chemicals are precipitated in the upper layers to form DURICRUSTS. See SOIL WATER. See also WILTING POINT.

**capillary water,** see SOIL WATER.

**capitalized value,** see LAND VALUE.

**carbon cycle,** *n.* the natural circulation of carbon in the biosphere. Separate but interconnected cycles exist for the circulation of carbon on land and in the sea. The cycles are connected at the junction of the ocean and the atmosphere. The carbon cycle begins with the fixation of atmospheric carbon dioxide by the process of PHOTOSYNTHESIS, conducted by plants and certain microorganisms. Thereafter, the cycle extends to the animal kingdom, the animals eating the plants thereby respiring and releasing carbon dioxide back into the atmosphere.

Until 1860, the amount of carbon in circulation was approximately stable. However, following the Industrial Revolution and the dramatic increase in the burning of FOSSIL FUELS, the amount of carbon in the atmosphere has rapidly increased (see GREENHOUSE EFFECT). Compare HYDROLOGICAL CYCLE, NITROGEN CYCLE.

**carbon fixation,** *n.* a process occurring in PHOTOSYNTHESIS whereby atmospheric carbon dioxide gas is combined with hydrogen obtained from the splitting of water molecules. Carbon fixation results in the formation of an energy-rich three-carbon compound called *phosphoglyceraldehyde* (PGAL). PGAL becomes rearranged to form the sixth carbon sugar, glucose, which is traditionally regarded as the end point of photosynthesis.

**carbon-14 (¹⁴C),** see RADIO-CARBON DATING.

**carbonaceous,** *adj.* composed of or containing carbon.

**carbonation,** see CHEMICAL WEATHERING.

**Carboniferous period,** *n.* the geological PERIOD that followed the DEVONIAN PERIOD and that extended from 345 to 280 million years ago. In the USA, the Carboniferous is divided into two major sub-periods, the *Mississippian period* and the *Pennsylvanian period*. There was considerable volcanic and igneous activity during this period, with widespread GLACIATION in the southern hemisphere in the latter stages. Extensive swamps were common in many parts of the world during the Carboniferous and these have subsequently formed the COAL MEASURES, probably the most important commercial geological deposit yet discovered: the term *Carboniferous* itself means 'coal-bearing'. Other economic resources to be found in Carboniferous rocks include OIL, oil shale, iron, FIRE CLAY, chalk and limestone. A variety of invertebrate marine fossils are used to date and correlate Carboniferous rocks. See EARTH HISTORY.

**carcinogen,** *n.* any substance capable of causing cancers in animal tissues. The mechanism by which carcinogens cause cancers in living tissue is not fully understood; nevertheless, much is known about the substances which predispose to the disease. The first carcinogenic material to be discovered was arsenic, but since the 1930s beryllium, cadmium, cobalt, chromium and ASBESTOS have all been added to the list of carcinogenic materials. The range of suspected or proven carcinogens include tyre rubber, tar macadam and the synthetic PCBs, of which there are now over 200 different types used in everyday items such as plastics, electrical components and hydraulic fluids in motor vehicles. Many of the synthetic compounds used in industrial processes have been found to be carcinogenic, for example, many of the epoxy resins and adhesives.

The National Cancer Institute of the United States has estimated that in excess of 30% of all Americans may die from cancer caused by exposure, contact or ingestion of carcinogenic material before the age of 74.

**carnivore,** *n.* any animal which gains the majority of its food supply by consuming other animals. The term is often restricted to any mammal of the order Carnivora, which includes cats, dogs, bears, racoons, hyenas and weasels. The feeding habits of mammalian carnivores has resulted in the development of a distinct tooth design, featuring strong, pointed teeth (the canines) for tearing and ripping meat. Carnivores form the third TROPHIC LEVEL of the food chain; they are the secondary consumers.

**carrier solvent,** see AEROSOL.

**carrying capacity,** *n.* (Ecology) the optimum population size that a given HABITAT can support indefinitely under a given set of environmental conditions. The concept of the carrying capacity can be illustrated by reference to a simple organism, such as the water flea species *Daphnia* (see Fig. 15). The population of *Daphnia* increases steadily until a LIMITING FACTOR in the environment exerts an ENVIRONMENTAL RESISTANCE which restricts further population growth. The rate of increase becomes slower until the population fluctuates around a theoretical optimum size, the magnitude of which will vary over time due to the inherent variability of the environmental inputs. See also POPULATION CRASH.

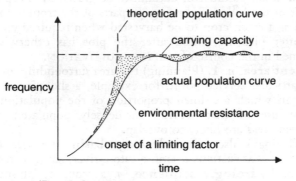

Fig. 15. **Carrying capacity.** See main entry.

**cash crop,** *n.* any crop grown for direct sale for a cash income. In developed countries where COMMERCIAL FARMING is predominant, almost all farm output is sold, and the term is usually employed to distinguish crops for sale from those fed to livestock on the farm. In developing countries, where a much greater proportion of farm output is consumed directly by the family the term is used for crops other than those for domestic subsistence.

**catalytic convertor,** *n.* a device fitted to the exhaust systems of motor vehicles which chemically changes the noxious hydrocarbons and carbon monoxide in AUTOMOBILE EMISSIONS to carbon dioxide and water vapour. The catalytic convertor may only be used with UNLEADED PETROL since lead actually destroys the convertor. The device has been a compulsory fitment in most states of the USA since 1976 and is also widely used in Japan, West Germany and Switzerland. They will become mandatory fitments to all new cars sold in the EC from 1993. However, because of the expense of fitting and renewing catalytic convertors and the reduction in vehicle performance which results from their use, they are unpopular with both car producers and users. Furthermore, the inability of catalytic convertors adequately to filter nitrogen oxides seriously limits their effectiveness in reducing motor vehicle pollutants. LEAN-BURN ENGINES are an effective alternative. See also PHOTOCHEMICAL SMOG.

**catch crop,** *n.* a minor crop (in terms of output) that is planted in the same year as, but immediately after, the main crop to 'catch' remaining soil moisture and other nutrients. Catch cropping is a form of MULTIPLE CROPPING. In temperate zones LEAFY CROPS may be sown in late summer or autumn to be grazed by sheep later in the year or the following spring. In parts of the tropics, cassava is planted as a catch crop to be harvested when required up to four years after planting, often after the plot has otherwise been abandoned in the course of SHIFTING CULTIVATION.

**catchment area,** *n.* **1.** (Planning) the area surrounding, or related to, a particular service point, for example, a shopping centre or school, in which a defined proportion of the population makes frequent use of that service. In densely populated regions, catchment areas are likely to overlap.

**2.** (Geology) also called **drainage basin.** the area of land bounded by WATERSHEDS draining into a river, basin or reservoir.

**catena** or **hydrologic sequence,** *n.* a sequence of soils in a particular locality which are derived from the same PARENT MATERIAL but have differing characteristics due to variations in topography and drainage. A catena may be apparent from soil profiles which traverse a hillslope (see Fig. 16).

**cation,** see ION.

**cation exchange capacity,** see CLAY-HUMUS COMPLEX.

**cave,** *n.* a natural underground chamber with an entrance from the surface. Caves form in a variety of ways including the MARINE EROSION of sea cliffs, the carbonation and/or solution weathering of limestone in KARST and chalk regions and occasionally, as a result of human excavation (see CHEMICAL WEATHERING).

Caves are presently used for a variety of purposes including SPELAEOLOGY, for the maturation of wine, as tourist attractions such as the caves of Drach on the island of Majorca, for the dumping of domestic and industrial wastes and, controversially, for

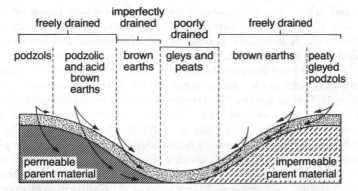

freely drained | imperfectly drained | poorly drained | freely drained

podzols | podzolic and acid brown earths | brown earths | gleys and peats | brown earths | peaty gleyed podzols

permeable parent material | impermeable parent material

Fig. 16. **Catena.** Cross-section of a slope showing the variety of soils found as drainage patterns (indicated by arrows) alter.

the storage of radioactive waste and stockpiling of military weapons and supplies.

**cavernous weathering,** see PHYSICAL WEATHERING.

**cavitation,** see FLUVIAL EROSION.

**CDA,** see COMPREHENSIVE DEVELOPMENT AREA.

**cementation,** *n.* the process by which unconsolidated sediments are bonded by siliceous, calcareous or ferruginous minerals to form SEDIMENTARY ROCK. The mineral cement then forms an integral part of the rock.

**Cenozoic era, Cainozoic era,** or **Kainozoic era,** *n.* the most recent geological ERA that commenced after the MESOZOIC ERA about 65 million years ago. The Cenozoic may be divided into the TERTIARY PERIOD and the QUATERNARY PERIOD. Lifeforms that developed during the Cenozoic are largely similar to those found today. See EARTH HISTORY.

**central area,** *n.* the central part of a city in which major shops and offices are concentrated and where public transport lines terminate; this is known in the USA as the *central business district* or *downtown*. The creation of RING ROADS around city centres have removed traffic from shopping streets, so allowing the development of PEDESTRIANIZED STREETS in many modern central areas.

**CEQ,** see COUNCIL ON ENVIRONMENTAL QUALITY.

**CFC,** see CHLOROFLUOROCARBON.

**chamaeophyte,** see RAUNKIAER'S LIFEFORM CLASSIFICATION

**chamaeophyte climate,** see BIOCHORE.

**chaparral,** *n.* the evergreen, SCLEROPHYLLOUS vegetation of the lower flanks of the Coastal and Santa Lucia ranges of southern California. It is usually known as *coastal chaparral* and has strong similarities with MAQUIS. The vegetation shows numerous adaptations to the long summer drought. Annual rainfall may be less than 380 mm, and in the driest areas the chaparral vegetation rarely exceeds 3 m in height. In damper areas where rainfall may reach 1000 mm per annum the chaparral becomes dominated by evergreen oaks, typically *Quercus agrifolia*, which may attain 20 m in height. The floral diversity is considerable with a wide variety of minor species. Chaparral species are notably resistant to fire damage, being able to produce vigourous new shoots from underground root systems. Many areas of the Californian chaparral are now under commercial development for irrigated agriculture.

**chemical fertilizer,** see FERTILIZER.

**chemical sedimentary rock,** see SEDIMENTARY ROCK.

**chemical weathering,** *n.* the loosening, dissolving and decomposition of rocks and their constituent minerals by the chemical action of water, oxygen, carbon dioxide and organic acids. Six types of chemical weathering have been recognized:

(a) *oxidation* results when atmospheric oxygen dissolved in rain and soil water combines with rock minerals to form oxides. Rocks and minerals containing iron compounds are particularly prone to oxidation which causes the yellow, red and brown colouration of many soils and rocks;

(b) *hydrolysis* entails a complex chemical reaction between water and rocks and is especially important in the weathering of silicate minerals, which are common in many igneous rocks. A variety of clay minerals may result after hydrolysis;

(c) *hydration* involves the incorporation of water within the chemical structure of certain rock minerals. The altered mineral increases in volume and may assist in the PHYSICAL WEATHERING of rock. An example of hydration is the conversion of anhydrite to gypsum;

(d) *carbonation* is the decomposition of rock and soils by carbonic acid. This acid forms either when carbon dioxide released by biological processes dissolves in SOIL WATER or when atmospheric carbon dioxide dissolves in rain water. Carbonate minerals are particularly affected and carbonation plays an important role in shaping the surface and underground topography of limestone areas;

(e) *organic acids* derived from the decay of plant and animal matter can cause the chemical weathering of rocks;

(f) *solution weathering* is largely confined to a few highly soluble minerals although all minerals are soluble to some extent. This process may dissolve the minerals that cement sedimentary rock particles, and is greatly accelerated in the presence of carbonic and organic acids.

Chemical weathering occurs under all climatic conditions but is more prevalent in the humid tropics where high temperatures and rainfall levels can cause the weathering of rock to depths of 45 m in extreme cases (see also VAN'T HOFF'S 'LAW'). Chemical weathering frequently operates in conjunction with PHYSICAL WEATHERING and BIOLOGICAL WEATHERING. The weathering of rocks can form new minerals which are of great importance such as BAUXITE and KAOLINITE. It is thought that 'concrete cancer,' which affects many man-made structures, may be a form of chemical weathering.

**chernozem, prairie soil, black earth** or **mollisol,** *n.* a black or dark brown, humus-rich soil found beneath the dry, mid-latitude grasslands of North America, the USSR, northern China, southeast Australia and Argentina. Chernozem soils characteristically display a deep A-HORIZON (up to 1.5 m in depth) with a well developed crumb structure, and a B-HORIZON rich in calcium carbonates, as the limited rainfall restricts the amount of LEACHING that occurs. They form excellent agricultural soils, being stone-free and located on extensive flat or gently sloping terrain. They are extensively used for cereal cultivation. The main management problem arises from wind-blown SOIL EROSION when incorrect ploughing regimes are used. See STEPPE.

**Chinook,** *n.* a warm dry ADIABATIC WIND which flows on the eastern side of the Rockies in the USA. The rapid rise in air temperatures associated with the Chinook causes the rapid melting of snow and can increase the likelihood of avalanches.

**chlorofluorocarbons (CFC),** *n.* a family of inert, relatively non-toxic gases mainly used in developed countries as propellants in AEROSOL spray cans, as refrigerant gases, solvent cleaners and for blowing foam plastics much used in the packaging associated with 'fast food' retailing. When 'discovered' by the American chemist, Thomas Midgley Jnr. in 1930, CFCs were hailed as a wonder chemical, non-flammable, non-toxic, non-corrosive, stable and with thermodynamic properties.

CFCs mainly comprise methyl chloroform and carbon tetrachloride. Prior to industrialization, the natural atmospheric level of chlorine was 2 parts per billion (ppb). By 1988 the figure had risen to 3 ppb and was sufficient to cause destruction of the atmospheric OZONE layer above the Antarctic. The function of the ozone layer is to protect the earth from the sun's ultraviolet radiation.

Under the Montreal Protocol, agreed in September 1987 by the world's leading industrial nations, the use of CFCs was to be held at the 1986 level until 1990. Thereafter, a 50% reduction in consumption was planned by the year 1999. However, so serious was the rate of accumulation of CFCs in the atmosphere that in March, 1989 the member states of the EC decided to work towards a 100% reduction by 2000. The USA has endorsed this decision. Unless THIRD WORLD countries can be encouraged not to use CFCs in their future industrialization programmes then atmospheric levels of chlorine will double to 6 ppb by the year 2000. In 1989, for example, India consumed 5000 tonnes of CFCs out of a world total production of 1 million tonnes p.a. By 1999, the Indian level of consumption will rise to 18 000 tonnes.

Alternatives to CFCs will be both expensive and difficult to find. Research in the US and the UK centres around HFC 134a as a replacement for the much used CFC 12. By 1989 a sum of £60 million had been invested in two experimental production lines in the UK. HFC 134a contains no chlorine and does not deplete the ozone layer but it may well have other unknown side effects upon the BIOSPHERE. Extensive testing of the compound will be necessary before its use can be sanctioned. If widely adopted, its cost will be 5 times that of CFCs. Its use will necessitate the development of new sealant materials as existing substances shrink upon contact with HFC 134a. New synthetic lubricants will also be required to enable electric refrigerator pumps to operate with the new substance. See GREENHOUSE EFFECT.

**C-horizon,** *n.* the layer lying below the B-HORIZON and above the PARENT MATERIAL in a well-developed mineral soil (see Fig. 55). The C-horizon is transitional between the weathered material of the A-HORIZON and B-horizon, and the underlying parent material. The occurrence of soil fauna and flora is minimal.

**chronosequence,** *n.* a sequence of related soils which differ in certain characteristics due to variations in their age. Well-developed chronosequences can be found along TRANSECTS taken across sand dunes, SPITS and successional MORAINES. See also CLIMOSEQUENCE, LITHOSEQUENCE, TOPOSEQUENCE.

**cinder cone,** see VOLCANO.

**circum-Pacific zone,** see EARTHQUAKE.

**cirque, corrie** or **cwm,** *n.* a steep-walled, semi-circular mountain hollow formed under glacial conditions (see Fig. 66). As snow accumulates, the hollow is enlarged through PHYSICAL WEATHERING, particularly as a result of freeze-thaw action on marginal and underlying rocks. The resulting debris is removed by rock falls, AVALANCHES, SOLIFLUCTION and meltwater. With time, a GLACIER may form and further enlarge the hollow. Lakes, or TARNS, are frequently found in cirques. Corries are often favoured locations for skiing developments as their sheltered nature provide some protection against winter storms as well as extending the season through the retention of snow long after it has melted on surrounding mountain slopes.

**cirrocumulus,** *n.* a high-altitude CLOUD consisting of thin layers of white globular masses. Cirrocumulus clouds usually occur in groups which form a rippled pattern known as a 'mackeral sky'.

**cirrostratus,** *n.* a thin, whitish veil of high-altitude CLOUD that produces solar or lunar halo phenomena. It is associated with the approach of the WARM FRONT of a DEPRESSION.

**cirrus,** *n.* a whitish, wispy high-altitude CLOUD associated with fair weather.

**CITES,** see CONVENTION ON INTERNATIONAL TRADE IN ENDANGERED SPECIES.

**city,** *n.* any large urban settlement. No strict population size has been set at which a town becomes a city; often it is the sociopolitical function of the urban area that determines city status. In Europe the presence of a cathedral is often used as an indicator of city status. However, this produces anomalous situations where, for example, St. David's in Dyfed, Wales with a population of 1800 is termed a city as well as the Italian capital Rome which has a population of over 3 million. Small cities may generally contain between 20 000 and 200 000 people and provide services to a HINTERLAND containing up to 1 million people; large cities of between 200 000 and 500 000 people may serve a hinterland of up to 3 million. When populations of this size are reached it is commonplace for several cities to exist in close proximity to each other, so creating a *megalopolis*, as occurs along the eastern seaboard of the USA. See CONURBATION.

**city climate,** see URBAN CLIMATE.

**city form,** *n.* the three-dimensional appearance of a city. A city's form is composed of its spatial layout and its fabric. The spatial layout is shown by the network of roads and railways, and the effect on these of topography and any rivers which may flow through the city. The massing, height and external character of the buildings together with the distribution of landscaped spaces relative to the buildings, presents an impression of the texture of the city's fabric. See BUILT ENVIRONMENT.

**Civic Amenities Act,** *n.* an enactment of 1967, which for the first time introduced the concept of CONSERVATION AREAS into the practice of town planning in the UK. The Act made it a duty of local planning authorities to define areas within their jurisdiction which had special qualities of architectural and historic interest, and to designate these as conservation areas. Once an area has been so designated, any proposals to alter the environment have to be judged on whether or not they will preserve and enhance the quality of the area. Conservation areas may comprise a whole village or small town, part of a town or city such as a street, market square, or group of buildings, or simply a unified terrace or crescent of buildings. See also LAND USE PLANNING, AMENITY.

**Civic Trust,** see AMENITY.

**class,** *n.* any of the taxonomic groups into which a PHYLUM is divided and which contains one or more ORDERS. For example, Amphibia, Reptilia and Mammalia are three classes of the phylum Chordata. See CLASSIFICATION HIERARCHY, LINNEAN CLASSIFICATION.

**classification hierarchy,** *n.* the ordering of organisms in a series of increasingly specialized groups (or *taxa*) because of similarites in structure, origin, etc., that indicate a common relationship. The major groups are kingdom, PHYLUM (in animals) or division (in plants), CLASS, ORDER, FAMILY, GENUS, and SPECIES. Each taxa is a collective unit containing one or more groups from the next lower level in the hierarchy whereby species are classified in a logical manner according to their evolutionary history. Over 1 million animals and 325 000 plants have been successfully included in the scheme.

All plants and animals are assigned a two-part name, the first part comprising the name of the genus to which the species belongs and second, a designation for that particular species. The naming of new species (or alteration of existing species names) is made according to the *International Rules of Botanical Nomenclature*, the *International Rules of Zoological Nomenclature*, and the *International Bacteriological Code of Nomenclature*.

While an individual plant or animal may be known by many different local or regional names it has only one scientific name which is used throughout the world. See LINNEAN CLASSIFICATION.

**classification of farming systems,** *n*. the grouping of agricultural enterprises into categories based on their organization and
production methods. Classification schemes have traditionally been
based on such criteria as crop and livestock associations, the
farming methods employed, the intensity of farming, the degree of
subsistence or commercial orientation, and the complex of farm
structures associated with the farm enterprise; other systems have
taken more account of the predominant physical environmental
conditions, levels of technology and population density. None
have successfully incorporated such important variables as LAND
TENURE or FARM SIZE and layout. Examples of more or less distinct
types of agricultural activity include MARKET GARDENING, MIXED
FARMING, large-scale grain production, PLANTATION agriculture,
WET-RICE CULTIVATION and livestock RANCHING.

Classifications of farming systems are now based on more
rigorous statistical assessments and include CROP COMBINATION
ANALYSES derived from acreage percentages, the use of farm-based
data on standard MAN-DAY units, and the identification system
advanced by the Commission on AGRICULTURAL TYPOLOGY. See
AGRICULTURAL REGION.

**clastic sedimentary rock,** see SEDIMENTARY ROCK.

**clay,** *n*. a sedimentary deposit composed of very fine grain mineral
material. It derives largely from the chemical weathering of rocks
containing felspar. Clay particles have a diameter of less than 0.002
mm in most soil classification systems (see Fig. 57) and are able to
absorb relatively large quantities of water. In a dry state, clay is hard
and apt to crack irregularly. Although porous when dry, wet clay
becomes plastic and virtually impermeable due to the swelling of
the clay particles and the retention of water in PORE SPACES as a
result of the surface tension exerted by the particles. Serious
waterlogging often occurs within clay-rich soils. The clay component of a soil affects the SOIL FERTILITY due to its role in the CLAY-
HUMUS COMPLEX and cation exchange capacity. Soils with a high
clay content are difficult to manage for agricultural use, being slow
to warm in spring and heavy to cultivate.

Clay is a valuable economic resource and is used in the
manufacture of paper, tiles, bricks, linoleum, pottery and cement.
Areas with an underlying clay geology are often suitable for the
construction of LANDFILL sites as the impermeability of the clay
restricts the movement of pollutants into the groundwater table.
Reservoirs are often located over clay-rich stratigraphy to prevent
water loss from seepage to the groundwater table.

**clay-humus complex,** *n.* the chemically active part of the soil formed by the intimate association between finely weathered CLAY mineral particles and the decomposed remains of plants and animals (in particular, MULL humus). Each clay-humus particle or *micelle* acts as a weakly charged *anion* (a negatively charged ION). Surrounding the micelle and attracted to it by the negative charge are numerous bases or *cations* (positively charged ions) such as calcium and magnesium. These adsorbed cations are capable of being exchanged between micelles, and between micelles and plant rootlets. The total amount of exchangeable ions is known as the *base exchange capacity* or *total exchange capacity* of the soil.

Cations can be easily replaced by the accumulation of hydrogen ions (which are them selves aggressive cations). After rainfall, the hydrogen ions in the percolating soil water cause the LEACHING of cations, leading to a temporary increase in acidity of the micelles. As the water drains from the soil so some of the hydrogen ions are removed, allowing a reduction in the soil acidity.

**claypan,** see HARDPAN.

**Clean Air Act,** *n.* the legislation of 1956 which empowered local authorities to establish SMOKE CONTROL ZONES within which it would be an offence to have dark smoke emitting from any chimney for more than 5 minutes in any hour. The Beaver Report of 1953 had shown that domestic fires were the main producers of particulate AIR POLLUTION: of the 200 million tonnes of coal consumed each year in the UK in the 1950s, domestic consumption accounted for 25% of this total while producing 50% of all smoke, ash and grit. Grants were subsequently made available to private residents for the conversion of household fireplaces to burn smokeless fuel. Since 1956, the incidence of smoke control orders has steadily increased to the extent that the total area of most towns and cities in the UK is now covered. The resulting clean air has had marked effects on public health and on the BUILT ENVIRONMENT, the latter being due to the opportunity which clean air gives for the successful refurbishment of the exterior of buildings. See SMOG, PARTICULATE MATTER.

**clear felling,** *n.* (Forest management) the widespread practice of completely felling and removing a STAND of trees. Usually, the cleared area is quickly replanted with a single species, an example of MONOCULTURE. In managed forests clear felling will usually occur when the curves of MEAN ANNUAL INCREMENT and CURRENT ANNUAL INCREMENT cross but in natural forest clear felling can occur at any stage of the life history.

**cleavage,** *n.* the property of certain rocks such as slate to split along planes of weakness which are often not related to the original bedding.

**climate,** *n.* the average weather characteristics of a particular area over an extended period of time. Climates are largely determined by latitude, topography, the distribution of land and sea, ocean currents, and the nature and influence of vegetation and soils. A climate can be described in terms of mean seasonal temperatures, precipitation, wind direction and speeds, and the nature and extent of cloud cover. See ARID CLIMATE, SEMI-ARID CLIMATE, MEDITERRANEAN CLIMATE.

**climatic change,** *n.* the largely natural short- and long-term alterations in climatic characteristics. Geological evidence indicates substantial climatic change during the earth's history. Since the last ICE AGE, there has been considerable climatic fluctuation. Temperatures throughout the world reached a maximum about 4000 BC whilst the period between 1550 and 1800 AD was known as the *Little Ice Age* in Europe due to a substantial decrease in mean annual temperatures. The average January temperature may have been 2.5°C lower than that of the present day. The decline in temperatures seriously affected agricultural productivity and caused the abandonment of land in northern Europe.

More recently, declining rainfall totals in the SAHEL region of Africa has caused an expansion of desertification and an increased occurrence of crop failure and famine.

The reasons behind climatic change are unclear but it is thought that short-term change may result from variations in solar radiation, natural alterations in the earth's atmospheric circulation system and modifications in the atmosphere's chemical composition, largely due to AIR POLLUTION incidents such as the depletion of the ozone layer by CHLOROFLUOROCARBONS. Longer term changes may be caused by movements in the earth's rotation and orbit around the sun and geological factors such as PLATE TECTONICS and CONTINENTAL DRIFT. Compare CLIMATIC MODIFICATION.

**climatic classification,** *n.* the classification of climates according to factors such as temperature, precipitation and wind patterns. Several classification systems have been devised but none are completely satisfactory:

(a) W. Köppen's climatic classification delineates five major climatic groups based on mean monthly and annual temperature and precipitation patterns;

(b) C.W. Thornwaite's classification system is based on moisture budget patterns and recognizes five climatic groups;

(c) H.Flöhn's classification is often regarded as one of the most satisfactory systems. It is based on precipitation characteristics and global WIND BELTS and has six categories.

**climatic climax,** *n.* a CLIMAX COMMUNITY controlled by climate, as opposed to other factors such as soil, fire, etc.

**climatic modification,** *n.* the artificial alteration of climatic characteristics as a result of human activity. AIR POLLUTION, HEAT ISLANDS and CLOUD SEEDING can cause climatic modification. Compare CLIMATIC CHANGE.

**climatic region,** *n.* any geographic area that displays the same broad climatic characteristics throughout, such as the Mediterranean region.

**climax community,** *n.* the final phase of PRIMARY SUCCESSION within which a fully developed and mature ECOSYSTEM attains equilibrium with the environment. BIOMASS increases to a maximum and FOOD CHAINS become more complex with a maximum diversity of plant and animal species.

When the model of climax communities was originally developed in the 1920s, they were considered the inevitable, stable endpoint for vegetation succession, with climate forming the main controlling influence over the rate and luxuriance of development. This simplistic model subsequently came to be known as the *monoclimax theory* with its vegetation succession said to have attained a CLIMATIC CLIMAX. However, in some regions, climate cannot form the controlling factor. Instead, soil limitations may result in an *edaphic climax*, or grazing pressure may result in a *biotic climax*. Vegetation development may, therefore, be under the control of several variables and the term *polyclimax theory* can be applied in these situations.

For many centuries the dominant influence on vegetation has been human activity. As a result of burning, grazing, deforestation, urbanization and pollution, an *anthropogenic climax* has resulted.

Climax theory is no longer considered relevant to explain the development of vegetation communities. It has been shown that it is the exception for vegetation to pass through a series of successional stages from pioneer community to climax. Plant communities are complex probablistic systems and as such their succession does not follow a predictable pattern.

**climosequence,** *n.* a sequence of related soils which differ in certain characteristics due to variations in the climate of the region as each layer of soil was being formed. See also CHRONOSEQUENCE, LITHOSEQUENCE, TOPOSEQUENCE.

**clone,** *n.* an individual new plant which bears an exact resemblance to its parent. Clones can be produced naturally by vegetative reproduction (by suckers, adventitious roots, layering, runners, etc.) or artificially by the cultivation of meristem tissue in a culture medium in a laboratory.

Clones display four distinct advantages over new individuals produced by sexual reproductive methods:

(a) if the parent plant is ideally adapted to its environment then the clone will show a similar advantage;

(b) if the parent plant has a commercially desirably characteristic (size, rate of growth, colour, shape) then the clone will also display this feature;

(c) naturally produced clones gain protection and sometimes nourishment from the parent plant;

(d) artificially produced clones can be replicated in large numbers on a production line basis throughout the year, independent of the growing season conditions.

The main disadvantage of a clone is that is does not show the genetic variability of an individual produced by sexual reproduction and cannot readily adapt itself to meet changes in its environment.

Cloning of commercially valuable species is now widespread. Flowers, vegetables and even trees are produced by this method. Research in the UK has shown that clones produced from tissue taken from fast-growing Sitka spruce trees exhibit a 20% increase in productivity over seed produced by sexual reproduction.

Total reliance on stock produced by cloning is undesirable. The genetic resource base becomes restricted and in the event of a disease epidemic or the emergence of a natural predator then total crop loss may result.

Cloning is the end result of a long line of technical applications by humans in their efforts to increase agricultural output and thereby reducing the numbers of inadequately fed humans. See GENETIC ENGINEERING.

**closed community,** *n.* a COMMUNITY of plants that has totally colonized an area of ground. Old closed communities display the greatest number and variety of species. LAYERING can frequently be observed. Apart from when very severe climatic or very poor soil conditions prevail, closed plant community conditions can be assumed to be the norm.

Humans can also disrupt a closed community by repeatedly removing plants of economic value. In some circumstances, for example, in TROPICAL RAIN FORESTS, removal of the dominant vegetation forms in a community can so disrupt the system that substantial erosion of the soil takes place making the re-establishment of a closed community a precarious process. Compare OPEN COMMUNITY.

**cloud,** *n.* a visible mass of particles of water or ice suspended in the atmosphere, and formed by the condensation of WATER VAPOUR on HYGROSCOPIC NUCLEI. This condensation is caused by the upward movement and rapid cooling of masses of air resulting from conditions of INSTABILITY, CONVECTION currents, orographic processes or the passage of FRONTS.

Various types of cloud are recognized and these are usually classified by their height, with subdivisions according to form. Within the three height groups, there are 10 major forms of clouds:

(a) Low-altitude clouds
   (i) CUMULUS
   (ii) CUMULONIMBUS
   (iii) STRATUS
   (iv) NIMBOSTRATUS
   (v) STRATOCUMULUS
(b) Medium-altitude clouds
   (i) ALTOSTRATUS
   (ii) ALTOCUMULUS
(c) High-altitude clouds
   (i) CIRRUS
   (ii) CIRROSTRATUS
   (iii) CIRROCUMULUS

The heights at which the three groups occur varies considerably with latitude (see Fig. 17). Cloud which forms just above the ground surface is known as FOG. Persistent cloud layers severely reduce the amount of SOLAR RADIATION at ground level and have an adverse effect on the development of natural vegetation and agriculture.

**cloudburst,** *n.* a sudden and usually short-lived shower of heavy RAIN associated with strong convection cells and the formation of THUNDERSTORMS. Cloudbursts can cause severe SOIL EROSION and flooding and may result in the damage of crops and loss of livestock.

| Cloud group | Tropics (metres) | Mid-latitudes (metres) | High latitudes (metres) |
|---|---|---|---|
| Low | Under 2000 | Under 2000 | Under 2000 |
| Medium | 2000–7500 | 2000–7000 | 2000–4000 |
| High | Above 6000 | Above 5000 | Above 3000 |

Fig. 17. **Cloud.** Variations in the altitude of cloud occurrence with latitude

**cloudseeding,** *n.* the artificial introduction of chemicals into the atmosphere, usually from aircraft and rockets, to induce the formation of rain-bearing clouds. Its success rate is erratic. It is thought that seeding agents, such as solid carbon dioxide, silver iodide and salt act as HYGROSCOPIC NUCLEI around which raindrops can form. Cloudseeding is used as a means to relieve drought but it is only successful if large, unstable masses of moisture-laden air are present. It has also been noted that seeding may reduce the destructive force of hurricanes by utilizing large amounts of energy in the condensation of water vapour.

**Club of Rome,** *n.* a multinational association founded in 1968 by the Italian industrialist Aurelio Peccei to examine the predicament of the human state in a world of finite resources and to suggest alternative policy options for meeting critical needs. Its members are drawn from business, politics, and the social and environmental sciences.

With funding from various agencies and foundations, the Club of Rome has commissioned research into the problems of pollution, urban planning, inflation, unemployment, and the widening gap between developed and developing countries. The Club has also contributed to the call for a NEW INTERNATIONAL ECONOMIC ORDER. See LIMITS TO GROWTH.

**CND,** *abbrev. for* Campaign for Nuclear Disarmament, a PRESSURE GROUP founded in the UK in 1958 by the philosopher Bertrand Russell and Canon L. John Collins to oppose the development and use of nuclear weapons. After attracting popular support in the early 1960s, the movement fell into decline until 1979 when a NATO decision to station intermediate-range cruise missiles in Europe (and particularly at Greenham Common in the UK) gave it a new impetus which it has sustained throughout most of the 1980s.

**coal,** *n.* a brown or black carbonaceous SEDIMENTARY ROCK formed by the ANAEROBIC decomposition of plant material. Coal results from the LITHIFICATION of peat, and depending upon the degree of compaction and heating during this process, deposits of coal may occur as LIGNITE, BITUMINOUS COAL or ANTHRACITE.

Coal seams may range in thickness from a few centimetres to tens of metres. Occasionally, coal may be found in rocks of the Devonian period but the largest deposits are associated with the Carboniferous, Permian and Cretaceous periods. Coal deposits are found throughout the world, with the most extensive reserves located in the USA, Western Europe, the USSR, Japan, China, India and Australia. The reserves of coal are the greatest of all the non-renewable FOSSIL FUELS. DEEP MINING and OPENCAST MINING techniques are used to exploit coal deposits depending on the depths at which they occur in the earth.

The history of coal is lost in antiquity although archaeological evidence suggests its use as a fuel in the Bronze Age. Coal was also used by the North American Indians and was a common domestic fuel in throughout medieval Europe. The large-scale exploitation of coal deposits followed the Industrial Revolution. Nowadays, coal is not only used in the generation of electricity and as a domestic fuel but is a raw material for a host of products such as petroleum, coal tar and coke.

Coal is a poorer quality and less clean fossil fuel than oil and natural gas, and over the past 30 years its popularity has waned in comparison to these other fuels. Despite this, the vast reserves of coal throughout the world, combined with the use of improved technology to increase burning efficiency and minimize pollutants, and the diseconomics of oil production may bring about a reversal of this trend.

There are a number of environmental problems associated with the exploitation and subsequent use of coal. In many countries, the collapse of abandoned underground mine workings has caused subsidence which has led to considerable damage to property, roads and railways, while ACID MINE DRAINAGE can cause the pollution of groundwater supplies. The spontaneous combustion of coal seams and spoil heaps causing the release of toxic fumes is a further problem. However, perhaps the greatest environmental problem stems from the AIR POLLUTION resulting from the large-scale burning of coal.

**coalification,** *n.* the LITHIFICATION of organically rich sediments to form COAL.

**coal measures,** *n.* a sequence of rocks from the CARBONIFEROUS PERIOD which often, but not always, includes coal seams of commercially exploitable thickness.

**coal tar,** *n.* a dark viscid substance obtained from the destructive distillation of BITUMINOUS COAL and used in the manufacture of plastics, pesticides, drugs and dyes.

**coast,** *n.* a poorly defined term for the land that borders the SHORE. It includes all land surfaces influenced by wave action, both at present and in the recent past. See also SHORELINE.

**coastal chaparral,** see CHAPARRAL.

**coastal classification,** *n.* the classification of a COAST and its features according to their origin. Several classification systems have been proposed, although it is unlikely that any can be fully accepted as the dynamic nature of coastal evolution has meant that most coasts are of compound origin. Three classifications that are commonly used.

(a) D.W. Johnson's classification (1919) recognized four types of shoreline: *shorelines of emergence*; *shorelines of submergence*; *neutral shorelines* due to factors other than earth movements or changing sea level; and *compound shorelines* whose origins were varied and complex and could incorporate at least two types of the preceding types;

(b) F.P. Shepard's classification (1937) suggested that coastlines were either *primary* and shaped by non-marine forces or were *mature* and largely influenced by marine processes. This classification was then applied to glaciated coasts, areas with broad coastal plains, coasts with young mountains that is, those formed in recent geological time), and coasts with old mountains;

(c) J.L. Davies' classification is more recent and is largely based on the comparative influence of MARINE EROSION and MARINE DEPOSITION on coasts.

**coastal climate,** *n.* the MICROCLIMATE developed over coastal regions and usually characterized by greater wind speeds and a reduced temperature range compared to inland areas. LAND BREEZES and SEA BREEZES are part of an ameliorating oceanic influence which influences coastal regions.

**coastal protection,** *n.* the protection of coastlines against MARINE EROSION and the transportation of sediment by LONGSHORE DRIFT. Coastal protection measures are usually only employed to prevent the damage and destruction of harbours and coastal settlements, and to limit the loss of sand from beaches. Coasts may be protected in several ways including the use of seawalls, BREAKWATERS and GROYNES.

**coastal reclamation,** *n.* the reclamation of shallow sections of the CONTINENTAL SHELF by the dumping of rubble and refuse, or the construction of BREAKWATERS and SEAWALLS and the drainage of the enclosed areas by pumping.

Reclamation schemes are found throughout the world and the reclaimed land is used for a variety of purposes including settlement, agriculture and industry. The Zuider Zee scheme in the Netherlands is one of the best-known reclamation projects and has provided 2300 km² of new land. However, once reclaimed, this low-lying coastal land is susceptible to flooding which may cause the temporary loss of productive agricultural land; in addition it may take up to five years to flush sodium chloride out of a soil following seawater flooding. Coastal reclamation can cause the loss of shellfish beds and inshore fishing grounds as well as damaging local ECOSYSTEMS. See also POLDER

**coastline,** see SHORELINE.

**coke,** *n.* a solid-fuel product comprising approximately 80% carbon produced by the distillation of coal to drive off its volatile impurities. Coke is used in the manufacture of steel.

**col, pass** or **saddle,** *n.* a depression or gap in a line of hills or mountains. Cols are formed in several ways including RIVER CAPTURE and GLACIAL EROSION. Lines of communication often utilize cols as they allow relatively easy access through otherwise difficult terrain.

**cold desert,** see DESERT.

**cold front,** *n.* the boundary between a warm air mass which is being undercut by an advancing cold air mass (see Fig. 18). This

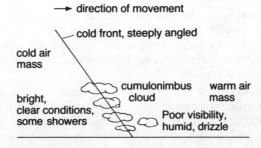

Fig. 18. **Cold front.** Cross-section through a cold front.

usually occurs at the rear of the WARM SECTOR in a DEPRESSION. The gradient of a cold front is steeper than that of a WARM FRONT. The rapid upward movement of moist air along the cold front results in condensation and the formation of CUMULONIMBUS cloud. Short, heavy rain showers are associated with the passage of a cold front which also causes a drop in temperature, a rise in air pressure, and the wind to veer from southwesterly to northwesterly in the northern hemisphere and vice-versa in the southern hemisphere. See FRONT. See also OCCLUDED FRONT.

**cold occlusion,** see OCCLUDED FRONT.

**collective effective dose equivalent,** see RADIATION DOSE.

**collective farm,** *n.* a type of state-owned farm originating in the USSR and now common in many Communist countries. Unlike KIBBUTZIM that operate without interference from the state, the collective farm or *kolkhoz* in the Soviet Union operates on land owned by the state and leased permanently to the workers on the farm. Prior to 1966 workers received a share, based on their labour contribution, of the product of the kolkhoz remaining after fulfillment of state delivery obligations. This has since been replaced by a regular wage payment system, one of several changes that has made collectives increasingly similar to STATE FARMS.

Collective farms are not confined solely to the Soviet Union and other communist regimes, but have been introduced in Italy, Mexico, India and elsewhere.

**collectivist tenure,** see LAND TENURE.

**commensalism,** *n.* the close association of different species where at least one species benefits from the other, but not at its expense. Compare PARASITISM, SYMBIOSIS.

**commercial farming,** *n.* any form of AGRICULTURE geared to the production of CROPS and ANIMAL HUSBANDRY for a cash return rather than for the satisfaction of domestic subsistence needs.

Commercial farming is a common feature in the advanced agricultural systems of most DEVELOPED COUNTRIES where the bulk of farm output is sold, and where farmers have a keen awareness of financial profit and loss. It is further characterized by the employment of wage labour, and the increased application of farm inputs, such as AGROCHEMICALS and MECHANIZATION. In much of the THIRD WORLD, commercial farming is less easily defined. The sale of incidental surpluses of staple foodstuffs by farmers otherwise engaged in SUBSISTENCE FARMING should not be regarded as evidence of the commercialization of agriculture. See also AGRIBUSINESS, AGRO-INDUSTRY.

**commodity,** *n.* any resource that is traded, especially on the international market. Commodities may be classified as soft or hard, according to whether or not they are renewable. Soft commodities include PERMANENT CROPS of trees and shrubs, and tillage crops (GRAIN CROPS and LEAFY CROPS) which with shorter production cycles are able to respond more quickly to shifting commodity prices. Hard commodities may be broken down into ferrous and non-ferrous metals, non-metallic minerals, energy minerals, and a group of precious minerals which are distinguished by their scarcity and value. Non-metallic minerals may be subdivided into widely available or ubiquitous commodities, and localized ones with a generally lower volume and higher value. A number of commodities, irrespective of their broad grouping may be further categorized as strategic minerals on account of their significance in the economies, and especially the defence sectors, of many DEVELOPED COUNTRIES, and the concentration of their production in politically sensitive areas of the THIRD WORLD, South Africa and the USSR.

**Common Agricultural Policy (CAP),** *n.* the set of measures forming the common approach of the members of the EC towards the regulation of agricultural markets and the promotion of STRUCTURAL REFORM within the community. Established in 1962, and extended by many subsequent directives, the objectives of CAP are formally stated as being: to improve agricultural productivity; to ensure a fair standard of living for the agricultural community, in particular by increasing farm incomes; to stabilize markets; to ensure the availability of supplies; and to ensure that supplies reach consumers at reasonable prices. The degree of market support is unusually large and accounted for 63% of the entire EC budget in 1986, and as much as 75% throughout the 1970s.

By far the greater part of the fund is spent on a complex and wide-ranging set of price support mechanisms (the Guarantee Section) while policies aimed at improving the structure of agriculture (the Guidance Section) have been much overshadowed. For most products, a target or guide price is set each year. If prices on the open market fall below this, the Community is responsible for *intervention buying*, that is, for supporting the market by purchasing and storing produce for subsequent sale. EC farmers are also protected from cheap imports by the setting of fixed threshold prices below which foreign produce cannot be sold. Intervention surpluses may be sold on the international market with the assistance of export subsidies known as restitution payments, that is, they make good to the trader the difference between EC purchase prices and world market prices.

To some extent, the Guarantee and Guidance Sections of the CAP are incompatible in that support of farm incomes has allowed many inefficient farmers on small, fragmented holdings to remain in production. A more intractable problem has been the build-up of farm surpluses as a result of intervention buying, and the inability of the system to discriminate in support between the large and small farm sectors. How to dispose of the so-called mountains and lakes of surplus produce, and how to reform the CAP to prevent further overproduction remain the most critical issues facing the EC at the present time.

**common land,** *n.* any land in England and Wales, subject to the 'rights of common', provided that the land has been registered as common land under the Commons Registration Act, 1965. Common land once formed an important feature of the English landscape.

Many commons were over-used and suffered from severe over-grazing, DEFORESTATION and SOIL EROSION. The effect of the application of animal manures, however, was to result in a gradual improvement in the quality of the common land. The better areas of common land were gradually incorporated into farm land, a process which culminated in the General Enclosure Act of 1845 (see ENCLOSURE). The poorer areas of common land have remained as tracts of under-utilized land and as such form important habitats for wildlife. These areas typically comprise heather-dominated moorland, bracken-infested land or birch scrub overlying infertile, sandy soils.

Nowadays, common land often provides recreational space in an otherwise urbanized landscape, for example, Wimbledon Common in south west London or the Town Moor in Newcastle-upon-Tyne. See also RIGHT OF WAY.

**commons attitude, the,** *n.* an attitude to certain kinds of resources such as the atmosphere, clean water, oceans, and fish in the sea that holds them to be inexhaustible common property, freely available to all, but without any individual or group having the incentive or responsibility to conserve them or to prevent any adverse outcome of that use, such as POLLUTION, over-fishing, DESERTIFICATION, DEFORESTATION, SOIL EROSION, etc. The 'tragedy of the commons' is that individuals continue to maximize their use of a common resource until the cumulative effect of several individuals each pursuing the same rational strategy depletes the supply to the point where it can no longer be exploited economically by anyone. Similarly, recreational pressures on the countryside, especially NATIONAL PARKS, can destroy the very attributes and values the visitors seek. Sea FISHING, WHALING, and ocean-bed mineral exploitation are among the last preserves of the commons attitude, and have defied a lasting solution to international exploitation despite efforts by the United Nations to establish a LAW OF THE SEA.

**communication network,** see TRANSPORT NETWORK.

**community,** *n.* **1.** any collection of plants or animals living and growing together and which possesses a certain unity and individuality. The community comprises a typical species composition that has resulted from the interaction of populations over time. See OPEN COMMUNITY, CLOSED COMMUNITY.

**2.** an assemblage of people linked together by some common interest, allegiance or service provision. In popular usage, the term refers to the residents of a definable geographic area, the assumption being made that such residents are drawn together socially by shared local loyalties or problems. See CITY, GHETTO, RESIDENTIAL SEGREGATION.

**commuter,** *n.* any person who travels an appreciable distance on a regular basis between their home and their workplace. The act of travel, by suburban railway, UNDERGROUND, bus or car is termed *commuting* after the first commuters, who travelled to work by train and used a season ticket for which a commuted charge was made. See JOURNEY TO WORK, PARK-AND-RIDE SYSTEM, COMMUTER VILLAGE.

**commuter village,** *n.* any village, usually on the fringe of a metropolitan region, in which a high proportion of the workforce commutes substantial distances to jobs elsewhere in the region. See COMMUTER, COMMUNITY.

**compaction,** *n.* the consolidation of sediment by earth movements or the compression of overlying material to form SEDIMENTARY ROCK.

**composite cone,** see VOLCANO.

**comprehensive development area (CDA),** *n.* an area scheduled for total renewal within a programmed plan of development for a city. Within the CDA, the local planning authority had powers of compulsory purchase that permitted it to acquire land and the buildings thereon for the purposes of clearance prior to total rebuilding. Families were moved out successively according to a progamme and resettled in other areas of PUBLIC SECTOR HOUSING while the clearance and rebuilding took place. Some returned to the newly built houses although in view of the high densities of population living in the CDA originally, it was seldom possible to return the same numbers to the site when modern building standards were observed. Those unable or unwilling to return to the original locality became part of the city's OVERSPILL total, which had to be permanently relocated.

CDA fell into disfavour in advance of the establishment of STRUCTURE PLANS and LOCAL PLANS by the Town and Country Planning Act (1969) as they were thought to be too crude an approach to renewing the environment and were certainly damaging to the social networks of the residents who were relocated.

**condensation,** *n.* the process by which energy is released from a gas whereupon it undergoes a change of state to become a liquid. The condensation of atmospheric water vapour is an important process in the HYDROLOGICAL CYCLE and is caused when air is cooled below its DEWPOINT. Condensation is responsible for the formation of CLOUDS, PRECIPITATION, FOG and HOAR FROST. See also HYGROSCOPIC NUCLEI.

**conducting tissue** or **vascular tissue,** *n.* cells in higher plants (GYMNOSPERMS and ANGIOSPERMS) that function as tubes or ducts through which water, dissolved nutrients and manufactured sugars move from one part of the plant body to another. There are two types of conducting tissue, the *xylem* which conducts water and salts from the roots upwards through the plant and the *phloem*, through which fluids can move either upwards or down. The phloem is responsible for distributing manufactured foodstuffs to all parts of the plant.

**conduction,** *n.* the direct transfer of heat from one medium to another. Much of the heat from the earth's surface is lost to the atmosphere through the conduction of outgoing longwave radiation despite air being a poor conductor. See INSOLATION, ALBEDO.

**conformer organism,** *n.* any organism whose internal rate of metabolism is controlled by the external environmental conditions. Most plants are conformer organisms as are most animals apart from the warm blooded mammals. External temperature exerts a major control on conformer species with most life forms being active over a fairly small temperature range 0-36°C. Compare REGULATOR ORGANISM.

**conglomerate,** *n.* a SEDIMENTARY ROCK consisting of rounded fragments of rock debris cemented within a matrix of finer material and formed by the LITHIFICATION of certain coastal and riverine sediments.

**coniferous tree,** see GYMNOSPERM.

**connate water** or **fossil water,** *n.* the water trapped in SEDIMENTARY ROCKS during their formation. Connate water is an important source of GROUNDWATER and is tapped by wells for domestic, industrial and agricultural purposes, especially in arid regions.

**conservation,** *n.* the management, protection and preservation of the earth's natural resources and environment. It has been suggested that a 'total' conservation policy must incorporate a conservation ethic into the everyday life style of human society (see LAND ETHIC). Such an approach involves using biospheric resources for the purpose of maximizing the aesthetic, educational, recreational and economic benefits of those resources to society. A successful conservation policy would help ensure a BIOSPHERE which showed minimal evidence of DISTURBANCE to natural material cycles and energy flows; plant and animal diversity would be enhanced, physical erosion of soils would be minimized and the recycling of materials would prolong the life of existing and future resources.

No nation currently operates a total conservation policy. Instead, conservation measures are applied only when severe signs of biosphere degradation become apparent, for example, the *'Thirty Percent Club'* of European nations, dedicated to the reduction of sulphur dioxide in the atmosphere by at least 30% by the 1990s. To varying extents, governments recognize the importance of conservation and accordingly have financed the establishment of scientific bodies, such as the NATURE CONSERVANCY COUNCIL, or passed legislation such as the NATIONAL ENVIRONMENTAL POLICY ACT of 1970. See GREEN POLITICS.

**conservation area,** *n.* any urban area in which neighbourhoods containing varied and valuable items of TOWNSCAPE are protected against ad hoc replacement. In the UK, the CIVIC AMENITIES ACT (1967) forced planning authorities to establish conservation areas; consequently, planning applications to develop within such an area or to alter a component building are vetted with particular care and stringent conditions concerning both internal and external redevelopment can be attached to any permission which may be given. See also AMENITY, HISTORIC BUILDING, ANCIENT MONUMENT.

**conservative boundary,** see PLATE TECTONICS.

**consistency,** *n.* **1.** the degree of adhesion and cohesion demonstrated by a soil.

**2.** a measure of the resulting resistance of the soil to mechanical deformation.

**consolidation,** *n.* the amalgamation and reallocation of scattered plots of land within a region to produce farms whose fields have contiguous borders. Consolidation is undertaken to permit more rational and efficient farming, facilite MECHANIZATION, improve access to roads and other infrastructure, reduce labour costs, and create conditions favourable for land improvement.

Consolidation may be voluntary through a mutual exchange of parcels of land (as in Belgium), involve state financial assistance (as in France), or be undertaken in wholesale regional rationalization and resettlement schemes (as occurred in the Netherlands and West Germany). The progress of land consolidation in Europe has been variable, but with notable exceptions in Kenya, India and Taiwan, has not featured greatly in developing countries where the disadvantages of FRAGMENTATION may not, as yet, be so apparent.

The impact of consolidation on the rural landscape and settlement pattern can be dramatic, creating fewer, larger and more regular fields and a greater dispersal of farmsteads. Consolidation may produce a detrimental effect on local HABITATS, resulting from the removal of field boundaries, hedgerows and small woodlands.

**constructive boundary,** see PLATE TECTONICS.

**constructive wave,** see WAVE.

**contact metamorphism,** see METAMORPHISM.

**container port,** n. any large area of land, strategically located on, or adjacent to, a railway network, or alongside deep-water harbour frontage, where freight containers can be assembled into groups for loading onto trains or ships. At the terminal, large travelling cranes manoeuvre containers either directly on to their onward transport vehicle or into temporary storage clusters to await despatch. Access to a local road system is an important feature of their location but local traffic disruption can result from the movement, especially onto city streets, of very long and heavy container vehicles. Examples of container ports in the UK include Felixstowe and Seaforth (Liverpool). See also FREEPORT.

**continental divide,** see WATERSHED.

**continental drift,** *n.* the fragmentation of the supercontinent of PANGAEA following the CARBONIFEROUS PERIOD, and the gradual and continuing drifting of the continental landmasses to their present positions (see Fig. 19 overleaf). This widely accepted

## CONTINENTAL GLACIER

theory is supported by PLATE TECTONICS. Continental drift is thought to be the result of convection currents, established in the MANTLE through the accumulation of radioactive heat, causing the movement of LITHOSPHERIC PLATES on the partially molten ASTHENOSPHERE. Evidence to support the theory of continental drift includes the apparent fit of Africa's west coast and South America's east coast, the distribution of Carbopermian glacial landscapes in the southern hemisphere, and the distribution of certain living and fossil plant and animal species.

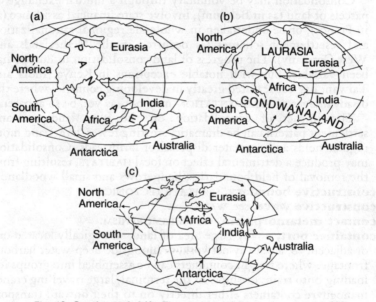

Fig. 19. **Continental drift.** The position of the landmasses (a) 200 million years ago (b) 180 million years ago (c) currently.

**continental glacier,** see GLACIER.

**continental shelf,** *n.* the gently sloping sea floor around the land masses. The continental shelf is terminated by the CONTINENTAL SLOPE at a depth of between 120 and 370 m (see Fig. 20). Shelves may only be marginal (as along the west coast of South America) or can extend up to 1200 km (as off Florida, USA). Most continental shelves appear to be marine PLANATION SURFACES that may have been subsequently built up by riverine and glacial deposition. The shallow waters of the continental shelves often contain important fishing grounds and have allowed the exploitation of numerous economic resources including sand, gravel, tin, gold, platinum, oil and gas.

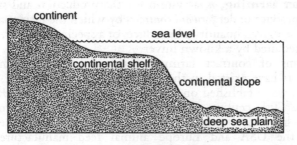

Fig. 20. **Continental shelf.**

**continental slope,** *n.* the steep slope at the edge of the CONTINENTAL SHELF that descends to the DEEP-SEA PLAIN (see Fig. 20). The angle of slope may vary from 2° to the 45° found off the coasts of Sri Lanka and Cuba. The irregular terrain of continental slopes is thought to be due to underwater FAULTS, the scouring action of TURBIDITY CURRENTS, and MARINE EROSION during a glacial period when the sea level was lowered.

**continuous cropping,** *n.* a system in which one crop is grown after the other for several seasons in succession, without recourse to seasonal FALLOW. The land may then be fallowed for a number of years. It is thus a feature of some, but by no means all, tropical farming systems employing techniques of SHIFTING CULTIVATION or ROTATIONAL BUSH FALLOW.

Continuous cropping may be achieved by sequential or relay planting (see MULTIPLE CROPPING). In the West African savanna, cycles operating over some eight seasons rotate grain crops such as millet and sorghum with root crops such as yams and cassava.

**continuum,** see VEGETATION CLASSIFICATION.

**contour ploughing,** *n.* a form of TILLAGE in which furrows are ploughed along the line of the natural contour of the land, rather than across it, in order to reduce the liklihood of SOIL EROSION from the unimpeded run-off of rainwater (see STRIP FARMING).

Contour ploughing has been encouraged in many of the drier areas of the USA and in tropical regions where the need for SOIL CONSERVATION is paramount. See CONTOUR RIDGING.

**contour ridging,** *n.* the construction of ridges along the line of the natural contour of the land to prevent the unimpeded runoff of rainwater and subsequent soil erosion. The technique is common in tropical farming systems employing hand tools rather than the plough. See CONTOUR PLOUGHING. See also MOUND CULTIVATION.

**contract farming,** *n.* a system for the production and supply of farm produce under *forward contracts* by which the farmer agrees to provide a given quantity of produce, of a specified quality, and at a time required by a known buyer.

Forms of contract farming between livestock farmers and butchers have existed in the UK since the 19th century, but the system was established on a more formal footing in the USA early in the 20th century when farmers signed contracts to supply particular kinds of meat to meat-packing companies. Since then in both the USA and Europe animal feed manufacturers have pioneered contract production of poultry, eggs and livestock. Contract farming also is common in the marketing of fruit, vegetables, cereals, oilseeds and cotton. It has contributed to the trend towards greater farm specialization, the concentration of production on larger holdings and increased capital-intensity in operation. For the farmer it means a reduction in uncertainty and financial risk, but at the loss of independence and control over production decisions. See also FACTORY FARMING, AGRIBUSINESS.

**contract rent,** see FARM RENT.

**contraflow,** *n.* a method of TRAFFIC MANAGEMENT employed on dual carriageway roads or MOTORWAYS when repairs necessitate the temporary closure of one carriageway. The adjoining carriageway is divided along its centre line by movable markers and traffic from the closed carriageway is diverted onto one side of it, and moves against the normal direction of traffic on that carriageway which continues to use the nearside lane. During the operation of such a contraflow system the carriageway in use becomes a conventional single carriageway road.

**controlled tipping, sanitary landfill** or **secured landfill,** *n.* the burying of refuse or waste materials including hazardous and/or TOXIC WASTES underground. The refuse is usually spread in layers between which soil or building rubble is laid in order to minimize the generation of spontaneous heat through decomposition and the accumulation of toxic gases such as METHANE and hydrogen sulphide. In the USA, some 82% of urban wastes are disposed of by controlled tipping, the remainder being burnt, recycled or composted. The sites of controlled tipping are carefully chosen so that the geological strata does not permit the transfer of the wastes via groundwater movement to areas away from the LANDFILL site.

**control rods** or **safety rods,** *n.* rods which can be inserted into the core of a NUCLEAR REACTOR in order to control the rate of NUCLEAR FISSION. Normally, the reactor operates without the rods in place but in the event of an emergency or breakdown sufficient rods can be inserted into the core of the reactor in order to slow down the fission chain reaction. Control rods are usually made from graphite, boron, hafnium or cadmium, all of which have high neutron capture capability. See ADVANCED GAS-COOLED REACTOR.

**conurbation,** *n.* a continuously built up area, often consisting of a central city and a number of adjacent towns. Originally, the latter may have been separate from the central city but expansion at the outer edges of the urban areas has resulted in a coalesced agglomeration. In the UK conurbations have traditionally been administered as a group of autonomous local authorities. The reform of local government in the mid-1970s had, as one of its main objectives, the rationalizing of this fragmented administration by the creation of supra-authorities at a geographical scale large enough to control conurbations as single administrative units. In England and Wales these authorities were called metropolitan counties until their abolition in 1986. Elsewhere in Western Europe most conurbations have some form of metropolitan government with area-wide functions including strategic land use and transportation planning.

**convection,** *n.* the vertical movement of air or water which follows the transfer of heat from a warmer body via CONDUCTION.

A *convection cell* is established in the atmosphere when the layer of air above the earth's surface, which has been warmed by INSOLATION, is also warmed, causing its density to be reduced; the mass of heated air then rises to be replaced by colder, denser descending air which in turn is heated so maintaining the circulation. Convection cells can generate atmospheric INSTABILITY and lead to the formation of CLOUDS and associated convectional rainfall. THUNDERSTORMS are caused by severe convection currents.

**convectional rain,** see RAIN.

**convection cell,** see CONVECTION.

**Convention on International Trade in Endangered Species (CITES),** *n.* a treaty signed by over 80 countries since 1973 which prohibits international trade in the rarest 600 or so species of plants and animals, and requires a licence from the country of origin for export of about 200 other groups of species. Despite the undoubted success of CITES in stemming trade in endangered species, considerable illicit trade in species such as South American parrots, African 'big-game', birds eggs and rare orchids still occurs. The Far East is responsible for much of the illegal trade which was estimated to be worth $100 million in 1986.

**convergent boundary,** see PLATE TECTONICS.

**coombe ridge,** see ARÊTE.

**coppicing,** *n.* the periodic felling of mature trees and the subsequent regrowth of new side shoots from the old tree stump. Coppicing was widely used throughout Europe between 1500 and 1800 as the FOREST MANAGEMENT technique best suited for the provision of medium-sized timber (10-20 cm log diametre). Fast-growing species such as willow, alder, poplar and birch were coppiced every 15-30 years. Occasional trees (*standards*) were left uncut to provide seeds for replanting new areas of forest. Because of its labour-intensive nature, coppicing has been superceded in cost-effectiveness by other forest management techniques and is now rarely used in developed countries. However, new interest has been shown in coppicing as a means of producing BIOMASS fuel. In the tropics, the management of forest by coppicing every 3-5 years can supply an abundant source of fuelwood.

**coral,** *n.* a small primitive marine animal with a calcium carbonate skeleton found in tropical and equatorial waters. Coral may be solitary or colonial and grows only in clear, sunlit salt water up to a depth of 55 m and where water temperature exceeds 20°C. Colonial forms of coral may develop into extensive formations known as CORAL REEFS.

**coral reef,** *n.* a limestone ridge situated at or near the surface of the sea and formed from the calcareous skeletons of reef-building CORALS and other marine organisms. Coral reefs occur in tropical and equatorial regions between the latitudes 30°N and 25°S, especially on the eastern sides of landmasses where warm ocean currents occur. Reefs do not develop on western coasts in these latitudes due to the flow of cold currents. Three types of coral reef have been recognized:

(a) *fringing reefs* consist of a platform of coral attached to the shore, and may only be visible at low tide. A shallow LAGOON may lie between the shore and the outer edge of the reef. Fringing reefs may extend up to 2 km seaward and are usually near the mouths of rivers. The river channels allow the passage of shipping in an otherwise impenetrable coral coast,

(b) *barrier reefs* can be situated up to 300 km offshore and are separated from the coasts by a lagoon. Breaks or passes in the reef allows the movement of shipping. Fringing reefs can form within barrier reefs. The Great Barrier Reef off the north-east coast of Australia is the largest coral structure in the world,

(c) *atolls* are low-lying, ring-shaped islands that enclose a lagoon. Changing sea levels during and after the last glacial period and the subsidence of land masses are thought to explain the formation and development of the different types of coral reef.

**core,** *n.* the central part of the earth's interior lying below the GUTENBERG DISCONTINUITY. The core is a high density metallic mass largely composed of nickel and iron and is about 3475 km in diameter. It consists of a less dense fluid outer section and a very dense solid inner core of 1200 km diameter. Temperature in the core is thought to be about 3000°C and is under a confining pressure of 1.5 to 3 million atmospheres.

**core area,** see BIOSPHERE RESERVE.

**Coriolis force,** *n.* the force generated by the earth's rotation which causes the deflection of winds to the right in the northern hemisphere and to the left in the southern hemisphere. Named after the French civil engineer, Gaspard Coriolis (1792-1843).

**corrasion,** see FLUVIAL EROSION.

**corrie,** see CIRQUE.

**corrosion,** see FLUVIAL EROSION.

**Cosmos 1870,** *n.* an 18-tonne Soviet space platform, eight times larger than any previous earth resources satellite. The platform was launched in July 1987 and circles the earth every 90 minutes providing information on earth surface land use at a scale hitherto unachieved by civilian satellites. Objects as small as 5 m² can be identified. Unlike the earlier American and French earth resources satellites, Cosmos does not transmit information back to earth in digital signal form; instead it takes pictures and sends the film back to earth in re-entry modules. The films are retrieved and processed every month. An equivalent civilian American earth observation satellite is not planned until 1996. See LANDSAT, EARTH RESOURCES TECHNOLOGY SATELLITE, SPOT.

**cost-benefit analysis,** *n.* a technique used to evaluate and distinguish between alternative investment proposals. It differs from most other forms of financial appraisal in costing, in monetary terms, all the factors involved, commercial, social and environmental, regardless of to whom the costs and benefits accrue. It is less widely but more correctly known as *benefit-cost analysis*.

Cost-benefit analysis proceeds by:

(a) identifying the alternative proposals;

(b) quantifying the gains or benefits and losses or costs of each alternative;

(c) discounting the benefits and costs in order to obtain one figure to represent the current value of all the present and future benefits and another to represent the value of all present and future costs;

(d) employing these figures to calculate benefit-cost ratios. Proposals where the ratio is greater than 1 are considered worthwhile.

(e) selecting the proposal with the greatest net benefit, that is, the one with the greatest margin between benefits and costs.

Cost-benefit analysis has been used in decisions over infrastructural provision, for example, the choice between sites for a new airport or dam, and routes for a new road. The technique has been widely employed by the the WORLD BANK in the THIRD WORLD, and in situations where policy goals are relatively simple. However, there are a number of fundamental problems. In practice, it may not be possible to quantify in money terms the value of human lives likely to be saved through flood control, the loss of wilderness area or a particular AMENITY, the effect of change on wildlife, or the gains and losses in mental and physical effort. There is a contentious subjective element in whom decides which costs and benefits are to be considered, and whether or not they represent particular interest groups; for example, a dam project may provide cheap hydro-electric power for a distant urban area and an increase in industrial employment opportunities, but cause settlement relocation, a loss of productive farmland, and a rise in water-related diseases for local inhabitants. The difficulty in resolving many of these issues has led in many countries to the adoption of more comprehensive procedures of project appraisal, for example, the use of ENVIRONMENTAL IMPACT ASSESSMENT favoured under the US NATIONAL ENVIRONMENTAL POLICY ACT.

**Council on Environmental Quality,** *n.* a US Presidential committee formed under the terms of the NATIONAL ENVIRONMEN-TAL POLICY ACT (1969) and responsible for conducting detailed analyses of the cost of environmental protection and the effect this would have on economic activity. It was calculated that in the decade 1972-82 $274 billion would be required to maintain environmental quality levels; this figure was about 2% of the American GNP. However, the cost of *not* protecting the environment is considerably greater. In 1968, air pollution damage alone in the USA was calculated to amount to $16 billion, mainly due to reduced agricultural yields, loss of labour through pollution-induced illness, hospitalization costs, delays to aircraft through poor visibility, etc. By comparison, the direct costs of air pollution in the UK in the early 1970s amounted annually to £360 million.

**country park,** *n.* an area of scenic countryside adjacent to a large urban area, created in the UK under the provision of the Countryside Act of 1968. Its purpose is to give urban dwellers increased access to outdoor recreation on sites which can be managed to this end; it was hoped that country parks would reduce the level of destruction caused in rural areas by the indescriminate pursuit of leisure activities. The parks offer walks, picnic areas and water-borne leisure activities; some have a ranger service to provide guidance and wildlife interpretation for visitors. See also AREA OF OUTSTANDING NATURAL BEAUTY, AMENITY.

**country rock** or **host rock,** *n.* an existing rock into which IGNEOUS ROCK masses or mineral veins intrude.

**crater,** *n.* **1.** the bowl-shaped depression around the vent at the summit of a VOLCANO.

**2.** a depression formed in the earth's surface by the impact of a METEORITE.

**crater lake,** see LAKE.

**creep,** *n.* the slow, almost imperceptible downslope movement of soil and rock debris under the influence of gravity. Creep is most effective in upper soil layers and is usually absent below depths of 90 cm. Rates of movement rarely exceed 2.5 cm per year. Creep is initiated by several processes including raindrop impact, the wetting and drying of soil, freeze-thaw action (see PHYSICAL WEATHERING), the heating and cooling of soil and rock, root growth and the activity of animals on and below a slope. The economic cost of creep can be considerable due to the repairs required for such things as tilting telephone poles, bulging and broken walls and stress fractures in road surfaces. See SOLIFLUC-TION. Compare MUDFLOW, SLUMP, EARTHFLOW.

**Cretaceous period,** *n.* the geological PERIOD that followed the JURASSIC PERIOD and extended from 136 to about 65 million years ago, when the TERTIARY PERIOD began. The Cretaceous is the final period of the MESOZOIC ERA. There was a major phase of mountain building and associated volcanic and igneous activity in both North and South America. During this period the dinosaurs rapidly disappeared and were replaced by the mammals, which made significant evolutionary advances. The plant kingdom also made significant gains during this period, with ANGIOSPERMS appearing for the first time in significant numbers. By the end of this period modern flora and fauna were well established. See EARTH HISTORY.

**critical factor,** see TRIGGER FACTOR.

**crofting,** *n.* a method of land use and a way of life based on the occupation of a croft, a small individual lot created in the Highlands and Islands of Scotland in the early decades of the 19th century, and subsequently regulated by law.

Agriculturally, crofting has involved continuous cropping of inbye or infield land and the common grazing of PASTURE, shared with all the other crofters in the township. A central feature of crofting is its essentially PART-TIME FARMING character, with most crofters both in the past and in the present relying for most of their livelihood on other sectors of economic activity, either being self-employed as in the provision of tourist accommodation, craft production and fishing, or in part-time wage employment, such as FORESTRY, FISH FARMING.

**crop,** *n.* any useful plant grown on a farm, except PASTURE. Crops can be classified in several ways, according to:

(a) botanical family. Over 80% of crops are grasses (Gramineae), legumes (Leguminosae) and crucifers (Cruciferae).

(b) the length of soil occupance. ANNUALS are plants that grow, flower, produce seed and die in one growing season (for example, GRAIN CROPS). They may occupy the soil for a few weeks only, or for up to 18 months in tropical areas where seasons are weakly differentiated. If the growing season is long, MULTIPLE CROPPING may be possible. BIENNIALS have a vegetation cycle of 18-30 months, growing in one season to flower in the next (for example, sugar beet and turnip). PERENNIALS occupy the soil for more than 30 months and yield annually for some years. They may be further differentiated as herbacious or perennial field crops that require cultivation and are generally replaced after 3-12 years (for example, bananas, sisal and sugar cane), and woody stemmed perennials or PERMANENT CROPS of trees and shrubs. Not all perennials are utilized in the manner of their natural productive cycle. The potato is a perennial herb but in normal agricultural practice the tubers planted in the spring are lifted in the autumn. Cassava (manioc) is a woody-stemmed shrub but the swollen tuberous roots can be harvested after eight months, and for up to four years after planting.

(c) their primary use as a FOOD CROP, a FODDER CROP or a non-food crop in which the product is used industrially, such as cotton and rubber. A crop that is grown in one region principally as a food crop may elsewhere be a green fodder crop or grown for industrial purposes. Maize, for example, is an important food crop in Africa, yet in northern temperate zones is fed green to animals or made into SILAGE. Its industrial value is recognized in such activities as starch manufacture and whisky distilling.

(d) their dominant economic function for the cultivator as the basis of SUBSISTENCE FARMING, or as a source of income or CASH CROP.

(e) their specific agronomic function, for example BREAK CROPS and CATCH CROPS.

(f) their broad agricultural subdivision as PERMANENT CROPS of trees and shrubs, GRAIN CROPS or LEAFY CROPS.

**crop combination analysis,** *n.* a technique used to establish the boundaries of AGRICULTURAL REGIONS, based on the statistical comparison of acreages of different CROPS. Traditional AGRICULTU-RAL TYPOLOGIES have tended to emphasize the leading crop of a region, as in designations such as the corn belt or the cotton belt, when in practice these crops are grown in association with others. Following the work of the American geographer J.C. Weaver in the mid-1950s, various statistical techniques have been employed to identify the most representative combination of crops in a region. These may then be mapped either by shading each combination, or by shading the leading crop and overprinting with letters representing the other crops in the combination. Such a comparison of acreages makes no allowance for the intensity of production and takes no account of areas devoted to ANIMAL HUSBANDRY, a weakness that has led to the development of FARM ENTERPRISE COMBINATION ANALYSIS.

**cropland,** *n.* all land on which CROPS are grown, either in regular alternation, or in MONOCULTURE. In the organization of farmland it is usual to distinguish cropland or arable land fit for ploughing from PASTURES, except under short-term LEY FARMING systems, and land devoted to PERMANENT CROPS of trees and shrubs.

Arable cropland comprises 1.4 billion hectares of land or 30% of the world's utilized agricultural area (excluding woodland). Among the 12 member countries of the EC in 1986 the percentage of the utilized agricultural area devoted to arable land ranged from 19% in Ireland to 91% in Denmark with a community average of 53%. International comparisons include : USA (44%), Canada (60%), Australia (10%), New Zealand (3%), and USSR (38%).

**cropland share,** *n.* the areal extent of specific crop types expressed as a percentage of total CROPLAND. Thus, in the UK in 1986, the cropland share of wheat was 29%.

**crop rotation,** *n.* a short-term sequence of cropping on one field, in which the *rotational period* is the length of the rotation in years. Although MONOCULTURES of wheat, barley, or maize are practical in some areas, most ARABLE FARMING systems involve some method of crop rotation. Crop rotation is employed to maintain soil fertility, to introduce BREAK CROPS, to produce high-value but often high-cost crops as a means of increasing income, to broaden the economic base of the farm enterprise, and to spread the demands on labour more evenly through the year.

A rotation of economically more important crops or those grown nearer the farmyard is a *primary crop rotation*, in contrast to a *secondary crop rotation* of less important crops or those grown in more distant fields.

**cross-section,** *n.* a profile of a river valley taken transversely across a stream's course. The shape of the cross-section is determined by the rates of WEATHERING and MASS MOVEMENTS on the valley sides, the nature and strength of the underlying bedrock, the extent of FLUVIAL EROSION and FLUVIAL DEPOSITION, and the length of time these factors have been in operation. Compare RIVER PROFILE.

**crowding** or **overcrowding,** *n.* the situation that results when the density of population exceeds a critical threshold. The particular drawbacks of overcrowding were pointed out by the architect-planner Raymond Unwin in a pioneering essay entitled *Nothing Gained by Overcrowding*. In this he advocated housing densities of 12 dwellings per acre (30 per hectare), a standard that was adhered to in much of the suburban expansion that occurred in the UK between the World Wars. See also RESIDENTIAL DENSITY, ROOM OCCUPANCY, OVERSPILL, PERIPHERAL ESTATE.

**crude birth rate,** see BIRTH RATE.

**crude oil,** *n.* OIL or petroleum in its natural state before refining.

**crustacean,** *n.* any member of the primitive, mainly aquatic class Crustacea, typically having a carapace hardened with lime and highly developed appendages such as mandibles and pincers; examples include lobsters, crabs, shrimps, barnacles, copeopods, KRILL and water fleas. There are about 35 000 distinct species of crustaceans, a few of which have become adapted to fresh water; one suborder (the woodlice) lives on land.

Crustaceans play a major role in the marine FOOD CHAIN: they form part of the PLANKTON found in the upper layers of the oceans and as such, form an important source of food for other marine organisms, and increasingly for humans also. Some 2.5 million tonnes of crustaceans are now caught for human consumption, mainly in the Pacific Ocean, the west central Atlantic, and the western Indian Ocean.

**cryptophyte,** see RAUNKIAER'S LIFEFORM CLASSIFICATION.

**cultigen,** *n.* any plant grown under cultivation that is sufficiently different from any wild plant to be of uncertain origin, for example, rape (*Brassica napus*).

**cultivar,** *n.* any cultivated variety of a plant SPECIES, differing in some respect from the rest of the species, that is maintained by cultivation. Cultivars of most crops are the product of SELECTIVE BREEDING, and differ genetically by one or more agronomically important traits, such as disease resistance or higher yielding performance.

**cultivation,** *n.* the preparation and use of land for growing CROPS. A wide range of cultivation systems are in use throughout the world and may be classified according to several distinguishing features; these include:

(a) the type of field rotation, that is, the long-term alternation between various types of land use, including FALLOW periods. As a result of the rise in AGRICULTURAL INDUSTRIALIZATION in the northern hemisphere, field rotation has become a less necessary feature and is now associated more with agriculture in developing countries;

(b) the intensity of rotation within a FALLOW SYSTEM, particularly in tropical regions. The relationship between crop cultivation and fallowing within a cycle of land utilization differentiates between extensive SHIFTING CULTIVATION, ROTATIONAL BUSH FALLOW, and progressively more intensive systems leading to PERMANENT CROPPING and MULTIPLE CROPPING;

(c) the nature of water supply, in particular the broad distinction between DRY FARMING and systems dependent on IRRIGATION (see WET-RICE CULTIVATION);

(d) the cropping pattern and animal activities of the holdings whereby farms are grouped according to broadly similar combinations of activities (see ALPINE FARMING, MEDITERRANEAN FARMING);

(e) the tools used for cultivation (see MECHANIZATION OF AGRICULTURE);

(f) the degree of commercialization (see SUBSISTENCE FARMING, COMMERCIAL FARMING);

(g) the operation of national, or mutlinational agricultural policies (see COMMON AGRICULTURAL POLICY, QUOTA).

**culvert,** *n.* a pipe or other covered passage under a road or railway, which carries a stream or drainage ditch. In the centres of some cities, small rivers that were important in the early development of the city are frequently enclosed in culverts and built over.

**cumulonimbus, anvil cloud** or **thunderhead,** *n.* a heavy, dense, grey CUMULUS cloud usually of immense vertical development. The top of cumulonimbus often spreads outwards to form the shape of an anvil. Cumulonimbus is associated with the development of THUNDERSTORMS.

**cumulus,** *n.* a flat-based, dome-shaped CLOUD of considerable vertical development. White isolated cumulus is associated with warm, anticyclonic summer weather whilst heavy, grey cumulus often develops into CUMULONIMBUS.

**curie,** *n.* a unit of radioactivity equal to $3.7 \times 10^{10}$ disintegrations per gram per second. Symbol: Ci. Its value is based on early measurements of the activity of 1 g of radium-226. Until recently, the curie was the standard unit of radioactivity but the term has been replaced by the BECQUEREL. Named after the Polish-born French physicist and chemist, Marie Curie (1867-1934).

**current annual increment (CAI),** *n.* the rate of annual increase in the volume of timber within the trunk of a tree. The term is most frequently applied by British foresters when measuring the

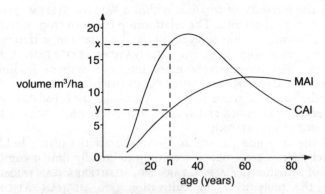

Fig. 21. **Current annual increment.** The pattern of volume increment in even-aged forests. After *n* years the annual volume increment at that time is *x* whilst the average annual volume increment from the time of planting to *n* years is *y*.

growth rate of even-aged coniferous STANDS of trees. Compare MEAN ANNUAL INCREMENT. Fig. 21 shows the relationship which exists between the CAI and MAI.

**current ground cover,** see LAND USE.

**cuspate delta,** see DELTA.

**cutoff,** see OXBOW LAKE.

**cwm,** see CIRQUE.

**cyclone,** see DEPRESSION.

**cyclone dust scrubber,** *n.* a device for extracting dust from industrial waste gases. The scrubber comprises an inverted cone into which the contaminated gas enters tangentially from the top. The gas is propelled down a helical pathway during which the dust particles are deposited by means of centrifugal force onto the wall of the scrubber. The particles fall to a hopper placed at the bottom of the scrubber while the cleansed gas escapes from an aperture at the top of the cone. When operating at their design capacity, cyclone dust scrubbers can effectively remove up to 95% of particulate load. They are best suited for removal of coarse particles, greater than 85 $\mu$m in size. See ELECTROSTATIC PRECIPITATOR, SCRUBBER.

**cyclonic rain,** see RAIN.

# D

**DALR,** see DRY ADIABATIC LAPSE RATE.

**dating techniques,** *n.* any of a number of methods whereby an object, natural phenomenon, or series of events can be dated to an acceptable degree of accuracy. Dating techniques can be divided into two types:

(a) *absolute dating* techniques that provide a specific date on a time scale. Absolute methods include RADIO-CARBON DATING, POTASSIUM-ARGON DATING, and DENDROCHRONOLOGY. Complete accuracy is rarely possible and it is necessary to apply a error factor; dates are given thus: 5000 years ±38, a date lying within the range 4962-5038 BEFORE PRESENT.

(b) *relative dating* techniques that provide a location for an event within a general chronological sequence. POLLEN ANALYSIS, diatom and Coleoptera analysis and varve sediments are widely used relative dating methods.

**day centre,** *n.* a non-residential meeting place for people, often the elderly or handicapped, that promotes social contact, or provides educational or recreational facilities as required. Day centres may be run by local authorities or voluntary organizations and make use of suitable premises within easy access of users' homes.

**DDT,** *n. Trademark.* 1,1,1-trichloro-2,2-di-(4-chlorophenyl) ethane, an organochlorine INSECTICIDE widely used between the early 1940s and 1960. It was first synthesized in 1874 by the German chemist Dr O. Zeidler, but no use for it was identified. The substance lay unused for 65 years until 'refound' in 1939, when it was discovered to be effective against fleas, lice, Colorado beetle, mosquitoes and flies. As with many other chemicals used in the battle with plant and insect pests, the precise action of DDT is not known; it affects the peripheral nervous system of insects causing characteristic tremors and convulsions.

It has been extensively used as a soil PESTICIDE, as a seed treatment, and against grasshoppers and cotton insects. When sprayed onto food crops it produces a distinctive taint. It was formerly used to control flies in milking parlours but DDT residues passed into the milk.

DDT became the most widely used insecticide in the world with annual production of 100 000 tonnes in the late 1950s. Since then, production has declined to about 30% of that figure. During its first decade of use, it has been estimated that DDT saved 5 million human lives and prevented 100 million serious illnesses due to malaria, typhus, dysentery and more than twenty other lesser known insect borne diseases.

By 1960, however, concern over DDT residues in soil and water stimulated new research, especially in Canada and the Scandinavian countries. The results of this work showed that organochlorides were not only stored in invertebrate and vertebrate tissue, but were also concentrated in the upper TROPHIC LEVELS of FOOD CHAINS. By 1965 stringent restrictions or complete bans on organochloride insecticides had been introduced in Europe and North America.

All organic tissue now contains traces of DDT and despite its ban in many countries the level of DDT residues in the biota and physical environment will continue to rise well into the next century.

DDT has been blamed for the catastrophic decline in the numbers of many species of wildlife, but in all probability changes in land and water development and usage have exerted a much greater influence than pesticides. It is probable that organochlorides make the greatest impact on aquatic organisms such as PLANKTON, KRILL and CRUSTACEANS.

**death rate** or **mortality rate,** *n.* the ratio of deaths in a specified area, group etc., to the population of that area, usually expressed per thousand per year. Compare BIRTH RATE.

**death zone** or **zone of intolerance,** *n.* a position at the extremity of the ENVIRONMENTAL GRADIENT at which a species cannot survive, for example, a fish in deoxygenated water or a plant in a soil devoid of water. See TOLERANCE.

**deceleration lane** or **slip road,** *n.* a lane on a carriageway, often at a widening of a major road, that permits a vehicle to detach itself smoothly from a line of fast-flowing traffic, and turn off the road. In the case of a MOTORWAY, deceleration lanes are also known as *off-ramps* since these lanes frequently drop or rise to a level where the ordinary road system will pass under or over the motorway.

**decibel,** *n.* the unit of measurement of sound intensity in watts per square metre, relative to the intensity of the quietest sound perceptible to the human ear. Symbol: dB. Several different decibel scales exist, for example, the flat or C-scale which is the unweighted range of sound from 20-10 000 Hz or the more common A-scale (dBA) which gives increased weighting to high-frequency notes which are of greatest annoyance to humans.

All sound scales are logarithmic so that each 10 dB rise represents a ten-fold increase in sound intensity. Thus an increase from 50 dB (normal conversation) to 100 dB (underground train) is equivalent to a $10^5$ increase in noise. (see Fig. 22). See NOISE POLLUTION.

| Example | Decibel (dBA) | Relative sound intensity | Effect with prolonged exposure |
|---|---|---|---|
| Jet aircraft (at take off) | 150 | $10^{15}$ | Eardrum ruptures |
| Thunderclap | 120 | $10^{12}$ | Pain threshold |
| Live rock music | 110 | $10^{11}$ | |
| Motorbike, farm tractor, blender, underground train | 100 | $10^{10}$ | Serious hearing damage (after exposure for 8 hours) |
| Dishwasher, noisy office | 80 | $10^8$ | |
| Typical suburb | 60 | $10^6$ | Intrusive |
| Quiet suburb | 50 | $10^5$ | Quiet |
| Library | 40 | $10^4$ | |
| Whisper | 20 | $10^2$ | Very quiet |
| Breathing | 10 | 10 | |

Fig. 22. **Decibel.** Some common noise levels.

**deciduous forest,** n. any forest comprising trees which shed their leaves or needles in response to an unfavourable climatic event, usually the onset of autumn when illumination and temperature levels become LIMITING FACTORS for growth. Deciduous forests were once extensive in north west, central and eastern Europe, in the eastern USA, northern China, Korea, Japan and in the far eastern part of the USSR and are still found mainly, although not exclusively, in the middle latitudes (40-55°) of the northern hemisphere.

Prior to 1500 AD forests showed a wide species variety (oak, ash, elm, beech, maple, hickory, birch, alder) with up to 40 different species per hectare. PRIMARY PRODUCTIVITY values averaged 1200 g/m²/year. The valuable HARDWOOD timbers have been exploited since the 1500s and today little original deciduous forest remains.

The annual deposition of leaves provides the soil beneath the deciduous forest with a rich supply of HUMUS and results in a fertile BROWN EARTH, a soil capable of producing high agricultural output with minimum loss of fertility and that is resistant to erosion. Compare EVERGREEN FOREST.

**deck access block,** *n.* a style of multistorey residential building in which the front doors of the individual homes at each level open onto a covered passageway which runs the length of the block. Such passageways are wholly contained within the building, but may have open balconies along their outer side. The slight similarity to a conventional street led to these passages to be described at one time as 'streets in the air'. Park Hill Flats, Sheffield, built in the 1950s are a particularly well-known example. Although intended to foster neighbourliness among the residents of each 'street', the design has proved to have some problems, notably that residents feel their privacy is infringed by having the common footway so close to the front windows.

**decommissioning,** *n.* the permanent shutdown and dismantling of a nuclear power plant. No nuclear reactors have yet been fully decommissioned although many of the MAGNOX REACTORS in the UK will have reached the limit of their working lives early in the 1990s. Considerable knowledge has been gained on the problems of making safe a nuclear reactor from the Chernobyl disaster. Three distinct stages have been formulated for the decommissioning process:

(a) *mothballing*, a temporary interval during which RADIATION within the power plant is allowed to subside naturally;

(b) *entombment*, the permanent covering of all radioactive components of a nuclear power plant with reinforced concrete, with the addition of a stainless steel, lead, sand or clay lining;

(c) *dismantlement*, the complete removal of all contaminated parts of the reactor involving extensive and thorough cleaning of components by water, steam and/or chemical means. Disposal of cleansing fluids must be carefully controlled.

See NUCLEAR REACTOR, NUCLEAR POWER, NUCLEAR WASTE.

**decomposer** or **detrivore,** *n.* any organism in an ECOSYSTEM that feeds on plant and animal protoplasm, breaking it down into its constituent parts, so bringing about decay. This action releases energy and allows soil nutrients to reach a high state of ENTROPY. All organisms can be said to be decomposers, although the term is usually confined to the SAPROPHAGES. Saprophages are responsible for up to 90% of the transfer of energy and nutrients within an ecosystem. Their operation is essential for the renewal of soil fertility and the maintenance of the mineral NUTRIENT CYCLES. See DETRITAL FOOD CHAIN.

**deep mining,** *n.* the extraction of deep-lying mineral resources from subsurface workings. Deep mining usually involves the driving of a vertical shaft from the surface down to the level of the mineral deposits which are then accessed by means of horizontal tunnels.

Many different mineral resources are exploited by deep-mining techniques including metalliferrous ores, building stone, coal, precious stones, sulphur, potash and rock salt. South Africa's gold and diamond mines are amongst the deepest in the world, reaching depths of several kilometres.

Deep mining is a dangerous occupation despite the stringent safety precautions taken in most countries. Underground fires, explosions and the build-up of toxic gases are a constant threat in many mines and can result in the loss of life, expensive machinery and productive workings. Miners can be afflicted by a number of dust-related pulmonary diseases including pneumoconiosis, silicosis, and asbestosis whilst those working in uranium mines have been found to have unusually high incidence rates of cancers.

The rock waste from deep mines, known as *tailings*, is usually dumped on the surface to form unsightly spoil heaps. The percolation of precipitation and groundwater seepage through spoil heaps can lead to the problem of ACID MINE DRAINAGE.

In areas of abandoned deep mines, subsidence often occurs due to the collapse of underground workings. This may cause serious damage to existing engineering structures although subsidence can usually be minimized by pumping concrete into the former workings. If records of former workings exist, careful land use planning can prevent the development of land susceptible to subsidence and avoid the need for expensive remedial work. Mined-out workings are occasionally put to other uses including storage and for military installations. Compare OPENCAST MINING

**deep-sea deposit, abyssal deposit** or **ooze,** *n.* the sediment found on the DEEP-SEA PLAIN. Two types of deposit can be distinguished:

(a) biological deposits, consisting of the remains of various marine organisms and material deposited from suspension, such as CLAYS. To depths of 3900 m, calcareous skeletons dominate biological deposits. Between 3900 m and 5000 m, calcium carbonate dissolves under the intense pressure and the remains of siliceous marine microorganisms are found. Biological oozes are usually named after the most abundant organism,

(b) non-biological deposits, which are found below depths of 5000 m where even silica is soluble. These red clay deposits are coloured by iron and manganese compounds and result from the sedimentation of particulate matter from the atmosphere such as volcanic ash and dust.

Manganese nodules are sometimes found in association with deep-sea deposits and may be of economic importance in the future.

**deep-sea plain** or **abyssal plain,** *n.* the extensive and relatively flat area extending from the base of the CONTINENTAL SLOPE. Deep-sea plains are, on average, 4000 m below sea level and are dissected by other submarine features such as DEEP-SEA TRENCHES, SEA-MOUNTS and GUYOTS. Deep-sea exploration has indicated the presence of substantial deposits of economic resources, especially manganese nodules, which may be exploited in the future. See ABYSSAL ENVIRONMENT.

**deep-sea trench,** *n.* a long, narrow and steep-sided V-shaped trough that dissects the DEEP-SEA PLAIN. At least 57 trenches have been identified, the deepest of which is the Marianas Trench near the island of Guam in the Pacific Ocean which is 11 033 m deep. Like MID-OCEANIC RIDGES and ISLAND ARCS, deep sea trenches form zones of intense volcanic activity and their origin is closely linked to the movement of LITHOSPHERIC PLATES.

**deep-sea zone,** see BENTHIC ZONE.

**deferred grazing,** see GRAZING MANAGEMENT.

**deflation,** see WIND EROSION.

**defoliant,** *n.* **1.** a HERBICIDE which causes the premature removal of leaves from plants. Modern defoliants such as 2,4-D and 2,4,5-T are *systemic* and attack the plant hormone system. They can be formulated to attack specific species, for example, allowing weeds to be removed from beneath a valuable forest crop. Defoliants were also widely used by the US forces during the Vietnam conflict to destroy the food sources and rainforest cover of their North Vietnamese opponents. AGENT ORANGE was the most notorious of these and the human and environmental repercussions of its use are still being felt today.

**2.** any animal, fungus or disease which feeds upon or infects the leaves of plants causing the removal of foliage. Caterpillars and locusts are notorious natural defoliators.

**deforestation,** *n.* the permanent clearing of forest land and its conversion to non-forest uses. Deforestation has been claimed by the WORLD RESOURCES INSTITUTE to be the world's most pressing land use problem. The precise extent of deforestation cannot be accurately assessed due to the remoteness of many areas from which forest is removed, the lack of written records for deforestation and the counteracting effects of AFFORESTATION.

Forest removal has continued, unabated, from the earliest days of human settlement. Attempts to estimate the former extent of the earth's forest cover suggest that $6 \times 10^9$ ha of land was forested some 5000 years before present but by 1954 this figure had been reduced to $4 \times 10^9$ ha.

Many developed countries recognize that forests are valuable RENEWABLE RESOURCES and tree removal is commonly followed by immediate AFFORESTATION. However, in the DEVELOPING COUNTRIES, sale of tropical hardwoods is a major source of foreign currency; in Africa and South America replanting schemes are, at best, only practised in a minority of instances. During the early 1980s the average annual rate of deforestation in tropical countries was estimated at 3.8 million ha. The FAO has estimated that 150 million ha (or 12% of remaining closed TROPICAL RAIN FOREST) will be deforested by the year 2000. Major losses have occurred in the Ivory Coast, Nigeria, Liberia, Guinea and Ghana where the rate of forest loss is seven times the world average.

Deforestation has been driven by population growth, from pressure to clear land for farming, for land speculation, for commercial ranching and by ruthless economic exploitation for profit. In its wake, deforestation has brought SOIL EROSION; DESERTIFICATION; sedimentation of water courses, lakes and dams; alteration of local climates through disruption of the energy balance and HYDROLOGICAL CYCLE, and massive EXTINCTION of plant and animal species which were dependent upon the forest habitat for their survival. However, perhaps most controversial of all consequences of deforestation are the changes to the atmospheric oxygen and carbon dioxide balance, which alter the ALBEDO and accelerate the GREENHOUSE EFFECT.

**deglaciation,** *n.* the retreat and erosion of an ice sheet or GLACIER. During the late PLEISTOCENE EPOCH, deglaciation led to the disappearance of several continental ice sheets, including the

Scandinavian Ice Sheet that centred on the Baltic Sea, and the Laurentide Ice Sheet that centred on the Hudson Bay in Canada.

**delta,** *n.* the flat alluvial area at the mouth of some rivers where the main stream divides into several distributaries. Deltas occur where there is an accumulation of river sediment deposited in the sea or a lake. Wave action and currents may operate simultaneously with river deposition to shape a delta. Three main types are recognized:

(a) the fan-shaped *arcuate delta*, of which the Nile Delta is an example;

(b) the lobate *bird's foot delta*, such as the Mississippi Delta;

(c) the tooth-like *cuspate delta*, such as the Tiber Delta in Italy.

Deltas can grow at considerable rates ranging from around 3 m per year for the Nile to 60 m per year for the Po in Italy. Some cities formerly at river mouths can now be some distance inland. With such rapid growth, flat deltaic plains result and are similar to FLOOD PLAINS. Such features in tropical areas are vulnerable to the flooding associated with tropical storms. In 1887, the Huang He delta in northern China was flooded with the loss of hundreds of thousands of lives and even today, innundations of the densely populated deltaic lands around the Bay of Bengal can cause considerable loss of life and damage to property.

**demersal fish,** *n.* any fish which lives near or at the bottom of a sea or lake. For example, plaice, cod and haddock.

**demographic transition,** *n.* the changes observed in the fertility and mortality of human populations, specifically where high BIRTH RATES and DEATH RATES alter to that of low birth and death rates. Demographic transition is expressed as a four-stage population MODEL intended to reflect the experience of countries developing and undergoing urbanization and industrialization; the stages are as follows:

(a) the *high stationary stage* in which disease, famine and war are common and birth and particularly death rates are high. The overall population level shows little variation and remains constantly low;

(b) the *early expanding stage* in which increased political stability and improved medicine and social conditions lowers the death rate while the birth rate continues at a high level. Total population thus expands at an increasing rate;

(c) the *late expanding stage* in which increasing urbanization and industrialization brings stability to the death rate and a drop in the birth rate. Total population continues to expand but at a reduced rate;

(d) the *low stationary stage* by which time both birth and death rates stabilize at a low level and the total population level is once again maintained at a constant, albeit higher, level.

Whereas the DEVELOPED COUNTRIES have long since completed the process of demographic transition, many DEVELOPING COUNTRIES are still in the throes of readjusting to rapidly reducing death rates.

**dendritic drainage,** see DRAINAGE PATTERN.

**dendrochronology,** *n.* the science of reconstructing and dating past bioclimatic events by means of studying the annual growth rings in tree trunks. Temperate forest species produce distinctive summer and winter accumulations of CONDUCTING TISSUE. It is possible to establish visual and statistical relationships between the rate of tissue growth and the prevailing climate. The technique requires that a living tree be cut down and the pattern of tissue growth to be carefully plotted. The sequence of growth over the age span of the tree is then compared and correlated with climatic data (average temperature, wetness, dryness, wind patterns, etc.) and a statistical model established between growth rate and climate. This model can then be tested with other trees of the same species growing in the same area. If proved valid, the model can be applied to dead trees of the same species, or to timber beams in old houses and the climatic conditions inferred from the relationship already established.

Many tree species have been successfully used in dendrochronology work, in particular those which live to great age, such as the oak and the yew. Without doubt the most useful and spectacular species has been the bristlecone pine (*Pinus aristata*) which grows at high elevation in the Sierra Nevada range of North America. This species can live for up to 4600 years while dead trunks have been dated back 8200 years. Correlation of past climatic events using the bristlecone pine and RADIO-CARBON DATING of materials has shown that the incorporation of $^{14}$C in living tissue has not been at a constant rate as had previously been thought. A correction factor of 1000 years at 6000 years BEFORE PRESENT must be applied. See Fig. 50.

**density dependence,** *n.* (Animal ecology) the control of population size by environmental factors whose effectiveness varies with population density. When high population densities are reached, the factors become more effective at restricting the continued growth of populations; at low population density the factors are less effective. Density-dependent factors include availability of food, living space, nesting sites as well as incidence of disease and competition between species. Several factors may operate at the same time, each contributing to the regulation of population. See also LIMITING FACTOR.

**denudation,** *n.* the lowering of the earth's surface by the processes of WEATHERING, MASS MOVEMENT, EROSION and TRANSPORTATION.

**Department of the Environment (DoE),** *n.* the department of the British government which deals with planning and environmental matters in England and Wales; the equivalent department for Scotland is the Scottish Development Department. The DoE is presided over by a senior minister who has a seat in the cabinet. The main functions of this large department of state include oversight of LAND USE PLANNING, housing, local government, the protection of HISTORIC BUILDINGS and SITES OF SPECIAL SCIENTIFIC INTEREST, in addition to general control over the physical amenities of the nation.

**deposition,** *n.* the laying down of soil and rock fragments transported by the action of rivers, glaciers, sea and wind. See SEDIMENTATION, FLUVIAL DEPOSITION, GLACIAL DEPOSITION, WIND DEPOSITION, MARINE DEPOSITION.

**depression, frontal depression, cyclone** or **low,** *n.* a region of low air pressure, usually ranging between 950 and 1020 millibars, and which originates along POLAR FRONTS. It is thought that depressions develop as small unstable wave-like irregularities along the polar front, which cause a localized drop in air pressure. These irregularities grow as warm air south of the polar front advances north. Cold air then sweeps in behind the WARM FRONT to generate a WARM SECTOR. Depressions frequently develop over the oceans in temperate latitudes and travel eastwards to bring cloud and rain to the western edges of the continental landmasses. A depression may range in size from 150 km to 3000 km and may travel at up to 1000 km/day.

In the northern hemisphere, winds circulate in an anticlockwise direction around the centre of a depression whilst in the southern hemisphere, the movement is clockwise.

Depressions are represented on weather maps by a series of oval or concentric closed ISOBARS. A deep depression has a strong declining outward PRESSURE GRADIENT and is indicated by many, closely spaced isobars. A shallow depression is characterized by only light winds, a weak pressure gradient and few isobars. See NON-FRONTAL DEPRESSION.

**deprivation, area of,** *n.* an area, usually within a city containing a concentration of people experiencing social, economic and environmental problems. Such areas are generally defined so that remedial measures aimed at improving the living conditions of the deprived population can be more precisely targeted. In the UK, deprivation areas were first designated in the late 1960s by education authorities; *educational priority areas* were defined around schools which were deemed to have a concentration of pupils disadvantaged in their school work by reason of family poverty, ill-health or a poor educational inheritance, and thus requiring a special input of resources. Town planners adopted the concept of areas of deprivation from the early 1970s onwards, based on wider criteria which included high rates of unemployment, infant mortality, delinquency, family poverty, low standards of education, chronic sickness, together with inaccessibility to services such as shops, health and social support services. In addition, housing quality, overcrowding and the lack of standard amenities is often taken into consideration, as is the proportion of elderly and one-parent families.

Such areas have become foci for the selective input of extra resources within many large cities, through policies of *positive discrimination*: priority is accorded to such areas under city-wide programmes for community regeneration and renewal, at the expense of other areas whose population enjoys a better quality of life. See also PRIORITY TREATMENT AREA, INNER CITY, CROWDING.

**derelict land,** *n.* any land left standing without a present use as a result of a past activity which has physically despoiled it or aesthetically disfigured it. Physical despoilation may be the outcome of past mineral workings, especially opencast workings, while past chemical manufacture may leave soils polluted or poisoned so that natural vegetation is slow to regenerate and premature re-use poses a health hazard to potential occupiers of the site. Sometimes the dereliction is a characteristic of the ruins of buildings, rather than of the physical land surface itself.

Clearance and REHABILITATION of derelict land is a major objective of urban renewal in an effort to promote new industrial and residential development. See also LAND RECLAMATION.

**derncarbonate soil,** *n.* a RENDZINA-like soil as listed in the Soviet soil classification system.

**derris,** *n.* a natural INSECTICIDE developed in 1912 from a South American root plant. The extract had been previously used by the native population as a fish poison. The precise action of derris as an insecticide is not known but it causes the paralysis of the respiratory system and heart of the insect. Derris has low toxicity on most warm blooded mammals, but on edible crops at least one day should elapse between application and harvest. Derris should not be applied near water courses because if it enters the water it causes total destruction of the fish stock.

**desert,** *n.* an area charcterized by a climatic pattern in which evaporation exceeds precipitation. A precise climatological definition of a desert is impossible due to considerable local variations, but generally, in areas where the annual precipitation is less than 250 mm then desert conditions will prevail.

Desert regions can be subdivided into:

(a) *semi-arid regions* in which the ratio of precipitation to evaporation is less than one. On an annual basis these regions suffer a deficiency of rainfall although a short-lasting moist season usually exists. Annual rainfall is usually between 380-760 mm;

(b) *arid regions* (or *true deserts*) with annual precipiation between 125-380 mm and which falls as short torrential downpours separated by many months of complete drought;

(c) *extremely arid regions* where the interval between rain may be

| Desert | Area (million km²) | % of world desert area |
|---|---|---|
| Sahara desert | 9.07 | 41.7 |
| Australian desert | 3.37 | 15.5 |
| Arabian desert | 2.60 | 11.9 |
| Turkestan desert | 1.94 | 8.9 |
| North American desert | 1.29 | 5.9 |
| Patagonian desert | 0.67 | 3.1 |
| Thar-Sind desert | 0.60 | 2.7 |
| Kalahari/Namib desert | 0.57 | 2.6 |
| Taklimakan desert and Gobi desert | 0.77 | 3.6 |
| Iranian desert | 0.39 | 1.8 |
| Atacama desert | 0.36 | 1.7 |
| All other deserts | 0.13 | 0.6 |
| *Total area* | 21.76 | 100.0 |

Fig. 23. **Desert.** Desert areas of the world.

up to five years, but when rain does fall it may exceed 50 mm in amount.

The amount of 'climatic' true desert is small, and semi-arid desert and, increasingly, human-made desert prevail.

These areas can also be classified as *hot deserts* (for example, Sahara desert) or *cold deserts* (for example the Kunlun desert in the lee of the Himalayas). In the latter, a distinct winter cold season exists in which daily temperatures may not exceed 0°C. In hot deserts no cold season can be distinguished although night temperatures may fall below 0°C.

The absence of water in deserts has made plant and animal life extremely hazardous. It is rare to find a sharp transition from humid to desert life forms; instead a transition zone exists across which vegetation and soils show a progressive adaptation to drought. Soils become increasingly dominated by the upward movement of water through the soil profile, a process dependent upon the excess of evaporation over precipitation. See DESERT VEGETATION, DROUGHT, CLIMATE.

**desertification,** *n.* a process by which the biological productivity of the land is so reduced as to lead to the spread of DESERT-like conditions in arid and semi-arid regions. Characteristic features of desertification include impoverishment of vegetative cover, deterioration of the texture, structure, nutrient status and fertility of the soil, accelerated SOIL EROSION, reduced availability and quality of water, and encroachment of sand.

The causes of desertification are complex. Although there may be some evidence for long-term climatic change towards greater aridity, the occurrences of serious desertification suggests the dominant cause to be environmental stress induced by human pressure coupled with extreme, short-run fluctuations in climate. This view was endorsed by the United Nations Conference on Desertification held in 1977 following the severe SAHEL drought of 1968-74. Out of some 45 identified causes of desertification, 38 were traced to human mismanagement of soil, water, energy, flora and fauna, including too intensive a form of cultivation, overgrazing beyond the CARRYING CAPACITY of the land, SALINIZATION through poorly managed IRRIGATION, and DEFORESTATION, particularly the cutting of fuelwood.

Although the characteristics of desertification frequently lead to a self-perpetuating deterioration of the ecological system, the process is not irreversible. Remedial and preventative action against desertification can be effective, for example, irrigation, AFFORESTA-

TION, the use of SHELTERBELTS, the planting of trees, grasses and fences to stabilize sand dunes, the protection of existing vegetation, SOIL CONSERVATION, careful management of water resources, and education to motivate local people to accept and adopt such changes.

**desert vegetation,** *n.* the vegetation type which has developed in response to very low annual rainfall levels (see DESERT). Desert vegetation usually shows a marked transition in appearance and character along a TRANSECT taken from the more humid edges towards the arid centre of a desert. Typically, woody plants become more gnarled and assume a spreading form. The number of species and individuals becomes reduced, leaves become smaller in size and eventually form vestigial spines or blades. Evergreen plant life forms predominate except in the true desert areas where trees are replaced by varieties of spiny SCRUB and ANNUAL herbs that are XEROPHYTES. See MICROPHYLLOUS FOREST, SEMI-DESERT SCRUB, TROPICAL DESERT VEGETATION.

**design guide,** *n.* the guidance notes which specify the main objectives to be fulfilled and standards to be adhered to in the proposed overall plan for a town or city. Design guides are especially important for building projects which will be part of the large-scale REHABILITATION of a historic town. Guidelines are often aimed at the private developer in an effort to improve the quality of proposed developments. In the UK, a particularly influential design guide was prepared by Essex County Council in the early 1970s. It served as a model for numerous similar guides produced by local planning authorities throughout the country. See also DEVELOPMENT BRIEF, LOCAL PLAN.

**desilication,** see LATERIZATION.

**destructive boundary,** see PLATE TECTONICS.

**destructive wave,** see WAVE.

**detergent,** *n.* an inorganic substance used to remove dirt and grease from a surface. The earliest detergents were soaps made from alkaline salts and fatty acids. From 1918, detergents were 'improved' by the addition of synthetic chemicals that increased their effectiveness in removing grime. However, additives such as tetrapropolyene benzene sulphonate (TBS) consist of complex, multibranched carbon chains (*hard chains*) which fail to break down in water. Effluent containing hard-chain detergents substantially reduces the efficiency of sewage treatment plants. Furthermore, the release of partially treated effluent into rivers causes the hard-chain detergents to produce large quantities of foam, commonly called *detergent swans.* The foam, contaminated by the effluent, can be dispersed by the wind, thus spreading pathogenic bacteria and worm eggs that cause infection in humans and animals.

In the 1960s the hard-chain additives were replaced by simpler, BIODEGRADABLE *soft chain* substances. While avoiding the problem of foaming, the new phosphate-rich materials have led to an increase in the EUTROPHICATION of water and have been partly responsible for ALGAL BLOOMS.

As little as 0.1 parts per million (ppm) detergent in water can reduce the availability of oxygen by up to 50%. Some fresh water fish, notably the trout, can be asphyxiated by detergent concentrations as low as 1 ppm.

**detergent swan,** SEE DETERGENT.

**detrital food chain,** *n.* a structured feeding hierarchy in the early stages of which the primary producers (green plants) form DETRITUS on which DECOMPOSER organisms feed. Compare GRAZING FOOD CHAIN. See SAPROPHAGE, SAPROPHYTE.

**detritus,** *n.* any organic litter accumulating on the ground, typically of plant origin but which may also include the remains of small animals. See DECOMPOSER, DETRITAL FOOD CHAIN.

**detrivore,** SEE DECOMPOSER.

**developed country,** *n.* any country characterized, in both absolute terms and relative to DEVELOPING COUNTRIES, by high standards of living and material well-being, high per capita gross national product (GNP), high per capita energy consumption, low birth and death rates, high life expectancy at birth, good provision of educational and health care services, high rates of literacy, and high standards of nutrition. Developed economies have relatively small numbers of the economically active population employed in agriculture, an industrial manufacturing base, and a growing tertiary (service) sector.

Developed countries include both the industrial market economies of Western Europe, North America, Japan and Australasia (see Fig. 24) and the centrally planned industrial economies of Eastern Europe and the USSR. Collectively described as the North, counterposed to the DEVELOPING COUNTRIES of the South, they exert a powerful political and economic influence on the international scene. The industrial market economies exhibit considerable diversity, ranging in terms of per capita GNP in 1986 from Spain ($4860) to the USA ($16 480), at a weighted average of $12 960. This contrasts with the average for developing countries of just $610.

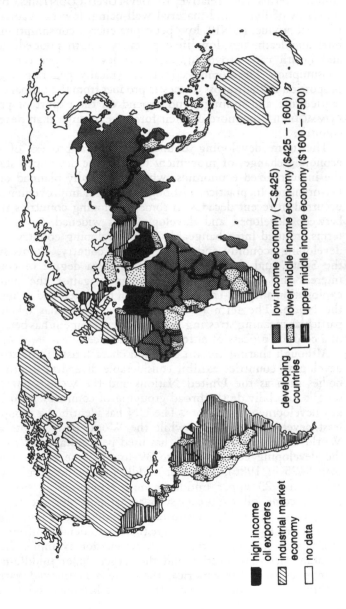

Fig. 24. **Developing country.** Developed and developing countries of the world.

high income
oil exporters

industrial market
economy

no data

low income economy (< $425)

lower middle income economy ($425 − 1600)

upper middle income economy ($1600 − 7500)

developing
countries

**developing country,** *n.* any country characterized, in both absolute terms and relative to DEVELOPED COUNTRIES, by low standards of living and material well-being, low per capita gross national product (GNP), low per capita energy consumption, high birth and death rates, low life expectancy at birth, poor educational and health care provision, and low levels of per capita food consumption. Developing countries typically generate a greater proportion of their gross domestic product from the primary sector (agriculture and mining), and depend on the export of primary COMMODITIES for more of their foreign exchange than developed countries.

The term developing country implies a process of socio-economic change, of movement towards the developed status of the industrialized economies, whether centrally planned or market-oriented. In practice, although very real improvements have occurred in recent decades in some developing countries the gap between developed and developing has widened. Several other terms are used interchangeably with developing country, such as developing economy and less-developed country. Terms such as the South, and the THIRD WORLD imply a degree of common interest and alignment in the struggle against the perceived exploitation and neo-colonial designs of the developed countries of the North. The term underdeveloped country has taken on a particular meaning stressing that underdevelopment has been a part of a colonial process of enforced dependence.

Although sharing certain common characteristics, the so-called developing countries exhibit considerable diversity. International bodies such as the United Nations and the WORLD BANK have sought to subdivide the broad grouping of countries according to key development indicators. The UN has identified a group of 31 least developed countries, while the World Bank, in its annual World Development Report, has used per capita GNP to divide the developing economies into 39 low-income economies (less than $425 in 1986), 35 lower-middle income economies ($460-1600), and 23 upper middle-income economies ($1600-7500). High income oil exporters such as Saudi Arabia are separately categorized, as are 35 countries with populations less than one million (see Fig. 24). Not only are there considerable differences in wealth and economic structure between, for example, the Sub-Saharan African countries and the larger upper middle-income economies of Latin America, there are also marked variations within countries, often greater than those between countries.

**development,** *n.* the process by which some system, place, object
or person is changed from one state into another; the term carries
the connotation that the change is in the direction of growth or
improvement. Many forms of development are invisible (such as
the cognitive development of an individual) but most developments become visible eventually through changes in the visible
environment which follows from invisible societal or personal
changes. Three main forms of development can be identified:

(a) *economic development,* involving the expansion and
diversification of the economic base of a region or nation, an
improvement in the work skills of the population, and the
fostering of technical innovation in production processes. Such
development is made visible through the demands made for new
and different styles of workplace, and the demands made upon
social facilities by the raised incomes of the population. Industrial,
agricultural and commercial developments are all encompassed by
economic development.

(b) *social development,* resulting from improvments in the quality
of life for people in particular geographical areas or social groups
(especially in the fields of health and education) and an increase in
the confidence and capability of a group to participate in decisions
about its own future. This development is made visible through
increased pressure on social and retail services and through rising
expectations of housing quality. Community development is a
component of social development.

(c) *physical environmental development,* in which there is planning
and provision for rising expectations and capabilities by ensuring
that the BUILT ENVIRONMENT is able to accommodate these new
expectations; altering the existing built structures and providing
new structures drawn together in a harmonious functional
relationship is the principal goal of LAND USE PLANNING. See also
DEVELOPMENT CONTROL, DEVELOPMENT PLAN.

**development brief,** *n.* a remit prepared by the owner or controller
of a site which specifies the constraints and opportunities for
DEVELOPMENT of that site, and which provides a framework within
which a prospective developer should prepare a detailed design.
Briefs prepared by the local planning authority will discuss topics
such as the size and physical configuration of the site, its physical
context relative to surrounding buildings, its social or economic
context relative to the uses of adjacent land and its planning
context relative to the existing LOCAL PLAN. A brief will also discuss
certain detailed considerations which will bear upon the design of

# DEVELOPMENT CONTROL

the development including the expected functional use (for example, housing), the number of units proposed, the constraints upon pedestrian and vehicular access, the provision for car parking, the arrangements necessary to provide security for the eventual occupants and the desirable height and massing of the prospective buildings. See DESIGN GUIDE, DEVELOPMENT PLAN.

**development control,** *n.* the process carried out by local authorities in the UK by which changes in land use are controlled. Control of DEVELOPMENT is carried out under powers bestowed by Act of Parliament and is therefore a statutory duty. The interpretation of what constitutes development is elaborated in the USE CLASSES ORDER and the GENERAL DEVELOPMENT ORDER which define respectively, those material changes in use that do not represent development but still require planning permission, and those alterations in land use that are exempted from the full rigours of development control procedures.

Applications by developers to change the use of land are made to local planning authorities. The planning authority, after technical consideration of the application and its conformity or conflict with the current DEVELOPMENT PLAN for the area, may approve the development, reject it, or approve it subject to conditions. Those refused usually have the right of appeal to the appropriate ministry in the central government. See also LOCAL PLAN, DEVELOPMENT BRIEF, DESIGN GUIDE.

**development plan,** *n.* a document produced by local, regional, county, provincial or national government setting out the principles by which DEVELOPMENT in an area will be permitted. In the UK, the term is used generally to describe the statutory plans produced by local authorities to guide land use development. Early development plans, prepared under the 1947 Town and Country Planning Act were map-based and the precise allocation of land use could be identified from the plan. Under the 1968 (and subsequently 1971) Planning Act, a two-tier system of development plans was introduced, comprising STRUCTURE PLANS (providing a strategic policy framework) and LOCAL PLANS (giving more information about detailed land-use proposals). In 1986, a system of *unitary development plans* was introduced in metropolitan areas of England and Wales. This system, which brings together strategic and local plan policies as two parts of a unified planning document, will eventually replace the present system of structure and local plans. In non-metropolitan areas, it is likely that structure plans and local plans will be replaced by *district development plans* prepared by district planning authorities, and set within the context of strategic planning guidance issued by county authorities. See DEVELOPMENT CONTROL, LAND USE PLANNING.

**Devonian period,** *n.* the geological PERIOD that followed the SILURIAN PERIOD, beginning about 395 million years ago and ending about 345 million years ago when the CARBONIFEROUS PERIOD began. A variety of marine invertebrates and fresh water fish fossils are used to date and correlate the Devonian. AMPHIBIANS appeared for the first time during this period. See EARTH HISTORY.

**dew,** *n.* water droplets which form on the ground, on plants and other cool surfaces following the CONDENSATION of atmospheric water vapour. Dew usually results from the nocturnal cooling of the lower levels of the atmosphere below DEWPOINT. Dew can also form when warm moist air passes over a cold ground surface. Dew fall is greatest under conditions of still, humid air. In arid and semi-arid regions, dewfall can form a significant part of annual precipitation and traps are made to collect the dew for domestic and agricultural purposes.

**dewpoint,** *n.* the temperature at which cooled air becomes saturated with water vapour and CONDENSATION occurs, leading to the formation of DEW. See also HYGROSCOPIC NUCLEI.

**diagenesis,** see LITHIFICATION.

**diatrophism,** see TECTONISM.

**dicotyledon,** *n.* the most numerous of the two sub-classes of flowering plants (the ANGIOSPERMS) in which the embryo is typified by two *cotyledons* (seed leaves). The leaves show a wide variety of shape although the veins within the leaves are invariably arranged in the form of an irregular net. The flower is arranged so that the petals, sepals and stamens are usually arranged in fours or fives or multiples of these numbers. Internally, the CONDUCTING TISSUE is similarly grouped in multiples of four or five.

The total number of dicotyledon species number in excess of 240 000 and contains most of HARDWOOD trees as well as shrubs and herbs. Compare MONOCOTYLEDON.

**dieldrin,** see ALDRIN.

**dietary energy supply,** see FOOD BALANCE SHEET.

**dike,** see DYKE.

**dilation,** see PHYSICAL WEATHERING.

**direct drilling,** see TILLAGE.

**direct recycling,** SEE RECYCLING.

**discharge,** see FLUVIAL TRANSPORTATION.

**dismantlement,** see DECOMMISSIONING.

**dissolved load,** see FLUVIAL TRANSPORTATION.

**district development plan,** see DEVELOPMENT PLAN.

**district heating,** *n.* a scheme whereby a group of houses receive their domestic heating from a centrally located heating plant. The plant may be purpose-built to the capacity required and may use any source of energy (geothermal, oil, electricity, gas or coal). In some cases use is made of excess hot water produced in the process of electricty generation in a combined heat and power scheme. District heating is little used in the UK but is common in other European countries and especially in Scandinavia.

**disturbance,** *n.* **1.** (Ecology) any change in an ECOSYSTEM as a result of environmental variation, for example, any atypical condition such as lack of snow-cover in an ALPINE ecosystem. Some ecosystems, notably TROPICAL RAIN FORESTS, appear unable to withstand any significant element of disturbance whereas northern coniferous forests require periodic disturbance by FIRE in order to stimulate regeneration.

The human disturbance of ecosystems now predominates and differs from natural disturbance in that it is more frequent, more severe and often deliberate in its application, for example the effects of DEFORESTATION and IRRIGATION.

**2.** (Meteorology) a small DEPRESSION.

**divergent boundary,** see PLATE TECTONICS.

**diversity,** *n.* a measure (either qualitatitive or quantitative) of the variety of species contained within a HABITAT. It can refer to either or both FLORA and FAUNA.

**divide,** see WATERSHED.

**division,** see PHYLUM.

**DNA,** *abbrev. for* deoxyribonucleic acid, the long-chain molecule that is the main constituent of the chromosomes of all organisms (except viruses) and within which are the genetic codes that permit the development and functioning of an individual organism. Each DNA molecule is a long two-stranded chain composed of *nucleotides* (combinations of phosphoric acid and monosaccharide deoxyribose) and one of four nitrogenous bases: thymine, adenine, guanine or cytosine. In 1953, F.H.Crick and J.D.Watson proposed that the strands were arranged in a complex double helix configuration, connected by hydrogen bonds. Within this coil specific base pairing occurs: adenine bonds only with thymine while guanine bonds only with cytosine. This specificity enables DNA to replicate itself with great accuracy, so ensuring that the genetic code can be passed unchanged from generation to generation. However, the total number of combinations of the nucleotides along the DNA chain is huge and permits a wide variation within a species type.

**DNOC,** *abbrev. for* 4,6-dinitro-o-cresol, a chemical first used as an INSECTICIDE in 1892, but from the 1930s widely used as a HERBICIDE. Its agricultural importance was originally confined to its ability to kill annual weeds in cereal crops without damaging the crop itself. DNOC has little effect on the persistent perennial weeds, as the substance is not translocated through the plant and underground organs are thus able to produce new leaves. DNOC compounds are rapidly broken down in the soil and do not appear to leave poisonous residues. As a result, DNOC is not transmitted through the FOOD CHAIN.

However, despite its use in cereal production, DNOC holds the unfortunate distinction of being one of the most harmful chemical sprays ever used in agriculture. It is exceedingly hazardous to mammals (including humans) either from inhalation or by absorption through the skin. Because of this, DNOC has been replaced by other selective herbicides, such as 2,4-D.

**DoE,** see DEPARTMENT OF THE ENVIRONMENT.

**doldrums,** *n.* an equatorial zone of low air pressure, developed over the oceans, characterized by high temperatures and humidity, and calm conditions or light indeterminate winds (see Fig. 70). Conversely, stormy weather may also occur in the doldrums due to the development of strong CONVECTION cells. The exact location and extent of the doldrums is largely determined by the season of the year.

**doline,** *n.* a steep-sided circular depression in KARST landscapes formed by solution weathering (see CHEMICAL WEATHERING) or from the collapse of underground caverns. Ancient dolines may be totally dry while dolines of more recent origin may be the sites at which OVERLAND FLOW disappears underground. See UVALA.

**domestication of plants and animals,** *n.* the process by which wild plants and wild animals are brought under human control, and are transformed by careful husbandry and SELECTIVE BREEDING into CROPS and livestock that can be systematically exploited for human use. The domestication of plants and animals enabled early HUNTER-GATHERERS to gain control over food production, permitting sedentary AGRICULTURE, higher population densities, the division of labour necessary for civilization, and, through further technological and political advances, the emergence of urban and literate society. Seeds from plants with favourable characteristics for cultivation, such as easy germination, uniform ripening without the seeds becoming detached from the plant, and good storage qualities were among the first to be domesticated in south-west Asia around 10 000 years ago. By deliberate cross-pollination early varieties of wheat and barley were created. The cultivation of plants was complemented by developments in animal exploitation whereby the wild ancestors of sheep, goats, cattle and pigs were isolated from their wild populations and herded for convenience to provide meat and later, milk and wool. Other animals such as the horse, ass, camel and water buffalo were domesticated primarily as beasts of burden, or as draught animals and so hastened the spread of plough cultivation. See AGRICULTURAL HEARTH, CULTIGEN.

**dominant,** *n.* (Ecology) any plant or animal species that competes most successfully for the essential requirements of life (food, light, territory space, a mate) and exerts a physical influence on its HABITAT so limiting the performance of other species which inhabit the same living area.

**doomsday syndrome,** *n.* the belief of some conservationists that all human activity leads toward ecological catastrophe. As a result of population growth, pollution, resource exhaustion, species extinction, increasing use of inorganic chemicals and soil erosion, it is believed that the natural stability of the BIOSPHERE is jeopardized. This attitude has been termed 'gloom-and-doom pessimism' and its advocates christened '*doomsters*'. Compare TECHNOLOGICAL OPTIMISM. See ECOCATASTROPHE.

**doomster,** see DOOMSDAY SYNDROME.

**dose equivalent,** see RADIATION DOSE.

**doubling time,** *n.* (Population dynamics) the amount of time necessary for a POPULATION to double its size. Doubling time will be dependent upon the number of breeding females in the population, upon the gestation period, on the proportion of multiple births, on survivorship and on DEATH RATES. As a result of these variables the population may increase (or decrease) and this can be expressed as a percentage increase (or decrease) per annum. The doubling time has been most widely applied to the growth of human population (see Fig. 25).

In an ideal environment, a population shows an EXPONENTIAL GROWTH pattern of population increase (see Fig 32).

**downland,** *n.* **1.** a grassland occuring on chalk outcrops, especially in the south of England. It consists of a closely grazed turf of *Festuca ovina* (Sheep's fescue grass) and *F. rubra* along with a number of herbs charcteristic of dry or highly calcareous soils.

| % increase per annum | Years to double population |
|---|---|
| 0.1 | 693 |
| 0.5 | 139 |
| 1.0 | 70 |
| 1.5 | 47 |
| 2.0 | 35 |
| 2.5 | 28 |
| 3.0 | 23 |
| 3.5 | 20 |
| 4.0 | 18 |

Fig. 25. **Doubling time.** The relationship between growth rates and doubling time.

Once widely used for sheep grazing, much downland has now been ploughed up and used for cereal production while the steeper land has been allowed to revert to forest SCRUB.

2. grassland-dominated areas in New Zealand and Australia and on which extensive sheep grazing occurs.

**downtown,** see CENTRAL AREA.

**down-valley sweep,** see MEANDER.

**drainage basin,** see CATCHMENT AREA.

**drainage basin management,** see WATERSHED MANAGEMENT.

**drainage pattern,** *n.* the arrangement of a river and its tributaries within a CATCHMENT AREA. The drainage pattern is influenced by

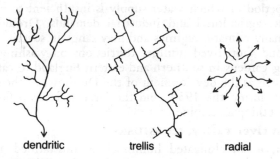

dendritic      trellis      radial

Fig. 26. **Drainage pattern.** Dendritic, trellis and radial drainage patterns.

the nature of the land surface, the type and structure of underlying bedrock and the climatic regime of the area. There are three main types:

(a) *dendritic drainage* which is characterized by the irregular branching of tributaries. It forms in areas where geological structure exerts little influence on the course of a river;

(b) *trellis drainage* which consists of a river network where stream junctions are at approximate right angles, and result from the marked geological control of drainage;

(c) *radial drainage* which occurs when streams radiate outwards from an elevated dome-shaped structure such as a volcanic cone.

ANTECEDENT DRAINAGE and SUPERIMPOSED DRAINAGE are recognized as being inherited drainage patterns.

**dredge,** *vb.* to remove sediment and ALLUVIUM, especially from rivers and ESTUARIES in order to maintain navigation channels. The task is performed by specially designed ships.

**drizzle,** *n.* a fine RAIN where the water droplets have a diameter of less than 0.5 mm.

**drought,** *n.* **1.** an extended and continuous period of very dry weather. The precise definition varies from country to country. In the UK, three types of drought are recognized:

(a) *absolute drought* is a period of 15 or more consecutive days with a rainfall of less than 0.2 mm;

(b) *partial drought* is a period of 29 consecutive days with an average rainfall of 0.2 mm or less per day;

(c) a *dry spell* is a period of 15 or more consecutive days, during which the rainfall does not exceed 1 mm per day.

In the USA, a dry spell is a period of 14 days without measurable rainfall.

**2.** a period in which water supply is insufficient to meet usual domestic, agricultural and industrial demands. Droughts occur under many climatic regimes and may range in severity from the minor and short-lived summer restrictions on washing cars and watering gardens in southern and eastern England to catastrophic events such as the development of the Dustbowl in the American Midwest during the 1930s and the large-scale crop failures and ensuing Ethiopian famine of 1985–86.

**drowned river valley,** see ESTUARY.

**drumlin,** *n.* an elongated hummock of unstratified glacial till deposited and moulded below an icesheet. Drumlins may be up to 3 km in length and 60 m in height and lie parallel to the direction of the former ice flow with their steeper blunt end facing upstream. Drumlins usually occur in clusters or *swarms*, giving rise to a 'basket-of-eggs' topography. The sands and gravels in drumlins are often extracted for construction and industrial purposes. See GLACIAL DEPOSITION.

**dry adiabatic lapse rate (DALR),** *n.* the rate at which heat is lost from an unsaturated air mass as it rises through the atmosphere. The DALR has been calculated at 1°C decline per 100 m rise, although some minor variations can occur in this value. Compare SATURATED ADIABATIC LAPSE RATE, ENVIRONMENTAL LAPSE RATE. See LAPSE RATE.

**dry farming,** *n.* a system of EXTENSIVE AGRICULTURE in which GRAIN CROPS are grown in semi-arid areas without IRRIGATION. In areas where annual rainfall may be less than 500 mm, the conservation of soil moisture is critical. This may be achieved by the introduction of a bare FALLOW every second year as bare ground loses less moisture through EVAPOTRANSPIRATION than do vegetated surfaces. Part of the rain falling during the fallow period is thus stored in the soil for the following year's crops. Exposing the soil surface to the elements increases the liklihood of SOIL EROSION although this may be lessened by the application of a MULCH, or, as in parts of the West African SAVANNA and SAHEL regions, by the practice of placing lines of stones along the contours of gentle slopes to impede run-off and allow moisture to infiltrate the soil. Frequent cultivation of the fallow land by cross-ploughing or pulverizing the soil into a fine tilth also aids water absorption and moisture conservation, although at some risk of WIND EROSION.

**dry spell,** see DROUGHT.

**dual carriageway,** *n.* a double-track highway consisting of two separate roads each of at least two lanes width, running alongside each other; each road carries traffic in one direction only. Some dual carriageways are limited access MOTORWAYS, while others have standard intersections with adjacent roads. Dual carriageways of both types are found as part of both rural and urban TRANSPORT NETWORKS. See also TURNPIKE, TRAFFIC LANE, BUS LANE.

**dune,** *n.* a low mound or hill of sand which has been deposited by the wind. Dunes are found throughout the world but tend to be associated with specific locations including sandy floodplains, sandy beaches along sea coasts or lakes and in deserts. Dunes form when an obstacle, such as a rock or vegetation, causes wind velocity to decrease and deposit any windblown material on the leeward side of the obstruction. Dunes vary greatly in size and shape according to the nature of the wind, the amount of sand available and the type and amount of vegetation. Various types of sand dune have been recognized including *parabolic dunes* (elongated dunes with horns pointing upwind), *longitudinal dunes* (long, narrow, symmetrical dunes running parallel with the prevailing wind direction) and *barchans* (crescent-shaped dunes with horns pointing downwind).

Most dunes are not static features and slowly migrate as wind blows sand up the gently sloping windward side over the crest and down the steep leeward side. Dune migration tends to be slow but some have been recorded moving over 30 m per year. Migration will continue until the dunes become vegetated. This can occur naturally or may be due to deliberate planting. In the Landes region of France and in north-east Scotland, agricultural land, roads and settlements have in the past been lost to invading dunes. Planting programmes were established in these locations to stabilize the dunes and prevent further loss. Similar schemes have also been necessary in the USA and Australia.

In coastal areas vehicular and foot traffic can cause the depletion of dune vegetation allowing erosion to set in. This has had serious consequences along parts of the north-eastern seaboards of the USA and the Netherlands where dunes protect low-lying land from inundation by the sea. Loss of life and damage to property can be considerable if the sea breaches the protective dunes.

**dung fuel,** *n.* dried animal dung used in DEVELOPING COUNTRIES as an alternative source of fuel. In the absence of firewood dung forms an important energy source and is used mainly for cooking purposes. In Africa and Asia at least 400 million tonnes of dung are burned each year. It has been estimated that if this quantity of organic matter was added to the soil as a manure it could help produce an additional 20 million tonnes of grain per year in those regions of the world which experience major food shortages. See also FUELWOOD CRISIS, BIOGAS, ALTERNATIVE ENERGY.

**duricrust,** *n.* a hard, compact, cemented layer in the upper horizons of some soils formed by the evaporation of mineral solutions; the concentration of the deposited minerals accounts for the hardness of the layer. These solutions are usually ferruginous, siliceous, calcareous, aluminous or magnesian in content; this high mineral content sometimes allows commercial exploitation, as occurs with certain LATERITES in Brazil. Duricrusts may be up to several metres thick and are most commonly found in semi-arid regions. See also CEMENTATION, HARDPAN.

**dust bowl,** *n.* a semi-arid region where soil has been or is being removed by deflation (see WIND EROSION). Dust bowls are associated with cyclical droughts and may be initiated by insensitive cultivation techniques or by the overgrazing of introduced livestock. The term is now in wide use but may refer specifically to the mid-west regions of the USA during the early 1930s. Improved farming techniques such as CONTOUR PLOUGHING, GREEN FALLOW and SHELTER BELTS can minimize dust bowl formation. See also SOIL CONSERVATION

**dust storm,** *n.* a dust-laden wind that may extend to great height. Dust storms are common in arid and overgrazed areas and may, in severe examples, contain so much dust as to reduce visibility substantially and blot out the sun. Dust storms take the form either of an advancing 'wall' or a WHIRLWIND and are usually short-lasting, although some examples have lasted for up to 12 hours. When accompanied by strong winds they can cause severe sandblasting of painted surfaces (notably of cars) and can cause the 'frosting' of glass. See also SANDSTORM.

**dyke** or **dike,** *n.* **1.** a discordant sheet-like mass of INTRUSIVE ROCK formed by the injection and solidification of MAGMA along weaknesses in the earth's crust, such as FAULTS. Dykes may range in thickness from a few centimetres to hundreds of metres. Dykes often occur in large numbers in parallel or radial arrangements (*swarms*) which may extend for hundreds of kilometres and are sometimes associated with large igneous bodies such as BATHOLITHS and LACCOLITHS. Differential erosion may result in outcropping sills forming distinct landscape ridges. Rock is often quarried from dykes for use as road metal.

**2.** an artificially constructed embankment to protect low-lying areas from river or coastal flooding. See POLDER.

**3.** a watercourse or ditch.

**dynamic metamorphism,** see METAMORPHISM.

**dynamic rejuvenation,** see REJUVENATION.

# E

**earthflow,** *n.* the rapid downslope movement of partially saturated soil and rock debris. Earthflows are more viscous than MUDFLOWS and often occur following spring thaws or heavy rainfalls. Earthflows may range in size from a few square metres up to several square kilometres.

Small earthflows may cause minor inconvenience by blocking roads and railways. They are often caused by the excavation of slope bases by river erosion and the artificial steepening of slopes that accompanies the construction of cuttings and embankments in association with the building of roads and railways.

Major earthflows are usually associated with areas of weak or impermeable underlying bedrock and may involve the movement of millions of tons of soil and rock. These movements can cause considerable loss of life and damage to property, as happened with the Nicolet earthflow in Quebec, Canada in 1955 and the Aberfan disaster in South Wales in 1966.

In areas prone to earthflows, movements can be limited by remedial action such as the drainage of slopes and their stabilization through planting schemes and the erection of retaining walls.

**earth history, geological column, stratigraphical column** or **geological timescale,** *n.* the chronology of the earth's geological past. Earth history can be divided into ERAS, PERIODS and EPOCHS. The use of the epoch and smaller time units are usually restricted as their definition can vary between countries. See Fig. 27.

**earthquake,** *n.* a naturally occuring and often detectable sequence of rapid earth tremors. Earthquakes result from tectonic movements (see PLATE TECTONICS) caused by the sudden release of energy due to the surpassing of rock resistance along FAULTS and LITHOSPHERIC PLATE boundaries after periods of deformation. The SEISMIC WAVES generated by these movements travel outward and upwards from the FOCUS to a point on the surface called the EPICENTRE. The magnitude of the earthquake, which is measured on the RICHTER SCALE, depends upon the location and scale of movement, the type and nature of rocks through which the seismic

| Era | Period | Epoch | Commencement date (millions of years BP) |
|---|---|---|---|
| Cenozoic | Quaternary | Holocene (Recent) | 0.01 |
| | | Pleistocene | 2 |
| | Tertiary | Pliocene | 7 |
| | | Miocene | 26 |
| | | Oligocene | 38 |
| | | Eocene | 53 |
| | | Paleocene | 65 |
| Mesozoic | Cretaceous | | 136 |
| | Jurassic | | 195 |
| | Triassic | | 225 |
| Palaeozoic | Permian | | 280 |
| | Carboniferous | | 345 |
| | Devonian | | 395 |
| | Silurian | | 440 |
| | Ordovician | | 500 |
| | Cambrian | | 600 |
| Precambrian | | | 4600 ? |

Fig. 27. **Earth history.** Table showing the geological timescale.

waves travel between the focus and the surface, and the duration of deformation. Earthquakes may also follow the sudden movement of MAGMA in the EARTH'S CRUST and volcanic explosions. Earthquakes may occur at depths of up to 700 km, although most originate within the crust.

The majority of earthquakes are confined to two principle zones:

(a) the *circum-Pacific zone* (also known as the *Ring of Fire* due to the associated frequent volcanic activity) which almost encompasses the Pacific, extending from the western coast of South and North America through the Aleutians, Japan, the Philippines and Indonesia to New Zealand.

(b) the *Mediterranean and trans-Asiatic zone*, which extends from Spain through the Mediterranean and the Middle East to the Himalayas.

Many thousands of earthquakes are recorded each year throughout the world but only a few are of any magnitude. Major earthquakes can cause catastrophic loss of life and damage to property: the worst examples include the Tokyo earthquake of 1923 when 300 000 lives were lost, and the 1988 Armenian earthquake when almost 25 000 people died. The destruction caused by earthquakes results initially from the violent shaking of the ground, although the secondary effects are also a major hazard; these include LANDSLIDES, MUDFLOWS, the release of toxic gases, subsidence, uplift, the collapse of buildings, fires caused by fractured gas mains or damaged electrical installations, and flooding from damaged dams. Submarine earthquakes may cause a TSUNAMI.

In earthquake zones, attempts are made to minimize the effects of earthquakes through careful LAND USE PLANNING and the construction of earthquake-resistant buildings. In many countries even outside earthquake zones, large engineering projects such as dams and nuclear power stations are built to withstand a major earthquake.

It is sometimes possible to predict the occurrence of earthquakes. Several methods are employed, including the use of seismographs and other sensitive instruments to monitor the build-up of stress, and the chemical and physical changes in crustal rocks prior to tremors; the establishment of major tremor frequency; the measurement of topographical changes; and the examination of animal behaviour.

Research is currently being carried out to investigate the feasibility of earthquake control through the initiation of frequent minor tremors using explosives to prevent the build-up of stress that would result in a major earthquake. Two other methods are also being examined as possible means to control earthquakes: the use of nuclear explosions to fracture rocks which would then lessen the effects of shock waves; and the removal of groundwater to increase frictional resistance. The long-term consequences of effective control methods are unclear as the limiting of stress in one area may lead to an increased build-up elsewhere. Earth tremors may also be artificially produced in a variety of ways, such as heavy traffic, blasting operations and nuclear tests.

**Earth Resources Technology Satellite (ERTS),** *n.* any of a series of orbiting satellites designed to relay information on civilian land use and forest, mineral, energy and water resource use.

The original ERTS programme comprised two satellites placed in a near-polar orbit at an altitude of between 880 and 940 km. The satellites crossed the equator every 103 minutes, thus making 14 orbits per 24 hours. In 18 days one satellite covered nearly the entire surface of the Earth. ERTS-1 was launched in July, 1972 by the National Aeronautics and Space Administration (NASA) of the USA. and ceased to operate in January 1978. ERTS-2 was launched in January 1975 on an identical orbit to ERTS-1 but separated by a 9-day distance interval.

ERTS was renamed LANDSAT in January, 1975 to distinguish it from a satellite called Seasat designed to survey only the oceans of the world.

**earth's crust,** *n.* the outer section of the LITHOSPHERE lying above the MOHOROVIČIĆ DISCONTINUITY. Below the oceans the crust may be only 5 km thick, whilst under the continental masses it may be up to 70 km in depth. Two layers can be differentiated within the crust: the discontinuous SIAL is the upper layer and is found only under the continental masses whilst the SIMA is the lower layer and is found under both the continents and the oceans. See Fig. 28.

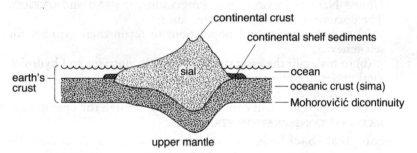

Fig. 28. **Earth's crust.** A cross-section through the earth's crust and upper mantle.

**easement,** *n.* the legal right of an individual to limit the use of land (or a resource) by another person. In practice, easement is usually confined to ensuring local issues such as confirmation of RIGHTS OF WAY, although it is conceivable that national or international easement may be considered necessary if a country proposed using land or resources in a way that would cause hardship to a third-party region or nation.

**ecocatastrophe,** *n.* any major catastrophe which results in a decline in biospheric quality and diversity. While natural disasters such as major volcanic activity can cause ecocatastrophes, they are usually the result of human activity. The use of HERBICIDES and PESTICIDES in agriculture, the release of pollutants into the air, land and oceans, the increasing use of NUCLEAR POWER and perhaps most fundamental of all, the EXPONENTIAL GROWTH of the human population are all possible contributing factors to potential ecocatastrophes. Specific examples of ecocatastrophes can be seen in the decimation of the Lapp reindeer herds following the nuclear accident at CHERNOBYL in 1986 or in the effects on the FRAGILE ECOSYSTEM of the Gulf of Alaska following the oil spillage from the supertanker *Exxon Valdez* in 1989. ECODEVELOPMENT would be considered by many conservationists to provide an antidote to potential ecocatastrophes. See DOOMSDAY SYNDROME.

**ecodevelopment,** *n.* a development approach to conservation based upon long-term optimization of BIOSPHERE resources. The ecodevelopment concept was introduced in the WORLD CONSERVATION STRATEGY document of 1980, a publication produced on behalf of INTERNATIONAL UNION FOR CONSERVATION OF NATURE AND NATURAL RESOURCES, the WORLD WIDE FUND FOR NATURE, the United Nations Environment Programme, the FAO and UNESCO. The document had three declared aims:

(a) to help species and populations to retain their capacity for self-renewal,

(b) to maintain the essential soil, climatic, nutrient and hydrological cycles in order that life itself can continue,

(c) to maintain genetic diversity.

The ecodevelopment concept is applied through the operation of NATIONAL CONSERVATION STRATEGIES.

**ecological backlash,** *n.* an often dramatic and undesirable response by components of an ECOSYSTEM to a disruptive influence. Examples would include the rapid decline of animal populations due to decimation of food sources and/or HABITATS, or soil erosion resulting from unsustainable agricultural practices. See also ECOCATASTROPHE, DISTURBANCE, FRAGILE ECOSYSTEM.

**ecological balance,** *n.* a rarely attained state in which all ECOSYSTEM inputs are equal to the system outputs. In the early years of this century many researchers assumed that most mature ecosystems reached a state of balance. However, it has been recently shown that ecological balance is the exception and not the norm and that a state of balance is attained only in theory and not in practice. See CLIMAX COMMUNITY, HOMEOSTASIS.

**ecological equivalent,** *n.* any unrelated or distantly related species in widely different parts of the world that occupy similar ECOLOGICAL NICHES. For example, prior to 1750 the grassland grazing niche in North America was filled by the bison whereas in

New Zealand the giant flightless bird, the moa, occupied a similar role.

**ecological evaluation,** *n.* the use of individual plant and/or animal species or, more rarely, groups of closely interdependent species to indicate the quality of an ENVIRONMENT. Evaluation of ECOSYSTEMS usually involves the identification of *indicator species* which have critical environmental requirements; for example, the presence of members of the Erica genus of heaths indicate a difficult environment with high soil acidity, pH 4.0 or lower, a shallow, nutrient-deficient soil and an accumulation of peaty material on the soil surface. Ecological evaluation can form an important part in an ENVIRONMENTAL IMPACT ASSESSMENT of a site.

**ecological niche,** *n.* **1.** the status or place of a plant or animal within its biotic community which determines its activities and relationships with other organisms. No two species can occupy the same niche in a community when in each other's presence.

**2.** *functional niche.* the role played by an organism within its community.

**ecological pyramid,** *n.* a graphic representation of the relationships which exist between successive TROPHIC LEVELS. Ecological pyramids can be constructed to show the relationships between trophic levels when the primary producer is large, as with a tree (Fig. 29a) or to show the reduction in number of organisms in each trophic level when the primary producers are small, for example, PHYTOPLANKTON (Fig. 29b). Inverted pyramids can also be found, for example, among parasitic organisms (Fig. 29c). See also FOOD CHAIN.

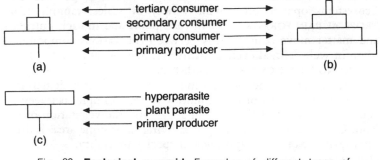

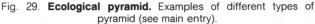

Fig. 29. **Ecological pyramid.** Examples of different types of pyramid (see main entry).

# ECOLOGY

**ecology,** *n.* the study of the relationships between living organisms (the *biota*) and their physical environment (the *abiota*). In its broadest sense, ecology is the study of organisms as they exist in their natural environment. See AUTECOLOGY, SYNECOLOGY.

**economic rent,** *n.* **1.** a measure of value, being the excess of actual earnings of any factor of production over the earnings which are sufficient to keep it in its present use. Economic rent differs from net income in that the costs of inputs include not only the actual costs incurred but also an evaluation of the possible returns those inputs might have yielded in alternative uses. The term 'rent' is not to be confused with its meaning in ordinary speech as a payment for someone else's property. The basic rent of a factor of production, say land for office space, might be £10. This represents the cost of bringing it into production. If demand for land increases the ensuing scarcity enables not only new landlords to benefit from higher rental values, say £15, but the existing landlords too. Thus, the first set of landlords benefit by £5 over the basic amount necessary to keep the property in office use. This surplus or additional value is economic rent. It serves as an allocative indicator in LAND USE COMPETITION, for only those prepared to 'bid' up to the amount including the economic rent will acquire the land. For a farmer anxious to maximize his returns, the land use that yields the highest economic rent, that is greater than all alternative uses, will be selected.

**2.** the return or net income that can be obtained from a given land use above that which could be realized at the margin of cultivation. See also MARGINAL LAND.

**ecosystem,** *n.* any system in which there is an interdependence upon and interaction between living organisms and their immediate physical, chemical and biological environment (see Fig. 30). Ecosystems operate on a wide variety of scales, ranging from TROPICAL RAIN FORESTS to small rock pools and in each nutrients and matter move continuously between the various components, often in well-defined cyclical pathways. See TROPHIC LEVEL, FOOD CHAIN, NUTRIENT CYCLE, COMMUNITY.

**ecotone,** *n.* any zone of transition between clearly demarcated groups of organisms or communities as, for example, between an aquatic vegetation at the edge of a lake and the vegetation of dry land. Competition between species living in ecotone areas is often intense and leads to a high level of species mortality.

**ectoparasite,** see PARASITISM.

**edaphic climax,** see CLIMAX COMMUNITY.

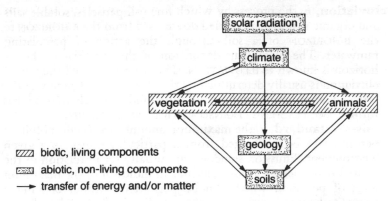

Fig. 30. **Ecosystem.** The relationship between the main components of an ecosystem.

**EDF,** see ENVIRONMENTAL DEFENSE FUND

**educational priority area,** see DEPRIVATION, AREA OF.

**effective dose equivalent,** see RADIATION DOSE.

**effluent,** *n.* any contaminating substance, usually a liquid, that enters the environment via an industrial, agricultural or sewage plant outlet. Effluents are usually harmful as a result of their chemical composition, temperature, pH, oxygen-absorbing capacity in fresh water, or in extreme cases, as a result of their radioactivity. See POLLUTION.

**E-horizon,** see A-HORIZON.

**EIA,** see ENVIRONMENTAL IMPACT ASSESSMENT.

**EIS,** see ENVIRONMENTAL INFORMATION SYSTEM.

**elbow of capture,** *n.* a sudden turn in a river course due to RIVER CAPTURE. Many local examples exist of elbows of capture forming easy routeways for rail and road links. See Fig. 54.

**electrostatic precipitator,** *n.* apparatus used to remove fine PARTICULATE MATTER from exhaust gases in chimneys such as those at coal-fired power stations, cement factories or in iron and steel works.

Electrostatic precipitators trap fine dust particles usually in the size range of 30-60 $\mu$m. When the dust-laden exhaust gas is passed between two electrodes across which a high electrical charge is passed, the particles become charged and are attracted to the oppositely charged electrode. See also CYCLONE DUST SCRUBBER, SCRUBBER.

**eluviation,** *n.* the process by which fine soil particles, soluble salts and organic material are carried downward from the A-HORIZON to the B-HORIZON of a soil through the action of percolating rainwater. The subsequent deposition of this material in the B-horizon is known as ILLUVIATION. The occurrence and the rate of eluviation is usually determined by the texture and structure of the soil and upon the climatic conditions under which the soil has formed. See LEACHING, TRANSLOCATION, PERCOLATION.

**emisson standard,** *n.* the maximum amount of a specific pollutant permitted to be discharged from a particular source in a given environment. Emission standards are usually set in relation to the desired *environmental quality standard*, an indicator of the maximum level of pollution which can be permitted without creating a hazard to human health or causing a decline in the visual quality of the environment.

Emission standards are usually set and enforced by public health departments or agencies. They may sometimes be substantiated by national legislation, for example, the CLEAN AIR ACT (1956) in the UK, or by local laws relating to acceptable standards, for example the discharge of EFFLUENT to a river as controlled by a river management authority. Penalties for exceeding an emission standard usually result in a fine. Persistant disregard for a standard can result in an industry being forced to install new pollution control equipment, or in extreme cases, being compelled to close. See BEST PRACTICAL MEANS.

**emplacement,** *n.* the formation of IGNEOUS ROCK within existing COUNTRY ROCK. See INTRUSIVE ROCK, BATHOLITH.

**enclosure,** *n.* **1.** the process in England and Wales, whereby land was enclosed by the establishment of permanent hedges or walls for the purpose of improving the quality of agriculture. Many types of land were enclosed: open field strips, waste and forest land, water meadows and, most controversial of all, COMMON LAND.

Enclosure of land prevented the indiscriminate wandering of domesticated farm animals, with their consequent trampling and eating of FIELD CROPS. GRAZING MANAGEMENT of PASTURE also became possible.

The enclosure movement was responsible for a major change in the visual appearance of the landscape. In place of small scale, disorganized farmsteads, a regular field pattern with larger collections of farm buildings appeared. The impact of enclosure on the native fauna and flora was catastrophic. Woodlands were cleared, wet areas drained and old grasslands ploughed. The loss of habitats along with organized hunting resulted in the extinction of a few species (wild boar, wolf) and an increasing rarity of many others.

**2.** any area of ground from which animals are excluded by means of a fence or cage. Enclosures are usually constructed during experiments to find the amount of vegetation eaten by HERBIVORES. Enclosures can be temporary, as in the case of most experiments, or of a more permanent form whenever grazing pressures are to be controlled in an effort to permit for example, tree regeneration.

**endangered species,** *n.* any plant or animal species which can no longer be relied upon to reproduce itself in numbers which ensure its survival. A species may become endangered due to natural environmental change or, more commonly, due to human activity (the giant flightless birds such as the dodo and the moa were made extinct by hunting). Changes in land use have removed many of the natural HABITATS of wild species while the increasing use of HERBICIDES and PESTICIDES have eliminated other species. For every one species which becomes extinct an average of 30 other species which were dependent on that species move into the endangered category. See also EXTINCTION, RED DATA BOOKS.

**endemic species,** *n.* **1.** any plant or animal species confined to, or exclusive to, an area. Remote oceanic islands or isolated mountain peaks often contain a large proportion of endemic species. Compare EXOTIC SPECIES.

**2.** any pest or disease-producing species occurring continuously in a given area.

**end moraine,** see MORAINE.

**endoparasite,** see PARASITISM.

**energy,** *n.* **1.** the capacity of a body or system to do work.

**2.** a measure of this capacity, expressed as the work that it does in changing to some specified reference state. It is measured in joules (SI units).

The planet Earth can be considered a single great energy system which receives SOLAR ENERGY as an input while it reflects light energy and radiates heat energy as an output. The flow of energy constitutes a major RENEWABLE RESOURCE. Within the system many transformations occur between the different types of energy. Over time the Earth neither gains nor loses energy; it exists in a state of energy balance or HOMEOSTASIS.

Over the millennia humans have attempted to channel energy sources to suit their needs. The major source of energy, the Sun, cannot be controlled. Solar energy is used in agriculture but as a 'passive' energy source to stimulate PHOTOSYNTHESIS. Instead, many other sources of energy have been developed, most of which are based upon the combustion of wood, coal, natural gas or oil. These FOSSIL FUELS are NON-RENEWABLE RESOURCES and attempts are presently underway, albeit on a limited scale to find and utilize renewable, ALTERNATIVE ENERGY sources.

**energy farm,** *n.* any area of land or water in which plants are specifically grown for their ability to produce large amounts of BIOMASS rapidly which can be converted into a range of biofuels such as METHANE or ETHANOL. Aquatic algae and trees are widely used for such purposes, while sugar cane, manioc (cassava), molasses, maize, wheat and sugar beet have all been used with varying degrees of success. Brazil has made extensive use of energy farms to produce ethanol and all petroleum used in Brazil has at least a 20% ethanol content. However, economic assessments made in the USA have suggested that the produce of energy farms cannot compete with traditional FOSSIL FUELS on a direct cost basis. See GASOHOL, ALTERNATIVE ENERGY.

**englacial transportation,** see GLACIAL TRANSPORTATION.

**enhanced oil recovery,** see OIL.

**enterprise trust,** *n.* an association of local businessmen and public agencies (either local- or central government based), established to promote the industrial and/or commercial development of a town or area. Such trusts arrange expert advice and logistic support for local people trying to set up in business. In addition, they seek to attract outside entrepreneurs to establish businesses in their area.

**enterprise zone,** *n.* an area specially designated by a UK local authority with the consent of national government, within which potential developers receive financial incentives to develop employment-creating enterprises. Simplified planning procedures operate within enterprise zones to facilitate the speedy implementation of development proposals.

Enterprise zones adapted the model of FREE PORTS commonly used in other countries and were officially adopted by the UK government in 1980 to target private and public investment upon run-down city areas stricken by high unemployment following the closure of traditional local industries. However, it is still too early to judge the success or otherwise of the policy. Early criticism suggests that the rise in employment in the zones merely represents a subtraction of job possibilities, or even transfer of existing jobs, from other parts of the cities concerned.

**entombment,** see DECOMMISSIONING.

**entrenched meander,** or **intrenched meander,** *n.* an INCISED MEANDER formed by rapid vertical FLUVIAL EROSION. River valleys containing entrenched meanders are usually steep-sided and symmetrical in CROSS-SECTION. If the neck of an entrenched meander is breached then a small hill is produced, surrounded on three sides by the abandoned stream channel and on the fourth side by the present river channel. Such sites were often the location of fortified settlements in medieval Europe, such as the town of Verdun on the River Meuse in France.

**entropy,** *n.* a measure of the degree of molecular disorder within an organism or system. In the physical world, erosion and decompostion serve to decrease the orderliness of components and entropy is said to increase. In the biotic world, the accumulation of energy via PHOTOSYNTHESIS and metabolic activity counteracts the trend towards decomposition; orderliness is maintained or increased and entropy is said to decrease. See Fig. 31.

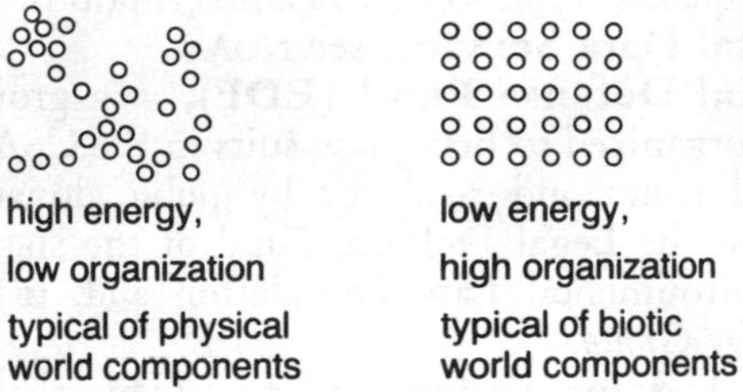

high energy,
low organization
typical of physical
world components

low energy,
high organization
typical of biotic
world components

Fig. 31. **Entropy.** High and low states of entropy.

**environment,** *n.* **1.** the combination of external conditions that influence the life of individual organisms. The external environment comprises the non-living, ABIOTIC components (physical and chemical) and the inter-relationships with other living, BIOTIC components.

**2.** the internal conditions, primarily chemical, which control the well-being of the individual plant or animal. The external environment can influence the internal environment, particularly in the case of the CONFORMER ORGANISMS. For example, external climate can influence the release of specific hormones within animals.

**environmental capacity,** *n.* **1.** (Planning) the safe level of use which may be placed upon an area whilst maintaining an acceptable environmental standard. For example, by limiting the capacity of car parks and/or restricting on-street parking of vehicles, it is possible to prevent CROWDING and a lowering of general comfort, convenience and aesthetic quality of an area. See AMENITY, DEVELOPMENT CONTROL.

**2.** (Pollution control) a level of pollution (gaseous, water or solids) which an environment can safely accommodate and which does not lead to a breaching of legally set limits or standards. See POLLUTION, EMISSION STANDARD.

See also CARRYING CAPACITY.

**environmental data base,** *n.* a collection of information pertaining to a specific environmental problem or area usually stored on computer disk as part of an GEOGRAPHICAL INFORMATION SYSTEM. Data such as geology and soil type, rainfall, elevation, vegetation and land use could be included in an environmental data base. Alternatively, data may be collected for specific species, for example, the RED DATA BOOKS provide scientific information on rare species. Data may also be collected on major natural events, for example earthquakes, typhoons, volcanic eruptions.

**Environmental Data Service,** see NOAA.

**Environmental Defense Fund (EDF),** *n.* a group of lawyers and scientists organized to bring law suits in the USA involving all environmental issues and paid for by pubic donations. Similar groups exist in the Legal Defence Fund of the SIERRA CLUB, the Canadian Environmental Law Foundation and, in the UK, the *Lawyers' Ecology Group.*

**environmental determinism,** *n.* the view that the natural environment controls the course of human actions, and determines the nature of economic activities, including settlement, and cultural development. The debate over the extent of the causal link between environmental conditions and human activity influenced the development of geography as an academic discipline in the late 19th and early 20th centuries. Many of the authors whose work has been regarded as highly deterministic, or environmentalist, were, on closer inspection, cautious to avoid the notion of environmental control. The debate led to many divisions, including the philosophy of *possibilism* in which it is held that the physical environment provides, not necessities in terms of human responses but the opportunity for a range of responses, with humans able to choose between the possibilities. It also gave rise to the intermediate positions of *probabilism*, in which it is argued that although the environment might not determine human activity, it nevertheless makes some responses more probable than others, and *stop-and-go* *determinism* which claims that people might determine the rate of an area's development but not the direction as indicated by the natural environment.

**environmental forecasting,** *n.* the prediction of environmental changes which might result from specific developments made either to the natural or built environments. Environmental forecasting is a widely used technique in ENVIRONMENTAL IMPACT ASSESSMENT. See BIOLOGICAL MONITORING.

**environmental geology,** *n.* **1.** the use of geological information to overcome problems created by the human exploitation of FOSSIL FUELS, minerals, soils and other lithological components. For example, the effectiveness of small irrigation ponds in Third World countries can be improved by locating the ponds on impervious materials thus minimizing water loss through groundwater seepage; also, rehabilitation of land formerly used for quarrying or dumping of industrial wastes can be assisted by the environmental geologist through knowledge of angles of stability for different sediment types, settlement rates, and the load bearing values of infilled ground.

**2.** the use of geological information to minimize the disturbance to the urban and industrial infrastructure; for example, the accurate mapping of fault lines, predicting the location and intensity of earthquakes and volcanic eruptions, predicting areas of subsidence in areas of former deep mining. This information can then be used by the planner, engineer and architect to construct buildings and communication networks that are resistant to geological hazards.

**3.** the use of geological information to locate commercial reserves of fuels and minerals and to develop extraction methods which will cause minimal disturbance to the environment.

**environmental gradient,** *n.* the normal curve of species distribution, showing the likely pattern of abundance or productivity of a species as environmental variables such as light, moisture or soil type alter (see Fig. 61). The tolerance of species to variations in environmental factors along the gradient results in variations in the frequency with which a species occurs. Species which can tolerate a wide range of environmental changes are termed *eurytypic* and generally occupy a wide geographic area; species which can tolerate only small altrations to their environment are termed *stenotypic* and are usually highly localized in their distribution. Each species usually functions most efficiently over a limited part of its environmental gradient, known as its *optimal range*. See also DISTURBANCE.

**environmental hazard,** *n.* a natural event occuring within the BIOSPHERE which impinges upon the well-being of humans, their property or their economic affairs. The prediction both of the location and timing of environmental hazards is often difficult to forecast. Events such as EARTHQUAKES, volcanic eruption and HURRICANES can occur with little warning and may result in massive loss of life particularly in undeveloped countries where the INFRASTRUCTURE to cope with post-disaster conditions may be poorly developed. For example, in November 1970, a hurricane killed 225 000 people and 280 000 cattle in Bangladesh in the space of 12 hours. Other environmental hazards may be slow to develop yet their effect over time is cumulative and equally devastating, such as the drought which has ravaged the SAHEL region of Africa.

The impact of environmental hazards can be minimized by quickly moving people away from the hazard area, although this is often impossible due to the speed with which the hazard strikes. The provision of financial AID and FOOD AID from DEVELOPED COUNTRIES then becomes vital. Technical solutions are sometimes possible as in Tokyo where earthquake damage has been minimized by the construction of buildings designed to withstand earthquakes and helped further by the planning of evacuation routes and education of the public in hazard behaviour. Along the Mississippi River flooding has been minimized by the artificial heightening of LEVEES.

The frequency of an environmental hazard is measured in terms of an event of specific magnitude occuring within a specified time period; in north west Scotland, for example, gales occur on 52 days each year while in southern England they often occur less than once per year.

**environmental impact assessment (EIA),** *n.* a method of analysis which attempts to predict the likely repercussions of a proposed major development (usually industrial) upon the social and physical environment of the surrounding area. In 1979 an EC draft directive was issued which proposed to make EIA mandatory for certain types of large-scale developments. All member states finally accepted the directive, after modifications, in 1985 and proceeded to incorporate its provisions into their planning procedures. In the USA, EIA became formalized in planning practice as a result of the NATIONAL ENVIRONMENTAL POLICY ACT of 1969, while Canada. Australia and New Zealand have adopted similar legislation.

The style of analysis adopted for EIA varies from country to country but common features are:

(a) a two-stage approach defining environmental effects during the construction stage (short term) and during the working life of the project (long term);

(b) an attempt to assess the effects on local employment, services and life style as well as the more directly visible effects on the physical environment, such as noise, AIR POLLUTION, visual intrusion, land degradation, and watercourse contamination. Impact assessments of this kind are 'before the event' predictions which are valuable in enabling a local planning authority to formulate a judgement about whether or not to grant planning permission for a proposed development, in the public interest.

**environmental information system (EIS),** *n.* the computerized collection, storage and manipulation of environmental data. This data may include a large proportion of geographical (spatial) information along with data specific to the description of plants, animals and their habitats. EIS are frequently derivatives of GEOGRAPHICAL INFORMATION SYSTEMS.

**environmental lapse rate,** *n.* the rate of change of atmospheric temperature with altitude at a specific place and time. On average, atmospheric temperature drops 0.6°C for every 100 m increase in height. Compare DRY ADIABATIC LAPSE RATE, SATURATED ADIABATIC LAPSE RATE.

**environmental perception,** *n.* the way in which individuals regard their ENVIRONMENT. An individual's perception of the external environment is formed and conditioned by the highly subjective interpretation of his or her sensory experiences of that environment. Thus, individuals' environmental perceptions even within the same locality are likely to vary considerably. For each individual this *perceived environment* is their reality and as such, it conditions attitudes towards life, and elicits a behavioural response in governing the way decisions are made.

There is generally a strong relationship between perceived environmental images and actual behaviour. DECISION-MAKING often reflects how individuals, rooted in their geographical, historical and cultural milieu, and influenced by their personal preferences and motives, view their environment and react to information about it. For example, in respect of agricultural decision-making it is now widely agreed that farmers rarely act rationally in the optimizing sense beloved of economic theorists;

yet their decision-making may be entirely rational within the way they perceive their environment.

**Environmental Protection Agency (EPA),** *n.* the US government agency with responsibility for federal efforts to control the pollution of air and water by solid wastes, pesticides and radiation hazards. It is also concerned with noise pollution and involved with research to examine the effects of pollution on ecosystems.

**environmental quality,** see AMENITY.

**environmental quality standard,** see EMISSION STANDARD.

**Environmental Research Laboratory,** see NOAA.

**environmental resistance,** *n.* the accumulated pressure of all environmental LIMITING FACTORS brought to bear on organisms to prevent them maximizing population growth.

As a result of environmental resistance, the size of the population is related to the CARRYING CAPACITY of the environment at a given point in time. It is unlikely that the carrying capacity will respond perfectly to the envirnmental resistance due to the operation of biological TIME LAGS.

**environmental science,** *n.* the relatively new interdisciplinary study which examines the interaction between mankind and the environment, with specific reference to the problems resulting from the emergence of the human species as the technological superorganism of the late 20th century. Environmental science includes many areas of traditional knowledge, notably, physical geography, ecology and geology along with aspects of the social sciences such as economics, politics and sociology. A training in environmental science permits a clearer understanding of how the current state of the BIOSPHERE has come about. It should also provide a training in the ways by which the repair and management of the biosphere and its resources can be achieved through the application of ECODEVELOPMENT principles.

**eolian process,** see AEOLIAN PROCESS.

**EPA,** see ENVIRONMENTAL PROTECTION AGENCY.

**epeirogenesis,** *n.* the formation and submergence of continents by broad, relatively slow displacements of the EARTH'S CRUST. The process involves continental uplift, subsidence, tilting and warping and results from the vertical movement of LITHOSPHERIC PLATES. Epeirogenic movements are gentle compared to the intense deformation associated with OROGENESIS. Epeirogenic uplift may lead to the formation of features such as plateaus and ENTRENCHED MEANDERS whilst subsidence may cause the formation of shallow continental seas, such as the Mediterranean Sea and North Sea.

**ephemeral,** *n.* any plant with a very short life cycle, often completed from seed to seed in 8 to 10 weeks. Under favourable conditions two or more generations of ephemerals can complete their life cycle in one GROWING SEASON. Many weed species are ephemerals, for example, chickweed.

**ephemeral stream,** *n.* any stream whose flow is intermittent. Most ephemeral streams occur in moisture-deficient environments such as deserts or in regions with permeable underlying rocks such as limestone. After heavy rainstorms, ephemeral streams may become raging torrents and cause considerable SOIL EROSION. As both the level of water and the speed of flow rapidly decrease then large quantities of debris are deposited haphazardly along the stream bed to await transport in the next flood.

**epicentre,** *n.* the point on the earth's surface directly above the FOCUS of an EARTHQUAKE.

**epilimnion,** see THERMOCLINE.

**epiphyte,** *n.* any herbaceous plant which grows upon another, usually a tree, but which does not feed upon its host; examples include the staghorn fern and many members of the orchid family. The epiphyte feeds on nutrients obtained in solution from rainwater and absorbed from the atmosphere. Epiphytes are most common in the humid rainforests of low latitudes.

**epoch,** *n.* a subdivision of EARTH HISTORY smaller than a PERIOD.

**equinox,** *n.* one of two occasions during each year when the apparent path of the sun bisects the plane of the equator. An equinox occurs on 21 March (the *vernal equinox*) and 23 September (the *autumnal equinox*). On these dates, day and night throughout the world are exactly the same length. After the 21 March, the southern hemisphere tilts away from the sun and following a period of twilight, the South Pole has 180 days of continuous darkness, whilst after 23 September, the northern hemisphere tilts away and the North Pole experiences increasing darkness.

**era,** *n.* the largest subdivision of EARTH HISTORY (see Fig. 31). The four eras recognized are: the PRECAMBRIAN, the PALAEOZOIC, the MESOZOIC and the CENOZOIC. See EPOCH, PERIOD.

**erosion,** *n.* the wearing down and removal of soil, rock fragments and bedrock through the action of rivers, glaciers, sea and wind. See SOIL EROSION, MARINE EROSION, WIND EROSION, GLACIAL EROSION, FLUVIAL EROSION.

**erosion surface,** see PLANATION SURFACE.

**erratic,** *n.* a fragment of rock transported by a GLACIER from its source and often deposited in an area of differing geology. Erratics can be very large; for example, the Maddison Boulder in New Hampshire, USA weighs an estimated 4600 tonnes. Erratics may also be transported considerable distances; some found on the east coast of England are of Scandinavian origin. Erratics may sometimes form long trails, providing an indication of the direction of ice movement. *Glacial striae* (scratch marks) can also be found on erratics. The depth and frequency of the striae indicate the intensity of erosion while the direction of the striae can sometime help determine the direction from which the erratic block has travelled. Erratics are sometimes positioned in such a fashion that they can be 'rocked'. On such occasions erratics are called *rocking stones* or *perched blocks.* Agriculture may be restricted if erratics are especially numerous.

**ERTS,** see EARTH RESOURCES TECHNOLOGY SATELLITE.

**escarpment** or **scarp,** *n.* a cliff or steep slope at the edge of an upland area. Escarpments are usually scenically spectacular but can form major obstacles for transport routes, for example the Drakensburg scarp which separates the South African high veld of the Transvaal from the coastal region of Natal.

**esker,** *n.* a long sinuous ridge of stratified glacial till originally deposited within or under a GLACIER by a meltwater stream flowing at approximately right angles to the ice margin. Some eskers may be over 240 km long and up to 1 km in width. In Canada, Finland and Sweden eskers often form causeways to islands in lakes, allowing the construction of roads and railways. Sand and gravel may be extracted from eskers for construction and industrial purposes. See GLACIAL DEPOSITION. See also KAME.

**estuary,** *n.* a semi-enclosed coastal body of water which has a free connection with the open sea, but within which the salinity level of the ocean is considerably diluted by the addition of fresh water brought in by a river system. Estuaries can be subdivided into four main groups:

(a) *drowned river valleys* which result from the submergence of a coastline, for example, Chesapeake Bay;

(b) FIORDS which form when the sea floods deep glacial valleys. These are found on most glaciated coastlines, including British Colombia, Southern Chile, Scotland, Norway, and New Zealand's South Island;

(c) *tectonic estuaries* which are caused by the down-faulting of small sections of the EARTH'S CRUST. Although generally uncommon, San Francisco Bay is a well known example;

(d) *bar-built estuaries* which are associated with the formation of baymouth bars. Examples include estuaries on the Gulf Coast of Texas, USA.

Human activities can often have detrimental effects on the estuarine environment. The rich and varied, but fragile, ecosystems in estuaries are particularly at risk from artificial changes in SEDIMENTATION and salinity caused by the upstream extraction of water or through engineering works on the estuary such as the construction and operation of flood-control barrages. In some estuaries, chemical pollution from domestic and industrial sources as well as from shipping can be a serious problem, as is THERMAL POLLUTION from the discharge of coolant water from power stations.

**ethanol, ethyl alcohol** or **grain alcohol,** *n.* a flammable organic compound ($C_2H_5OH$) formed during the fermentation of sugars.

**ethyl alcohol,** see ETHANOL.

**euphotic zone** or **photic zone,** *n.* the shallow surface zone in water through which sunlight can pass and in which PHOTOSYNTHESIS can occur. The euphotic zone effectively filters out the red, orange, yellow and green wavelengths in approximately the top 100 metres leaving only the blue light which provides the 'colour' of oceans and lakes.

PLANKTON is confined to the euphotic zone. NET PRIMARY PRODUCTION (NPP) for this zone show great variation, ranging from a low of 125 g/m²/year in open oceans to a maximum of 2500 g/m²/year in algae beds and reefs. The extent of the open oceans (332 million km²) means that despite a low NPP, their cumulative production levels of about 4150 million tonnes per year, make them the greatest single source of protoplasm production on the planet.

**eurytypic species,** see ENVIRONMENTAL GRADIENT.

**eustasy,** *n.* a worldwide rise or fall in sea level. The formation and subsequent decline of continental icesheets were the main cause of eustatic changes in sea level during the PLEISTOCENE EPOCH. However, for the last 20 million years PLATE TECTONICS and SEA-FLOOR SPREADING are believed to have been important factors in sea level changes.

**eutrophication,** *n.* the process of nutrient enrichment of an aquatic system. In water environments, occurrence of eutrophication has increased in frequency due to the accumulation of nutrient-rich EFFLUENT from agricultural production. The nutrients accumulate at a rate greater than can be recycled by decomposition or used in PHOTOSYNTHESIS. See ALGAL BLOOM.

**evaporation,** *n.* the process of energy absorption which allows a liquid to change state and become a gas. The evaporation of water from soil, rock and surface water bodies generates atmospheric water vapour and is an important process in the HYDROLOGICAL CYCLE. The rate of evaporation is dependent upon air temperature, the level of water vapour already in the atmosphere, the nature of the wind, and the surface from which the water is evaporated. High evaporation rates from surface water bodies in arid regions can often inhibit the development of human settlement and agricultural development of these areas. Particularly in dry areas, evaporation can be responsible for the upward movement of salts which would otherwise remain at depth in the soil profile. The salts may result in the contamination of domestic and agricultural water supplies (see SALINIZATION).

**evaporite,** *n.* a SEDIMENTARY ROCK derived from minerals precipitated from evaporating seawater, such as ANHYDRITE, GYPSUM, POTASH and ROCK SALT.

**evapotranspiration,** *n.* the total loss of water as a result of TRANSPIRATION from plants and the EVAPORATION of water from soil, rock and surface water bodies.

**evergreen forest,** *n.* any forest composed of tree species that retain their leaves or needles for more than one year. At high altitudes and latitudes where temperatures are low and growing seasons short, spruce and fir trees often retain their needles for up to 20 years, although elsewhere up to 5 years would be more common. The evergreen habitat is typical of:

(a) plants which grow in environments free from climatic adversity, typically the low latitude TROPICAL RAIN FORESTS. Recent work has shown that these habitats are not entirely free from cold or drought-prone spells and under these short lasting conditions the rain forest can shed its leaves.

(b) plants which grow in environments with very marked seasonal change, typically the MEDITERRANEAN and the BOREAL FOREST biomes. In these regions the evergreen habitat allows plants to benefit from more than one season of photosynthetic capacity from a particular set of leaves. Also photosynthesis can occur whenever chance favourable conditions exist. Compare DECIDUOUS FOREST.

**exclusive economic zone,** see TERRITORIAL WATERS.

**exfoliation,** see PHYSICAL WEATHERING.

**exosphere,** *n.* the uppermost layer of the ATMOSPHERE that occurs between heights of 500 km and 2000 km. Air molecules may escape from the exosphere into space.

**exotic species,** *n.* any plant or animal species that has been introduced to an area; a non-native species. Compare ENDEMIC SPECIES.

**exponential growth,** *n.* the rate of growth which follows a *geometric rate of increase* (where each increase differs from the preceding one by a constant ratio), as opposed to an *arithmetic rate of increase* (where each increase differs from the preceding one by a constant amount). See Fig. 32. Exponential growth is a characteristic of all life forms, including the human population. When plotted as a graph, an exponential curve shows a typical shape (see Fig. 33).

| Generation: | 1 | 2 | 3 | 4 | 5 | ... | 10 |
|---|---|---|---|---|---|---|---|
| Arithmetic rate of increase: | 2 | 4 | 6 | 8 | 10 | ... | 20 |
| Geometric rate of increase: | 2 | 4 | 8 | 16 | 32 | ... | 1024 |

Fig. 32. **Exponential growth.** Comparison of arithmetic and geometric rates of increase.

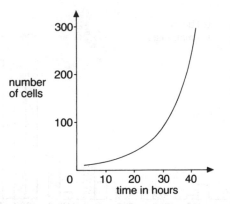

Fig. 33. **Exponential growth.** An exponential growth curve for yeast cell culture.

**extensive agriculture,** *n.* a method of farming in which there are relatively low inputs of capital and labour per unit area. On extensively farmed livestock enterprises, especially in HILL FARMING areas, the STOCKING RATE is low. In ARABLE FARMING, output in yield per hectare is generally low; however, this system is commonly practised on large-scale holdings where MECHANIZATION has displaced labour and consequently, total yields, yields per worker, and profits can be high, as on the cereal farms of North America, northern France and eastern England. However, extensive agriculture should not be regarded as synonymous with farming in large units. In many forms of tropical agriculture, for example, SHIFTING CULTIVATION, extensive farming may be practised on very small holdings. Extensive agriculture tends to be practised on fields furthest from the farmstead, and is generally associated with areas of low population density that are relatively distant from markets. Compare INTENSIVE AGRICULTURE.

**extinction,** *n.* the disappearance of plant and animals from the BIOTA. It has been claimed that of all known extinctions which have occurred since 1600 AD, 75% of the mammal extinctions and 66% of avian extinctions can be directly attributed to human activity. Hunting is the single most important cause of extinction closely followed by habitat modification. Figures for extinction rates refer mainly to the largest and hence visible plant and animal species (see Fig. 34). However, it is probable that the number of small and totally unknown species have suffered at least as great an extinction rate. For every one species which becomes extinct,

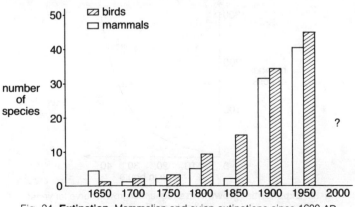

Fig. 34. **Extinction.** Mammalian and avian extinctions since 1600 AD.

approximately 30 other dependent species move into the 'at risk' category. See RED DATA BOOKS, CONVENTION ON INTERNATIONAL TRADE IN ENDANGERED SPECIES.

**extremely arid region,** see DESERT.

**extrusive rock** or **volcanic rock,** *n.* a type of IGNEOUS ROCK that forms following the cooling and solidification of MAGMA extruded onto the earth's surface. Extrusive rocks are usually fine-grained or glassy due to their rapid cooling. BASALTS are the commonest type of extrusive rock. Soils formed from weathered extrusive rock are often fertile, and consequently, are of great agricultural significance, such as those on the Deccan Plateau in India and on the plains surrounding Mount Etna in Italy. See LAVA.

# F

**facade,** *n.* the front exterior face of a building. See FRONTAGE.

**factor interaction,** *n.* the compensating effect which a high level of one ecological factor will make upon a second factor which may be present in limited amount. For example, in glasshouse production of tomatoes, a TRACE ELEMENT deficiency can be compensated for by an abundance of sunlight. See LIEBIG'S LAW OF THE MINIMUM.

**factory farming,** *n.* the intensive production of livestock in an essentially artificial and indoors environment. Factory farming is so called because of the adoption of management criteria and flow production methods similar to those employed by industrial enterprises.

Factory farms are characterized by highly specialized buildings for the close confinement at high density of often large numbers of animals, especially poultry. Temperature, ventilation, and lighting regimes are carefully regulated, and feeding is often automated and computer-controlled to achieve optimum production while reducing unit costs and the area of land required. Factory farming methods are most widely adopted in the poultry and pig-rearing sectors, and increasingly in the production of veal from calves fed on an enriched diet to promote rapid weight gain.

Although factory farming methods, in reducing costs, have widened the market for meat and eggs and thus contributed to improved diets, they are the cause of some concern and opposition. Animal welfare groups draw attention to the stress and boredom allegedly endured by animals in close confinement in artificial environments. Others argue that it is an efficient system only within a narrow economic definition of the term. Social consequences of the reduced labour demands may be damaging to rural communities. Moreover, because factory farming is dependent on expensive (sometimes imported) feed concentrates and greater energy inputs it is not only a high-cost production system but one that is vulnerable to external forces.

Factory farms are often part of the vertically integrated organization of AGRO-INDUSTRIES controlling selective breeding programmes, feed firms, food processing, marketing and distribution, or CONTRACT FARMING arrangements.

**factory fishing,** *n.* the use of highly mechanized and technical commercial fishing fleets to exploit deep-water fish stocks beyond the reach of smaller coastal fishing fleets. Shoals of fish are detected by sonar and may be netted, trawled or sucked on board by means of a giant pump before being transferred to a FACTORY SHIP with special refrigeration equipment to be gutted and processed at sea. So efficient is factory fishing that *over-fishing*, that is the exhaustion of fish stocks, has taken place in many of the world's traditional fishing grounds, for example, the North Sea and the Atlantic Ocean.

Factory fishing is very energy-intensive, requiring large quantities of fuel oil to travel the long distances to and from the fishing grounds, to remain at sea for long periods, and to power the freezing plants. It has been calculated that factory fishing has the lowest energy output (in terms of the energy value of human food) per energy input of any major food or fodder production. Japan and the USSR have been the main exponents of factory fishing methods, although many other national fishing fleets now use the technique.

**factory ship** or **mother ship,** *n.* any large purpose-built vessel which collects and processes its own and/or the accumulated catches of fish from auxilliary fishing vessels. The factory ship was a response to the need for fishing fleets to move further from their home waters as coastal fish stocks became depleted in the 1970s. As well as receiving the catches from the rest of the fleet, the factory ship serves as a supply point for the other ships, providing reserves of fuel, maintenance facilities, food, etc. The fish are processed on board the factory ship, frozen and stored for the duration of the voyage, which may last several weeks, before unloading at a suitable port. See FACTORY FISHING.

**fallout,** *n.* **1.** the deposition of particles formerly lodged in the atmosphere. The term is commonly applied to the rate of deposition of PARTICULATE MATTER, (soot, ash and grit) and also of radioactive particles which result from thermonuclear explosions or accidents in the atmosphere.

**2.** any solid particles that are so deposited.

**fallow,** *n.* any agricultural land left uncropped for one or more seasons in order to restore soil fertility. In developed countries the land may be ploughed and harrowed several times to aerate the soil, to destroy perennial weeds by desiccation, and to quicken the decomposition of crop residues which beneficially increases the nitrogen content of the soil. In tropical farming systems, such as SHIFTING CULTIVATION and ROTATIONAL BUSH FALLOW, the land is abandoned to recolonization by natural vegetation for a period of up to 30 years before being brought back into cultivation.

During the AGRICULTURAL REVOLUTION of the 18th and 19th centuries in western Europe the need for bare fallows was reduced by the introduction of CROP ROTATIONS. Between 1960 and 1982 the practise of fallowing land in Europe was generally discontinued as greater use of chemical FERTILIZERS permitted the continuous cropping of the land. However, over-production in EC countries has led to the development of a *set aside policy* in which farmers are effectively subsidized to place a proportion of their land into non-agricultural use for a period of 10 years. See FALLOW SYSTEM.

**fallow system,** *n.* a farming practice in which a period of cropping is followed by a period when the land is FALLOW. Fallow systems are characteristic of tropical regions and are made necessary by the rapid LEACHING of nutrients from the soil as soon as cultivation occurs. Different farming systems can be compared on the basis of the ratio of fallow years to crop years. Thus, SHIFTING CULTIVATION will have a fallow:cropland ratio of more than 10:1, while ROTATIONAL BUSH FALLOW systems are characterized by values of ranging between 4:1 and 10:1.

**family** or **taxon,** *n.* any of the taxonomic groups into which an ORDER is divided and which contains one or more related genera. For example, Felidae (cat family) and Canidae (dog family) are two families of the order Carnivora. See CLASSIFICATION HIERARCHY, LINNEAN CLASSIFICATION.

**false-colour imagery,** *n.* the display of images in colours that are not true to life, in order to enhance certain features contained in the images. The technique uses film emulsions responsive to wavelengths beyond the sensitivity of the human eye. The use of film that is sensitive to infrared wavelengths, which are not scattered by atmospheric haze, allows much sharper images to be rendered than with the use of visible-wavelength film emulsions (which record true colour and monochrome). Thus it is easier to distinguish vegetation (which records as red) from soil (which records as blue-green) using false-colour imagery than with conventional film.

False-colour imagery is much used in aerial photography, particularly for the plotting of vegetation and soil distributions, and for displaying electronic images relayed from earth-survey satellites such as LANDSAT.

# FAO

**family farm,** *n.* any agricultural holding capable of supporting a single family. Unlike the PEASANT farm, outside labour is employed in addition to family labour, but not to the extent of larger-scale farm enterprises which rely mainly or entirely on wage labour. The notion of the owner-occupied family farm remains enshrined throughout Western Europe and North America as an ideal, representing a scale of farming activity generally preferred to the inefficiencies of peasant smallholdings and the impersonal, institutional operations of commercial AGRIBUSINESSES.

**famine,** *n.* an acute shortage of foodstuffs in an area leading to MALNUTRITION and starvation. The root causes of famine are multiple and complex. The onset of famine may be rapid when crops or stocks are destroyed by natural hazards as in the Bangladesh floods in 1974, but is more generally a gradual process of deteriorating food availability. Environmental problems, such as persistant DROUGHT, DESERTIFICATION, or disease and pest depredations rarely lead to widespread famine on their own. Rather, they are compounded by other factors such as political unrest or civil war, as seen in Mozambique and the Ethiopian provinces of Eritrea and Tigre in the 1980s, by exhausting land resources through OVERPOPULATION or overgrazing, and by agricultural policies that depress production or fail to realise the land's full potential during the more favourable seasons.

**FAO,** *abbrev. for* Food and Agriculture Organization, a specialized agency of the United Nations, founded in 1945 to coordinate programmes of food, agriculture, forestry and fisheries development in order to improve the standards of living of rural populations and combat MALNUTRITION and hunger.

The FAO coordinates research and data collection, such as the WORLD FOOD SURVEYS, and analyses and disseminates information through regular statistical reports, commodity studies and occasional papers. It advises governments on policy and agricultural planning and provides technical assistance on such matters as nutrition and food management, improved food storage, marketing and distribution, FERTILIZER production, pest and disease control, SOIL CONSERVATION, and IRRIGATION. Since 1985, greater attention has been directed towards the problems of tropical forestry, with the introduction of measures to promote conservation and the increased use of AFFORESTATION in agricultural and industrial planning. Together with other United Nations agencies, the FAO administrators the WORLD FOOD PROGRAMME.

**farm enlargement,** *n*. the increase in the size of agricultural holdings, primarily through the physical amalgamation of farms and the aquisition of contiguous and non-contiguous plots of land. Farm enlargement is deemed necessary in order to reduce the number of holdings that are too small to provide more than PART-TIME FARMING opportunities or else are non-viable without external financial support. In many parts of Europe the process of farm enlargement and the reduction in the number of farms has been slow, particularly in areas with farm structures deficient in other respects (see FRAGMENTATION). To promote STRUCTURAL REFORM, many countries have provided aid in the form of farm amalgamation grants, or have established an intervention agency responsible for buying land as it comes on the market, with a view to selling it as part of a more rational programme of farm enlargement. In France, for example, such regional agencies have, since 1960, handled some 1.2 million ha of land in this way. The progress of farm enlargement is illustrated by the changing number of holdings in different size categories. In the EC between 1960 and 1985 the number of holdings with 1-5 ha of agricultural area fell from 4.0 million to 2.3 million, while those over 50 ha increased from 265 000 to 383 000.

**farm enterprise combination analysis,** *n*. a method of analysis used to identify the relative importance of different activities or enterprises on a farm holding which collectively define a particular type of farm, and may thus help delineate an AGRICULTURAL REGION. (An enterprise is here regarded as a single CROP or form of ANIMAL HUSBANDRY, or as a group of related activities, rather than the farm itself as a business enterprise.)

Unlike CROP COMBINATION ANALYSIS where data is compared using a relatively simple statistical examination of crop acreages, any comparison of the relative significance of different types of farm enterprise requires a common denominator to standardize data from a variety of sources, for example, the monetary value of farm outputs, or more commonly, the standard labour requirements applied to crop acreages and livestock numbers.

**farm forestry,** see AGROFORESTRY.

**farming,** see AGRICULTURE.

**farm rent** or **contract rent,** *n*. the monetary payment made by a tenant farmer for the right to occupy a farm holding. Periodic monetary payments often replace systems in which rent is paid in kind or services, as in SHARECROPPING. Farm rents may be taken as a crude measure of land value as they are determined by the location, quality and agricultural potential of the farm.

**farm size,** *n.* a measure of the profit-making or productive potential of a farm enterprise. Farm size is most commonly and conveniently measured by the land area of a farm. However, a holding may include areas of woodland and permanently or temporarily unproductive land not normally classified as part of the *utilized agricultural area* (UAA). Even where the agricultural land alone is measured it is not possible to differentiate land quality and potential.

Average farm sizes have increased throughout the developed countries. In the USA, for example, the average farm size rose from 61 to 174 ha between 1930 and 1981, while the number of farms fell from 6.8 million in 1935 to 2.4 million in 1980. In the UK the average size of the 258 000 holdings included in the 1985 Farm Census was 67 ha and against 38 ha for the 517 000 holdings listed in 1955.

**fault,** *n.* a fracture in the earth's crust along which there has been a displacement of strata. Faults are formed by compressional or tensional tectonic forces. Faults can range from a fraction of a centimetre to hundreds of kilometres in size. Various types of fault are recognized according to the direction and relative movements of strata along the fault plane, including normal faults, reverse faults, tear faults and thrust faults (see Fig. 35). Movement along faults may result in EARTHQUAKES.

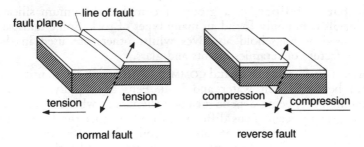

Fig. 35. **Fault.** Normal and reverse faults.

**fauna,** *n.* the total animal population that inhabits an area.

**faunal region,** see ANIMAL REALM.

**fecundity,** see BIRTH RATE.

**feeder service,** *n.* a bus service which connects an outlying residential or commercial area to a station on a railway network. Feeder services are scheduled to connect to particular train departures and are most heavily used in the mornings and evenings by COMMUTERS. See also TRANSPORT NETWORK, PARK-AND-RIDE SYSTEM.

**feedlot,** *n.* an area of land for the fattening or finishing of animals, generally beef cattle in which the *stocking densities* (the number of animals per unit area of land at a particular point in time) are very high. Fodder is not produced on the feedlot but is supplied in the form of concentrates and roughages prepared to promote rapid weight gain and lean meat. Cattle that have been restricted nutritionally earlier in their life benefit best from the feedlot regime, producing a high compensatory gain, often in excess of 1 kg per day, when put on a high-energy feed for three to five months.

Feedlots, first introduced in California in the inter-war years, are highly mechanized, with low labour demands in relation to the number of animals handled. The 1% of feedlots in the USA with capacities in excess of 1000 head account for more than half the fed cattle, and the very largest in Colorado and California may have an annual throughput of up to 100 000 cattle. The expansion of livestock feed and fattening industries owes much to the diffusion of HIGH-YIELDING CEREALS. Ever-increasing output has held grain prices down, enabling feedlots to satisfy the expanding North American market for beef, pork and poultry.

**feldspar** or **felspar,** *n.* a group of complex aluminium silicate minerals commonly found in many types of rock. Some feldspars are used as household abrasives whilst others are used in the manufacture of ceramics, paints and glass.

**fen,** *n.* a distinctive marshland COMMUNITY found in the transitional area between fresh water and land. Fens are rapidly evolving communities, being dependent upon the rate at which silting and peat growth occur. Fens differ from BOGS in that the former are nutrient-rich and have an alkaline soil (pH greater than 7.0). Fenland occupies upper parts of old estuaries as well as the edges of lakes. It is a particularly well developed feature of East Anglia, England where the rivers drain from inland, chalk-rich rocks. Much fenland has been reclaimed used as highly productive agriculture land.

**ferralization,** see LATERIZATION.

**ferrisol,** *n.* the term used in the FAO soil classification system for a LATOSOL-type soil.

**Fertile Crescent,** *n.* a crescent-shaped area of fertile plains and valleys in south-west Asia stretching from the alluvial lowlands of the Tigris and Euphrates rivers in Mesopotamia and the surrounding fringes of hill country to the east and north, through Syria and Lebanon to the south-east Mediterranean coast.

The Fertile Crescent was one of the world's major AGRICULTURAL HEARTHS. The early DOMESTICATION OF PLANTS AND ANIMALS on the higher ground led to the development of non-migratory systems of AGRICULTURE, population growth, and more elaborate settlements. The region saw the flowering of the Sumerian, Babylonian and Assyrian civilizations, and the emergence of city life and literate society in the period circa 3500-1600 BC.

**fertility rate,** see BIRTH RATE.

**fertilizer,** *n.* any material added to the soil to supply essential nutrients for crop growth. These may be by-products of natural (organic) materials, products manufactured for the purpose, or by-products from other chemical processing industries. Bulky organic substances such as compost, MANURE, and NIGHT SOIL are normally considered separately, and the term fertilizer is more strictly applied to substances derived from chemical processes or mining which are thus known variously as *artificial fertilizers, inorganic (mineral) fertilizers,* or *chemical fertilizers.*

Crops require varying quantities of six major elements – calcium, magnesium, sulphur, and especially nitrogen, phosphorus and potassium – as well as smaller amounts of TRACE ELEMENTS. Upwards of two-thirds of the nitrogen and phosphorus taken up by the plant is lost from the soil either in harvesting the crop or in the crop residues. It is the role of fertilizer to replenish these nutrients. Fertilizers generally provide nitrogen, phosphorus and potassium (potash) either individually (known as *straight fertilizer*) or as a mix of two or three of these elements, often with additional trace elements. Most fertilizers are of this *compound* or *mixed* type, and are produced in granulated form or as powders or liquids in combinations of nutrients to suit specific soil needs. When applied to the soil by hand, or by drilling, often along with seed, or as a top-dressing sprayed by aircraft, chemical fertilizers dissolve rapidly and are either absorbed by soil colloids or taken up by plant roots.

World fertilizer use has expanded rapidly from 14 million tonnes in 1950 to 131 million tonnes in 1986, leading to a rise in crop yields that has more than offset a one-third decline in grain area per capita. In the EC in 1984, the consumption of nitrogenous fertilizers stood at 8.0 million tonnes, phosphate fertilizers at 4.1 million tonnes and potassium fertilizers at 4.4 million tonnes. Application of nitrogenous fertilizers per hectare of utilized agricultural land ranged from 45 kg in Greece to 227 kg in the Netherlands with a Community average of 77 kg in 1982–83.

**fetch,** *n.* the distance travelled by a wind-generated wave before it arrives at a coastline. The fetch controls the height and strength of waves which in turn influences their ability to erode and deposit. See MARINE EROSION.

**field capacity,** *n.* the water content of a formerly saturated soil following the natural drainage by gravity of the excess water. It is retained around soil particles by surface tension and is readily available for plant growth. Field capacity is usually attained some 6-24 hours following the cessation of heavy rainfall and is largely determined by the SOIL TEXTURE. See also SOIL WATER.

**field crop,** *n.* any LEAFY CROP previously grown almost exclusively on small-scale MARKET GARDENING enterprises but now grown mostly in open fields, for example peas, beans, sprouts and cabbages. From the late 19th century many vegetable crops were adopted by arable farmers as additional cash crops or as part of a CROP ROTATION. With the post-war growth in demand for convenience foods and in CONTRACT FARMING an increasing proportion of vegetable output is now derived from field cropping. Growing vegetables on farms as field crops has several advantages over market gardens, for example easier MECHANIZATION requirements, a more efficient use of labour, and specialization. Field crops can also act as BREAK CROPS, and in the event of adverse price shifts can be ploughed back in as a beneficial green MANURE, thereby avoiding harvesting costs.

In some instances the term field crop is used more broadly to describe any CROP grown in rows other than PERMANENT CROPS of trees and shrubs, and small grain cereals such as wheat, barley, oats and rye. See also HORTICULTURE.

**field system,** *n.* the planned division of land into fields for agricultural use such as MEADOWS, PASTURES, and for ARABLE FARMING. The shape and utilization of fields varies considerably, in response to the physical conditions, the kinds of CROPS and livestock raised, and factors such as the system of LAND TENURE and inheritance, the level of agricultural technology or MECHANIZATION, and the density of population. In many parts of medieval

Europe, especially England, the predominant field system prior to ENCLOSURE, was one of *open fields*, where two or more large fields were held as COMMON LAND and worked by individuals in unenclosed and generally scattered strips (see STRIP FARMING). When lying FALLOW the open field was subject to common grazing rights. A two-field system operated where the common arable land was divided into two parts, one under cultivation, the other lying fallow. A three-field system involved a division into two cultivated areas and one fallow. Other types of field system include temporary land occupance or field-rotation, such as SHIFTING CULTIVATION, and the INFIELD-OUTFIELD system. See also CROP ROTATION.

**field work,** see field study.

**fiord** or **fjord,** *n.* a long, narrow and steep-sided coastal inlet formed by the drowning of a U-SHAPED VALLEY and resulting either from the submergence of a coast or by glacial erosion below sea level. Fiords may be over 1200 m deep although their seaward end is shallower and marked by a rock bar, formed by the depositing of terminal MORAINE or a decrease in a glacier's erosive power as it entered the sea. Fiords are found along the coasts of Scotland, Norway, Greenland, British Columbia, Patagonia and South Island, New Zealand. Fiords usually provide well protected deep water anchorages.

**fire clay,** *n.* fine-grain sedimentary deposits frequently found beneath the COAL MEASURES; the term is used primarily in British geological descriptions. Such clays are rich in silica and alumina and low in alkalis and lime, and are economically important because of their refractory properties (that is, they are heat-resistant and do not melt when placed in high temperature locations). They are used in the manufacture of fire bricks for lining flues, furnaces etc. See CLAY.

**fire (in agriculture),** *n.* the deliberate burning of natural vegetation and crop residues for agricultural management purposes. Fire was one of the first and most effective tools employed in early agriculture, and its use remains widespread throughout the world.

In tropical FALLOW SYSTEMS fire is part of the SLASH-AND-BURN method of land clearance, and has been subject to much debate on its contribution to soil fertility. The use of fire represents a convenient labour-saving device in initial plot preparation and in the destruction of weeds and grass seeds, reducing the need for later weeding. The burning of the bulk organic matter enriches the soil with an ash that contains greater concentrations of available nutrients, especially potassium and phosphorus, and ensures a rapid release of mineralized nitrogen when the first rains fall. Heating of the soil surface also reduces the numbers of insect pests. On the other hand, burning destroys HUMUS, and many nutrients are lost to the atmosphere. Burning offers the tropical farmer a means of regulating soil conditions so as to obtain an adequate crop yield, at least in the first year of cultivation; however, each time this expedient is used the means of building up long-term fertility is destroyed. See also PYROCLIMAX.

**firn,** see SNOWFIELD.

**fish farming,** *n.* a form of AQUACULTURE in which inland waters and marine environments are utilized to breed and rear fish for commercial purposes. Fish farming differs from FISHING in the greater control over the raising and harvesting of fish stocks, in freeing the fisherman from the uncertainties of hunting, and by producing much higher yields per unit of area.

Ponds or temporarily flooded areas, including rice PADDIES, have been fish-farmed for several thousands of years in parts of eastern Asia. Complex POLYCULTURES have evolved enabling several species to be raised through exploiting different ECOLOGICAL NICHES within otherwise limited pond areas. Freshwater fish farming makes a significant contribution to the total fisheries harvest in Asian countries such as China (40%), India (38%) and Indonesia (22%). The Japanese pioneered marine fish farming or *mariculture* at the turn of the century; elsewhere in the world, fish farming has developed significantly only in the last 30 years. In Africa, most progress has been made in rearing the Chinese carp and the Nile tilapia in freshwater ponds. In DEVELOPED COUNTRIES, the major advances have been in rearing freshwater trout in the entirely controlled environment of tanks, and salmon in cages lowered in coastal waters. Artificial fertilization has allowed fry of trout, and a species of catfish in Africa, to be raised in hatcheries before being transferred to larger ponds or enclosures. In several tropical countries where IRRIGATION and navigation channels have become blocked by aquatic vegetation, plant-eating species such as carp have been introduced both to clear vegetation and to provide fish stocks for human consumption.

Currently, inland and marine fish farming accounts for only about 9% of the total fish catch, or 7 million tonnes per annum. The goal of the FAO and the United Nations Development Programme is to raise the output of fish farms five-fold by the end of the century, concentrating on efforts in DEVELOPING COUNTRIES to integrate aquaculture with agriculture rather than on commercial enterprises which have tended to develop species for export and for more exclusive markets.

**fishing,** *n.* the harvesting of fish and shellfish from oceans and inland waters to provide food, industrial raw materials and as a form of recreation. Activities such as sealing, WHALING and the gathering of marine plants may be separately categorized.

The principal fishing nations in 1985 were Japan, accounting for 13% of the world catch, USSR (12%), and China (8%). More than three-quarters of the world catch is harvested in the northern hemisphere, and is heavily concentrated on coastal waters and the CONTINENTAL SHELF. Only about 12% of the catch is from inland waters. The most heavily fished marine area is the North-West Pacific.

The global catch has risen gradually, but individual fisheries have shown more dramatic growth rates, sometimes doubling each year as a particular species is exploited, for example, the Peruvian anchovy catch rose from less than 100 tonnes p.a. in the 1940s to 10 million tonnes in the late 1960s, before collapsing under the pressure of over-fishing to 94 000 tonnes in 1984. Although some species appear quite resilient to heavy exploitation, over-fishing remains a serious problem. See also AQUACULTURE, FISH FARMING, TRAWLING, SEINE NETTING, LINE FISHING, LAW OF THE SEA.

**fission reactor,** *n.* the most common type of NUCLEAR REACTOR in which atoms such as uranium-235 and plutonium-239 are split by neutrons thus releasing energy mainly in the form of heat. This process occurs in the reactor as a NUCLEAR FISSION chain reaction. The heat generated by the process is used to raise steam which in turn drives a turbine and generates electricity.

**fissure,** *n.* a weakness in the earth's crust, such as a FAULT, from which LAVA is extruded. Large-scale fissure eruptions over long periods may result in extensive lava plateaus, such as the Deccan Plateau in India which is around 700 000 km² and the Columbus-Snake River Plateau in the USA which is about 300 000 km². See also BASALT.

**flatted factory,** *n.* an industrial building where individual floors are subdivided into work areas for small industries, the products of which do not require mass production techniques or large-scale chemical handling. Such buildings were more frequently used in the 1950s and 1960s than today, when direct ground-level access is preferred, even by small businesses.

**flocculation,** *n.* the process whereby individual CLAY particles coagulate to form AGGREGATES. Flocculation usually occurs in alkaline soils when calcium and magnesium ions dominate the CLAY-HUMUS COMPLEX. Flocculation coarsens the texture of a soil, making it easier to work and can be encouraged through the application of lime to the soil.

**flood,** *n.* the overflowing of a river or the temporary rise in the level of the sea or a lake which results in the inundation of dry land. River and lake floods are caused by the inability of these water bodies to accomodate an increased supply of water. This increased supply may occur naturally as a result of, for example, unusually heavy rainfall or rapidly melting snow, or may be the result of human activity. DEFORESTATION and urbanization decrease INFIL-TRATION which causes more rapid runoff and an increase in peak discharge. Land drainage schemes can also disrupt the hydrological balance and generate flooding further down the drainage system. Flooding in low-lying coastal areas can result when a storm surge occurs at high tide. Floods can cause considerable loss of life and damage to property, such as the flooding which followed the exceptionally heavy MONSOON rains in northern India during July and August 1980.

A number of techniques are utilized to minimize the effect of floods. The maintenance of trees and other vegetation reduces rapid surface runoff and the marked peak in discharge which generates floods. The creation of small storage dams can also be used to regularize the flow of water through drainage systems. The construction of DYKES and LEVEES is a traditional form of flood control and is intended to prevent the inundation of low-lying areas by rivers and the sea. A sophisticated system of levees and associated *floodways* (subsidiary channels designed to reduce the discharge in the main river channel during very high floods) has been developed to prevent large-scale flooding in the Mississippi Basin. Flooding can also be averted by straightening river courses to increase water flow and the volume that can be accomodated at any particular time.

Flooding can occasionally have beneficial environmental conse-quences. The agricultural productivity of the Nile Basin in Egypt is largely maintained by the deposition of fertile sediment associated with annual flooding.

**flood basalt,** see BASALT.

**floodplain,** *n.* the floor of a valley over which a river may spread during short-lasting or seasonal floods. Floodplains can be created by rivers of all sizes. Other land forms are frequently associated with floodplains, for example, OX-BOW LAKES and marshy depressions indicating abandoned river channels.

Floodplains result from the downstream migration of MEANDER belts which cause a widening of the valley. During times of flood, water inundates the floodplain and a layer of ALLUVIUM is deposited as a result of the drop in river velocity. The deposits are thicker and coarser adjacent to the river channel and the resultant landform is known as a LEVEE. As floods recede and the river returns to its normal course, it is possible for sediment to be deposited in the river channel. If the rate of deposition in the channel is greater than the surrounding floodplain then the river can flow above the level of the flood plain due to the containment by the levees.

The agricultural importance of floodplains cannot be underestimated. Almost half the world's population lives in southern and south-eastern Asia, many of whom are directly dependent on farming the alluvial soils of the floodplains for their livelihood.

**flora,** *n.* the total vegetation assemblage that inhabits an area.

**flush,** *n.* an area of nutrient enrichment in the soil usually brought about by the inflow of enriched surface water to provide a *wet flush*, or more rarely as the gravitational movement of dry particles to a site thus forming a *dry flush*. Flush sites are frequently found along valley sides or in basin-like topography in upland areas.

**fluvent,** see ALLUVIAL SOIL.

**fluvial deposition,** *n.* the deposition of a river's load when its competence and capacity are decreased (see FLUVIAL TRANSPORTATION). There are several causes for these decreases including a loss of velocity, a decrease in volume, a reduction in stream gradient, the river freezing over, a widening of the stream bed or emptying into a slower moving body of water such as a lake. Debris deposited by a river is known as ALLUVIUM. As deposition commences, it is usual for the large particles in the load to be laid down first. Numerous landforms are associated with fluvial deposition, including ALLUVIAL FANS, FLOOD PLAINS, LEVEES and DELTAS.

**fluvial erosion,** *n.* the progressive removal of material from the bed and margins of a river channel. Erosion can take place in solid bedrock or in unconsolidated alluvial material and can be accomplished in four ways:

(a) the force of flowing water along a river channel (*hydraulic action*) can exert a dragging effect on highly fractured and poorly consolidated materials. A specific form of erosion caused by hydraulic action, *cavitation*, results from the explosive effects of water pressure caused by a very high flow velocity;

(b) erosion by solution is known as *corrosion* and is most effective in rivers which cross areas of soluble bedrock such as limestone;

(c) the most effective means of fluvial erosion is *corrasion* or *abrasion* which results from the physical impact of the river LOAD against the channel bed and margins. Corrasion is responsible for most of the downcutting in a river channel. Potholes in river beds are a common feature of rapid abrasion, as pebbles are rotated and ground into the bedrock by turbulent, high velocity flow;

(d) *attrition* is a subsidiary form of corrasion, as the particles in the load are broken and worn as they are rolled or carried along by the river.

The rate of fluvial erosion is dependent upon the stream size, the nature of the load, stream velocity and the gradient of the river valley. The regulation of river flow through storage in reservoirs and the periodic release of large amounts of water from dams and hydro-electric power stations may cause an artificial increase in downstream erosion. To prevent the localized loss of agricultural land and undermining of building foundations, bridges, roads and railways, the stream banks may be protected by rock pilings or concrete. The stabilization of river banks through the planting of trees may also reduce the effects of fluvial erosion.

**fluvial transportation,** *n.* the carrying of mineral material along the course of a river. There are three ways in which a stream can transport material:

(a) as a *dissolved load*, comprising materials in solution. This is dependent upon the solubility of the bedrock or ALLUVIUM over which the river flows;

(b) as a *suspended load*, comprising mainly light-weight materials, such as sand, silt and clay, and supported by the turbulent motion of river flow;

(c) as a *bedload*, comprising larger pieces of debris, such as rock fragments and pebbles, rolled or slid along the channel floor by the force of the river flow. Localized turbulence can also cause some of the bedload to move in a series of short jumps called *saltation*.

*Stream capacity* denotes the potential load that can be transported whilst *stream competence* refers to the largest size of particle it can carry. The ability of each river to carry debris largely depends upon the velocity and volume, or *discharge*, of the water. Thus, during periods of flood a river's competence and capacity can increase dramatically. For example, in 1952 the English seaside resort of Lynmouth was devastated when, after an intense downpour, the swollen West Lyn River, suddenly changed course and brought

40 000 tonnes of boulders, soil and masonry through the town, demolishing houses, roads, bridges and service cables, and causing 31 deaths.

**fluvioglacial deposits,** see GLACIAL DEPOSITION.

**fluvisol,** see ALLUVIAL SOIL.

**fly ash,** *n.* extremely fine particles of ash produced when coal is burnt efficiently in modern forced-draught furnaces, particularly those associated with modern coal-fired electricity generating stations. Because of their extremely low mass, fly ash particles are not easily removed from the atmosphere by gravitational fallout. Fly ash is, however, easily trapped by ELECTROSTATIC PRECIPITATORS and can be disposed of by dumping in quarries, made into concrete building blocks or dumped in estuaries to reclaim tidal flats.

**flyover,** *n.* a crossing of two roads where one is carried over the other by a bridge. When two MOTORWAYS merge or cross, a highly complex, multilevel bridging structure, or *interchange*, can result. On- and off-ramps are needed to allow traffic to change direction with minimum conflict or reduction in capacity of the road system. See also DUAL CARRIAGEWAY, UNDERPASS.

**focus,** *n.* the area in the EARTH'S CRUST where tectonic movement results in the sudden release of energy that generates seismic waves leading to an EARTHQUAKE.

**fodder crop,** *n.* any crop grown principally for animal food, and consumed either directly or in a preserved form. The most important cultivated fodder plants belong to three families, the *Gramineae* (grasses), *Leguminosae* (legumes) and *Cruciferae* (crucifers). Grasses forming natural PASTURE are not strictly fodder crops despite being foraged by grazing animals. Fodder is commonly produced from GRAIN CROPS, ley legumes (such as clover, lucerne, trefoil, sainfoin and vetch) often containing grass, root vegetables (such as turnip, swede, sugar beet and kohl rabi) and various plants cut or grazed as immature green fodder (such as maize, flax, linseed, fodder peas, broad beans, rape and kale).

Fodder crops may be cut, dried and stored as HAY, fresh cut and fed green as soilage, conserved as SILAGE, and fed off direct in the field in which grazing may be controlled by electric fencing (see GRAZING MANAGEMENT). Crops in which the bulk of the edible root is below ground have to be lifted and topped (as with turnip, swede and sugar beet). These are being replaced by kale, rape and kohl rabi which are both easier to cultivate and more accesssible to foraging animals.

Plants that are grown in some parts of the world as FOOD CROPS for human consumption are, elsewhere, grown largely for fodder. Maize, a STAPLE foodstuff in much of Africa and Latin America, is grown primarily as green fodder in temperate zones. In 1984, over 60% of the world's maize crop of 449 million tonnes, was fed to animals, much of it on FEEDLOTS. Some fodder crops may be grown to maturity for human consumption, or for industrial processing, for example, flax, linseed, rapeseed and sugar beet.

**fog,** *n.* a dense, cloud-like mass of water droplets suspended in the lowest layers of the atmosphere that reduces visibility to less than 1 km. Fog is formed by the cooling of the lower layers of the atmosphere and the resulting condensation of water vapour. Several types of fog are recognized according to the mechanism of cooling, including RADIATION FOG, ADVECTION FOG, and STEAM FOG.

**Föhn,** *n.* **1.** a warm dry ADIABATIC WIND which flows on the northern side of the Alps.

**2.** any adiabatic wind, such as the Chinook in the USA, the Nor'wester in New Zealand and the Samun in Iran.

**fold,** *n.* a symmetrical or asymmetrical bend in layered crustal rocks, usually formed by compressional tectonic forces. Small folds may sometimes result from the differential compaction or the displacement of deep-lying rocks. Folds are most commonly found in sedimentary rocks but can also occur in igneous and metamorphic rocks. See ANTICLINES, SYNCLINES.

**fold mountains,** *n.* a mountain chain formed by compressional tectonic forces folding a GEOSYNCLINE. The Himalayas and the Pyrenees are examples of fold mountain ranges. See FOLD.

**food aid,** *n.* the transfer of surplus agricultural commodities from DEVELOPED COUNTRIES to DEVELOPING COUNTRIES in order to promote economic and social development, to provide emergency relief, and to combat MALNUTRITION. There are three main forms of food aid. Emergency or FAMINE relief aid is the most visible form but is limited in scale compared to both *project aid*, in which food is allocated to specific development projects, and *programme aid*, in which food is donated to support a country's budget or balance of payments. Alternatively, food aid may be categorized as *food-for-cash*, in which the sale of low-cost food donations raises revenue for the recipient government that can then, in theory, be allocated to development projects; *food-for-nutrition*, in which food is distributed to especially vulnerable or deserving groups, such as mothers with infants; and *food-for-wages*, in which labour on projects is rewarded in part or full by a food ration.

About half of all food aid is donated by the USA, with a further 30% from the EC. In total, food aid amounts to only some 10% of all bilateral AID. Food is also donated to agencies such as the WORLD FOOD PROGRAMME (WFP), which now channels some 25% of all food aid, mostly into food-for-work projects and for the benefit of specified groups. At present, the WFP handles only 10-20% of total emergency food aid.

Food aid is controversial. Few would deny the humanitarian appeal of emergency food aid in famine relief, but long-run dependence on food aid can effect food consumption patterns, establishing a demand for commodities such as wheat and skimmed milk that are unlikely ever to figure significantly in domestic production. Many see food aid simply as a disguised means of disposing of burdensome surpluses produced by heavily protected farm sectors in developed countries. Food aid is often poorly targeted, and may benefit both poor and relatively affluent groups, while it is generally the recipient government that bears the financial costs of internal administration and distribution. Because food aid becomes part of the recipients' general food economy, it may reduce demand pressures on domestic sources, depressing prices, undermining farmers and removing their incentive to expand output. It is often welcomed by THIRD WORLD governments anxious to peg food prices at levels acceptable to urban consumers identified as politically more sensitive than rural interests.

Above all, food aid is an intensely political mechanism. The disposal of US surpluses is explicitly linked to foreign policy objectives, hence the large volume of food aid directed to South East Asia in the early 1970s, and the withholding of food aid to Chile during 1970-73 when a Marxist government was in power. Egypt is currently the world's largest recipient of food aid, receiving almost 2 million tonnes of cereals or their equivalent in 1984-85, mostly from the USA, or 16% of the developing countries' total. The year before, Egypt received 56% of all food aid to Africa, although this fell to 26% in 1984-85 as American and European donors responded to the famine elsewhere and released cereals, even to countries such as Mozambique and Ethiopia whose political regimes were otherwise out of favour.

**Food and Agriculture Organization,** see FAO.

**food balance sheet,** *n.* a measure of calories available per person per day in a country. Domestic food output, adjusted to account for imports and exports and changes in stocks, and excluding animal feed, seed requirements and estimated losses, is converted to its calorific equivalent or *dietary energy supply* and divided by the population.

National food balance sheets can be compared with the national minimum calorific requirements calculated by the World Health Organization, to indicate the percentage of minimum calorific requirements being met by domestic food supplies. The minimum daily Calorie requirement per person is calculated for each country as the energy necessary to meet the needs of an average healthy person, taking account the variations in body size, the age and sex structure of national populations, and climatic conditions. These range from 2160 kcal per person in parts of South East Asia with warm climates and a high proportion of children in the population, to nearly 2700 kcal in Scandinavia. Between 1982 and 1984 more than two-thirds of Sub-Saharan African countries were reported to have supplied less than their minimum calorific requirements, whereas several developed countries supplied 130% or more of their minimum requirements.

There are many problems inherent in the calculation, use and interpretation of food balance sheets including defining an adequate calorie intake for a healthy body, and the unreliability of population and food output data in THIRD WORLD countries. Nevertheless, they are widely and increasingly used as shorthand indicators of regional food supplies and of countries likely to have population groups suffering from MALNUTRITION.

**food chain,** *n.* a structured feeding hierarchy whereby energy in the form of food is passed from an organism in a lower TROPHIC LEVEL to one in a higher level. The first trophic level $(T_1)$ comprises the *primary producers* (plants); $T_2$, the *primary consumers* (herbivores or plant eaters); $T_3$, *secondary consumers* (carnivores or meat eaters); and $T_4$, *tertiary consumers* (meat eaters at the top of the food chain). Food chains are frequently arranged as complex interconnected networks call *food webs*. See DETRITAL FOOD CHAIN, GRAZING FOOD CHAIN.

**food crop,** *n.* any CROP grown primarily for human consumption. Crops cultivated in some areas almost exclusively for human consumption are elsewhere grown largely as FODDER CROPS or as raw materials for industrial processes. Maize is a STAPLE foodstuff in much of Latin America and Africa but in North America and Europe, it is grown mainly for feeding livestock. Palm oil is consumed in considerable quantities in West African cuisine but the greater part of the world's output is destined for soap-making and industrial purposes.

**food-for-cash,** see FOOD AID.

**food-for-nutrition,** see FOOD AID

**food-for-wages,** see FOOD AID.

**food policy,** *n.* any policy introduced by a government or international body to influence the food sector for the benefit of a group of producers, or consumers, or both. The most common forms of food policy include price support, the provision of grant aid and loans such as a subsidy on farm inputs and operating costs, mechanisms for the disposal of agricultural surpluses, and selective consumer subsidies.

In DEVELOPED COUNTRIES, most food policies have been oriented primarily to the producer through some from of market regulation for example, the COMMON AGRICULTURAL POLICY of the EC. Food policies in Europe and North America have been so successful in ensuring the security of domestic food supply and in encouraging production efficiency (albeit at steadily rising cost to the taxpayer) that it has been necessary to pay farmers to take land out of production, as occurs in the USA, and to subsidize overseas sales as FOOD AID. In communist countries such as the USSR almost the whole farm sector is controlled by centrally planned food and non-food policies which require COLLECTIVE FARMS and STATE FARMS to fulfill delivery obligations at state regulated prices.

Explicitly consumer-oriented food policies are not common in developed countries, although selective subsidies to benefit the urban poor have been implemented in the USA. Consumer subsidies are more widely employed in DEVELOPING COUNTRIES. In India, for example, grain surpluses accumulated by the state are sold through 'fair-price shops' at subsidized prices. Forms of indirect consumer subsidy include policies that control producer prices in order to secure low-priced food for urban consumers.

**food supply,** *n.* the production of food and the process by which it is made available to consumers. The vast majority of the world's population attain their food supply by means of SUBSISTANCE FARMING or by market purchase, although an increasing number of the THIRD WORLD population are becoming dependent on FOOD AID and other subsidized food sources. A decreasing proportion of food transactions, especially in DEVELOPED COUNTRIES, consists of direct open market sales of produce by the farmer to the ultimate consumer.

Indexes of agricultural production show a greater than 50%

increase in global food output since the mid-1960s, but given population growth since then this figure represents a real increase of only a 7% in per capita output (see FOOD BALANCE SHEET). The supply of GRAIN CROPS has increased much more rapidly than that of root crops and tubers, aided by price support mechanisms in many developed countries, and the spread of GREEN REVOLUTION technology elsewhere. Despite these advances, regional deficiencies in food supply continue to lead to MALNUTRITION and FAMINE. See also WORLD FOOD SURVEY.

**food web,** see FOOD CHAIN.

**footloose industry,** *n.* any manufacturing industry with no specific requirements or preconditions for its location. Such industries are neither resource- nor market-orientated and as such can enjoy a freedom to locate in a wide variety of areas. Footloose industries tend to be found in modern consumer societies in which MOTORWAYS provide good transport mobility and where there is an industrial trend towards assemblage of a product from components which are shipped in from a wide variety of sources.

**footpath,** *n.* any track intended solely for pedestrian traffic. A footpath may be consciously planned and paved, as within a city (and known as a *sidewalk* in North America) or consist of a naturally developed track in the countryside. Some rural footpaths follow historic trackways such as the cattle drove roads in the Scottish Highlands. Many pathways are legally protected RIGHTS OF WAY which are maintained in their open state by being walked over.

**forest climate,** *n.* the MICROCLIMATE developed over afforested areas and often characterized by variations in ALBEDO, diurnal temperatures, wind and humidity patterns compared to surrounding non-afforested areas.

**forest management,** *n.* the management practice applied to a forest to ensure the long-term and continuous supply of timber products and associated items. In northern industrialized countries such as West Germany and the UK, forest management embraces all stages of development and includes site drainage before planting, fencing, planting, weeding, fertilizing, herbicide/insecticide application, selective thinning and eventual felling. By careful choice of tree species and appreciation of landscape quality, the forest manager can enhance the appearance of the countryside, thus help to foster the tourist-recreation potential of his or her locality. The mature forest can provide facilities for a variety of uses, such as walking, camping, hunting and timber production.

By contrast, many forests in developing countries still suffer from lack of long-term management policies. Instead, commercial logging companies, often owned by northern hemisphere multinational companies, devastate vast areas of rain forest by CLEAR FELLING. Often, 75% of the forest can be destroyed in the quest to remove the remaining 25% of commercially valuable timber. Replanting schemes are not usually practised, the forest being allowed to regenerate as best it can.

Forests are RENEWABLE RESOURCES and as such can respond to sympathetic management policies. Forests can provide more products than timber alone and imaginative forest management can ensure a wide MULTIPLE LAND USE role. See SILVICULTURE.

**forest park,** *n.* an extensive area of mature forest, associated with landscapes of great scenic beauty, in which tourism, recreation and wildlife protection receive special management attention alongside the traditional pursuit of commercial timber production, as for example, in Tollymore Forest Park in Northern Ireland. On occasions the forest park may assume a role similar to that of a NATIONAL PARK, as with the Argyll Forest Park and the Queen Elizabeth Forest Park in Scotland, the Tsitsikama Forest Coastal National Park in South Africa and the Sherbrook Forest Reserve in Australia.

**forest reserve,** *n.* **1.** any area of forest, usually of limited extent (less than 100 ha) which contains luxuriant development and/or diverse flora and fauna and as such is considered worthy of protection from DEFORESTATION. These areas may be more formally conserved as SITES OF SPECIAL SCIENTIFIC INTEREST in the UK, or may form part of a much larger NATIONAL PARK, for example, the Redwoods National Park in California.

**2.** any area of forest land set aside from an area of currently managed forest, and retained for a variety of future uses, for example, as an economic source of timber, for conservation or as a gene bank. In the USA, extensive tracts of *primitive areas* (14.2 million acres in 1939, much of which was forested), were set aside and have subsequently been reclassified as WILDERNESS AREAS.

**forestry,** *n.* the practice of planting, tending and managing forests primarily for the exploitation of timber, whether commercially or for local subsistence needs. An increasingly important responsibility vested in forestry concerns, public and private, is the provision of recreational opportunities, and the management of landscape AMENITY.

The extent of global DEFORESTATION together with the escalating demand for timber products has emphasized the need for FOREST MANAGEMENT and the establishment of forestry agencies to promote AFFORESTATION and REAFFORESTATION. World output of commercial wood products in 1983 amounted to 175 million tonnes of paper and paper board (a 37% increase since 1970), 2.9 billion cubic metres of roundwood, that is, timber as it is felled (+22%), 448 million cubic metres of sawnwood, that is, timber in its first stage of processing (+11%), and 128 million tonnes of wood PULP (+22%). A quantity in excess of this is used for firewood or is converted to charcoal (see FUELWOOD CRISIS).

**formation,** *n.* a group of vegetation COMMUNITIES in a single geographical region or continent, of similar PHYSIOGNOMY and existing under related climatic and environmental conditions. For example, the tropical rain forest of the Indo-Malayan archipelago is a formation and within which are found detailed local and regional variations called ASSOCIATIONS.

**forward contract,** see CONTRACT FARMING.

**fossil,** *n.* a trace or the remains of a plant or animal from a past geological age which is usually preserved by natural processes in rock. To allow fossilization, the traces and remains of the organism must be quickly buried to avoid weathering and decomposition. Skeletal structures or other hard parts of the organism are the most commonly occurring form of fossilized remains although mineral casts or *trace fossils* showing trails, tracks or excrement are also found. Fossils are most commonly found in sedimentary rocks of marine origin although they may occasionally occur in igneous and metamorphic rocks. The oldest fossils are of simple unicellular algae and are an estimated 4000 billion years old.

Fossils are classified by the same nomenclature as used for modern plants and animals. They have been widely used to date and correlate rocks and to indicate former environmental conditions.

**fossil fuels,** *n.* any naturally occurring carbon or HYDROCARBON fuel derived from the anaerobic decomposition of organic material in the earth's crust. Examples include NATURAL GAS, OIL, COAL, PEAT and OIL SHALE. Fossil fuels form the world's dominant energy source and are a major raw material.

**fossil water,** see CONNATE WATER.

**Fourth World, the,** see THIRD WORLD, THE.

**fragile ecosystem,** *n.* those plant and animal COMMUNITIES which are particularly vulnerable to damage caused by human activity. All ecosystems can be damaged or destroyed by humans but those in slow-growing ALPINE regions and those in high-latitude in particular have suffered very considerable destruction due to the low powers of regeneration and recovery. Wetland communities (such as MARSH, FEN, BOG, and MANGROVE) are also considered fragile communities in that they require very specific conditions of wetness and pH for their survival. See TOLERANCE, ENVIRONMENTAL GRADIENT, ANTHROPOGENIC FACTOR.

**fragmentation,** *n.* the division of a farm holding into several non-contiguous plots. In Asia, Africa and many of the poorer regions of western Europe, holding patterns have assumed highly fragmented forms. Such fragmentation may have resulted from a fossilization of old open-FIELD SYSTEMS, but more often has arisen through the practice of *partible inheritance* by which land is divided equally among all direct heirs, a practice accentuated by *parcellement*, in which heirs receive an equal share of each existing plot.

Fragmentation is widely regarded as contributing to the emergence of poor agrarian practices. It is likely to lead to impaired labour efficiency, particularly in the journey to work, and in the movement of stock and equipment. There is increased scope for disputes over ownership and access to land, while small and scattered plots may prevent the enlargement of fields needed to facilitate MECHANIZATION, greater ease of access, and more efficient GRAZING MANAGEMENT. Problems arise in agricultural advisory work and land planning, and in the sharing of costs and benefits derived from improvements such as SOIL CONSERVATION, drainage, and IRRIGATION.

In many countries fragmentation has become excessive. By the early 1960s in Spain, for example, the average number of plots per holding was 14, and as high as 32 in the Galicia region.

Not all fragmentation should be seen as necessarily disadvantageous. In tropical and temperate agriculture, fragmentation may present a multiplicity of physical conditions and ECOLOGICAL NICHES for the farmer to exploit with different crops, yielding a more diverse farm output and minimizing the risk from localized climatic hazards such as frost or hail. Fragmentation, particularly across an altitudinal range, may permit staggered production regimes, and allow for a rational deployment of labour through seasonal peaks of demand. Excessive fragmentation may be tackled by schemes for FARM ENLARGEMENT and CONSOLIDATION.

**freeport,** *n.* an industrial development area with harbour facilities where raw materials are imported and processed, and from which finished goods are exported without being subjected to the host nation's customs and excise regulations. A freeport functions, for tax purposes, like an island separated from the rest of the country, while still being a source of employment for local people. The freeport concept has also been successfully applied to areas adjacent to airports, as for example at Shannon airport in the Republic of Ireland.

**free range,** see GRAZING MANAGEMENT.

**freeway,** see MOTORWAY.

**freeze-thaw action,** see PHYSICAL WEATHERING.

**freezing point,** *n.* the temperature at which a liquid changes state to become a solid. The most important freezing point with regard to the biosphere and its life forms is that of water which occurs at, or slightly below 0°C. See SUPERCOOLING.

**Friends of the Earth,** *n.* a non-party-political international pressure group dedicated to increasing public awareness of key contemporary issues concerning all environmental matters. Operating in 28 different countries the organization lobbys governments and politicians, provides information for inclusion in ENVIRONMENTAL IMPACT ASSESSMENTS and disseminates information through videos, lectures, and books. In recent years it has been particularly active in establishing adequate safety levels for existing NUCLEAR POWER stations, resisting the construction of new nuclear plants and in the preparation of a *Natural Heritage* document targeted at governments of developed nations. See also GREENPEACE.

**fringing reef,** see CORAL REEF.

**front,** *n.* the boundary between AIR MASSES of different temperatures and humidities. Fronts such as the POLAR FRONT may occur on a large scale and indicate the convergence of major air masses, whilst others are smaller and are usually associated with more localized DEPRESSIONS.

**frontage,** *n.* **1.** the facade of a building.

  **2.** the collective wall of adjacent facades along a street. See also TOWNSCAPE.

**frontal depression,** see DEPRESSION.

**frontal rain,** see RAIN.

**frost,** *n.* **1.** a form of PRECIPITATION consisting of the deposition of ice particles onto a surface, usually out-of-doors at night. The severity of a frost is dependent upon the moisture content of the atmosphere and the temperature of the air. Several types of frost are recognized according to the nature of formation, including HOAR FROST, RIME and GLAZED FROST.

**2.** also known as **air frost**. air whose temperature is at or below 0°C at STEVENSON SCREEN level.

**frost heaving,** see PHYSICAL WEATHERING.

**frost hollow,** *n.* a depression in the landscape into which cold air drains under gravity, causing a marked drop in temperature. The lowest temperatures are recorded on the sides of the frost hollow just above its base. Farmers and foresters must adapt their management systems to cope with frost hollows by planting frost-hardy crops or by providing shelters for animals which may mistakenly seek shelter in the hollow.

Road engineers and architects must also be aware of the occurrence of frost hollows as the increased risk of FROST can lead to black ice on roads and cause structural damage to property.

**frost wedging,** see PHYSICAL WEATHERING.

**fuel bundle,** see FUEL ROD.

**fuel rod,** *n.* a rod-like unit of nuclear fuel from which energy is released during a fission chain reaction in a NUCLEAR REACTOR. Fuel rods usually consist of uranium dioxide ($UO_2$) pellets comprising 3% fissionable uranium-235 and 97% non-fissionable uranium-238 held within a cladding tube made from zircaloy; this metal alloy optimizes the requirements for high neutron retention, reliable corrosion resistance in very hot water, and the strength required for cladding the uranium fuel used especially in light water reactors. The $UO_2$ pellets are manufactured by compacting and sintering $UO_2$ powder (*yellowcake*) into cylindrical pellets and grinding to size. This process is highly hazardous to the workers as it releases radioactive radon-222 gas. Also, the waste products from this process retain a low-grade radioactivity for thousands of years.

Fuel rods are usually grouped together into *fuel bundles*, typically of 63 rods, which are inserted into the reactor core. The rate at which the fuel rods release their energy is managed by inserting or removing CONTROL RODS located between the fuel bundles. After an interval of between 18 and 30 months the fuel bundles will be removed and sent to a nuclear fuel reprocessing plant for disposal or re-cycling.

**fuelwood crisis,** *n.* the increasing scarcity of wood and charcoal to meet the requirements of fuel for cooking and as a source of warmth in many parts of the THIRD WORLD. At least half of all the timber cut in the world is used as fuelwood. In Africa, 90% of the population use wood or charcoal for cooking. Even in urban areas where alternative energy sources are available fuelwood is favoured because of its cheapness relative to kerosene, electricity or gas, and despite the low fuel efficiency of most traditional wood burning stoves. Fuelwood is seen as reliable given the frequency of power cuts in urban areas and there is often a preference for the taste of food cooked over wood or charcoal. Long regarded as a 'free' resource (see COMMONS ATTITUDE), even rural areas some distance from urban centres are now finding that securing adequate supplies of fuelwood is costly. According to the United Nations Development Programme, 23 countries have already run out of wood. At present consumption rates, the estimated fuelwood deficit will double by the year 2000. Throughout the Third World some 1.5 billion people are unable to obtain sufficient fuelwood to meet their minimum needs or are forced to consume wood faster than it is being replenished. It has been estimated that the annual rate of consumption has exceeded the mean annual incremental growth of local tree stocks and forest reserves by 30% in the SAHEL, and by as much as 200% in Niger, 150% in Ethiopia, 75% in Nigeria, and 70% in Sudan, without taking account of other causes of DEFORESTATION.

The adverse consequences of the fuelwood crisis extend beyond accelerated deforestation and the attendant problems of SOIL EROSION and DESERTIFICATION, to include the increasing amounts of time, especially female labour, diverted from productive activities to the search for wood, and the use of other combustible materials such as dung and crop residues that might otherwise have enriched the soil. Despite the growing scarcity of readily accessible fuelwood supplies local perception of the problem and the response to AFFORESTATION has been poor. Fuelwood difficulties are generally seen as less pressing than food or water shortages. Solutions to the fuelwood crisis in rural areas are likely to lie less in large-scale programmes of afforestation or reforestation than in local AGROFORESTRY projects that can be shown to have tangible benefits for farming communities. See also DUNG FUEL.

**fumarole,** *n.* a small vent in the earth's surface from which steam and other volcanic gases escape. Fumarole activity is usually associated with dormant volcanoes. In Italy, some fumaroles are harnessed to produce electricity.

**functional land use,** see LAND USE.

**functional niche,** see ECOLOGICAL NICHE.

**fungicide,** *n.* any chemical substance used to destroy FUNGI, which are a common cause of plant diseases. Sulphur is one of the oldest fungicides (used from 1803), while Bordeaux mixture (copper sulphate, quicklime and water) may have been used to control fungus attacks on vines since Roman times. As far as can be assessed, fungicides have not caused any major environmental problems. Their use has been too localized to cause anything other than minor simplification of the DECOMPOSER group of organisms. See also HERBICIDE, INSECTICIDE, PESTICIDE.

**fungus,** *n.* any plant of the division Fungi, lacking chlorophyll, leaves, true stems and roots. Fungi reproduce by spores and live entirely as SAPROPHYTES or parasites. The group includes moulds, yeasts, mildews, rusts and mushrooms. Some can cause disease in plants and animals, including humans.

**fusion reactor,** *n.* a type of NUCLEAR REACTOR whose energy is derived from the fusion of two ATOMS (usually deuterium or tritium) to form one helium atom with the consequent release of energy. Fusion reactors are still at an experimental stage of development mainly due to the problems associated with the very high temperatures ($10^8$ °C) of fusion. Attempts have so far been concentrated on containing the fusion reaction within a magnetic field set within a ceramic-walled chamber.

**fynbos,** *n.* a low-growing SCLEROPHYLLOUS vegetation confined to the most southerly tip of South Africa. A wide variety of woody Protea plants often form a low shrub thicket within which a rich ground flora exists. In proportion to its size, fynbos has the greatest number of species of all the MEDITERRANEAN-type plant communities. Many of the plants are endemic to the region and as such are of considerable scientific importance. Unfortunately, little if any natural fynbos remains as a result of human activity.

# G

**Gaia concept,** *n.* a controversial hypothesis devised by the British scientist James Lovelock in 1979 concerning the role of living organisms in maintaining a climatic equilibrium on earth. The theory views the earth as a single complex organism which is both self-regulating and self-organizing. Biotic elements attempt to moderate their local environment which brings about an optimal chemical and physical environment for all life forms in which there is sufficient oxygen for animals and sufficient carbon dioxide for plants. The 'Gaia mechanism' moderates atmospheric and oceanic temperatures by a controlled GREENHOUSE EFFECT as well as regulating the major biogeochemical cycles essential for life, that is, water, oxygen, soil and rock.

Lovelock's theory has received only qualified support in scientific circles although it has been gaining limited respectability as further research uncovers evidence of links operating between biological systems and the environment in the way that the Gaia hypothesis suggests. (Named after *Gaia*, Greek goddess of the Earth.)

**gangue,** *n.* the valueless rock and minerals in an ORE deposit. The extraction and disposal of these materials add greatly to the cost of developing an ore deposit and waste tips remain an unsightly legacy on many industrialized landscapes. Improved ore extraction techniques have led to the reworking of many gangue deposits. See DEEP MINING.

**garden city,** *n.* a planned settlement in England which attempts to incorporate the best of both urban and rural environments. The garden city was a development of the 19th century philanthropic attitude of providing a 'model village' environment for work and residence. Ebenezer Howard is credited with developing the concept of the garden city in 1898. His model 'city' planned to accommodate 32 000 people on a site of 2429 ha. Population density together with space for roads, industry, shops, schools and parks was allocated according to a series of idealized standards. CROWDING was to be avoided and healthy living conditions were to be maintained. The garden city was to be self-sufficient in terms of providing employment for its inhabitants thus avoiding the need for commuting to a major urban area. As the optimum population for a garden city was reached then so new ones would be started, each to be located in a cluster around a central, older city and linked by rail and road.

The first garden city was constructed at Letchworth, 55 km north of London in 1903. The second and better known example was Welwyn Garden City (1920), also to the north of London. By modern planning standards garden cities suffer from a uniformity and monotony of style; however, the relatively low density of housing coupled with an abundance of open space contrasts favourably with the successors to the garden city, the NEW TOWNS. See also RESIDENTIAL DENSITY.

**garrigue,** *n.* a SCLEROPHYLLOUS shrub-dominated vegetation growing to a height of about 1 m and found around the Mediterranean Sea. Garrigue probably represents the most modified of the Mediterranean vegetation communities. Many generations of burning, grazing and deforestation have removed the natural mixed forest of the region and in its place is found a sparse vegetation in which geophytes and aromatic, herbaceous plants predominate. The growing period of the garrigue is unusual in that it is confined to the cooler and moister winter and spring seasons with the result that the garrigue often appears desolate in the long, hot and dry summers which characterize the Mediterranean region.

**gas-cooled nuclear reactor,** see MAGNOX NUCLEAR REACTOR.

**gaseous pollutants,** *n.* gases released into the atmosphere which act as primary or secondary pollutants. A variety of gaseous pollutants exist including nitrous oxides ($NO_x$ group), sulphur dioxide, and the PEROXYACETYL NITRATES. Gaseous pollutants are mainly produced as a result of human activity and are vital components in the formation of PHOTOCHEMICAL SMOG and ACID DEPOSITION, both of which can have a range of harmful effects including the deterioration of building materials, the degeneration of trees and other plants, and respiratory ailments in humans and animals. See PRIMARY POLLUTION, SECONDARY POLLUTION, AIR POLLUTION, AUTOMOBILE EMISSIONS.

**gasohol,** *n.* a US term for motor vehicle fuel comprising between 80-90% unleaded petroleum spirit and 10-20% ETHANOL. Gasohol production is uneconomic compared with conventional petrol (gas) production and often requires considerable political backing in the form of financial subsidy and legislation compelling its use in place of petrol. However, it is a 'clean' fuel producing very little pollution when burnt.

The greatest user of gasohol has been Brazil as its chronic balance of payments difficulties makes it impossible for it to purchase large quantities of crude oil on the world market; however, it does possess abundant supplies of organic material which can be used for gasohol production. The Brazilian National Alcohol Programme was responsible for the production of 3 billion litres of ethanol in 1981, derived mainly from sugar cane and manioc. By 1985, the gasohol production target was 8 billion litres with an increased ethanol content. Other countries are experimenting with gasohol.

In the USA, maize is used as the source material and a target production level of 5.7 billion litres by 1990 has been set. New Zealand, Canada and many THIRD WORLD countries are also producing smaller quantities of gasohol from waste wood products, molasses and waste animal products.

**Gatto report,** *n.* the *Report on Community Forest Policy* published in 1983 by the EC and which provided an assessment of forest production and timber product consumption until the year 2000.

**gene bank,** see GENE RESOURCE CENTRE.

**gene library,** see GENE RESOURCE CENTRE.

**gene pool,** *n.* the sum total of all genes and genetic information possessed by a given reproducing population.

**gene resource centre, gene library** or **gene bank,** *n.* a collection point for genetic material from which genetic engineers can draw a wide variety of wild genetic material for experimentation or for incorporation into existing agricultural HYBRID species in order to revitalize or improve the performance of species of economic value. The centres comprise collections of seeds of those plants threatened by EXTINCTION. They are held in the possibility that they may be used to restore the vigour of agricultural hybrid species or overcome some unforseen disease. The preservation of animal genetic material is more problematic but great advances are being made in the storage of ova and sperm in a deep frozen state for subsequent use. See GENETIC ENGINEERING.

**General Development Order,** *n.* a listing of types of DEVELOPMENT which in the UK do not require planning permission. If a proposed development relates to one of the listed classes, the developer need not apply to the local authority for permission. There are sometimes limitations imposed upon particular developments which thereby restrict the scope of the deemed permission, for example, limitations of size for a house extension beyond which planning permission must be sought in the normal way. Governments may tighten or loosen planning controls by means of changes to the General Development Order which may constrain or make freer the interpretation of what constitutes 'permitted development'. Compare USE CLASSES ORDER, DEVELOPMENT CONTROL.

**generally recognized as safe,** see GRAS.

**genetic conservation,** *n.* the retention of plants and animals for the purpose of safeguarding their genetic information. In practice this can be done by establishing collections of plant and animal protoplasm (see GENE RESOURCE CENTRE), by conserving those geographical areas which show the greatest natural plant and animal diversity or by selective breeding in captivity.

The Rio Palenque Research Station in Equador is an example of the second method of genetic conservation where 1025 different plant species along with animals dependent upon them are maintained in 170 ha of land. Despite its conservation status, however, DEFORESTATION is occuring due to local people invading the forest in search of FIREWOOD.

According to the Russian biologist Nikolay Vavilov it should be possible to identify areas of maximum genetic diversity by selecting a domesticated plant and then searching for the region in which its wild relatives show maximum diversity. The so-called 'Vavilov centres' (see Fig. 36 overleaf) have been proposed as the optimum sites for conservation in support of the gene pool although this method is directed mainly to existing agricultural crops. Of equal importance are the genetic codes held by organisms which have yet to be of proven value to humans. For example, biochemists are evaluating genetic structures of Australian rain forest trees in the expectation that an antidote for the AIDS virus may be found within one species. See GENETIC ENGINEERING.

**genetic engineering,** *n.* the manipulation and/or recombination of genetic materials to produce species of greater benefit to mankind. An early and simple form of genetic engineering was the practice of hybridization as studied by Mendol between 1856-68. It involved the crossing of genetic material between two closely related plants but which displayed detailed differences such as colour and smoothness of seeds. Crossing members with different visual characteristics produced a HYBRID.

Natural hybridization produces offspring with wide ranging characteristics and provides the mechanism whereby species can

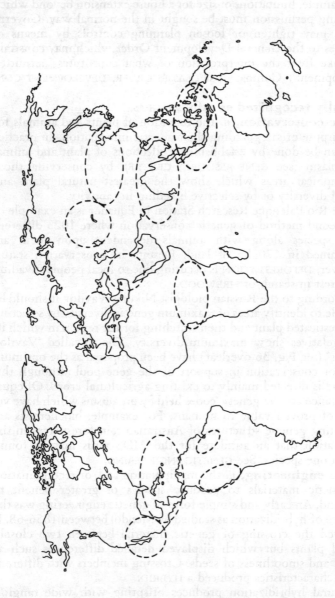

Fig. 36 **Genetic conservation.** The distribution of Vavilov centres, which represent regions of maximum natural plant and animal diversity.

migrate into new environments. Agriculturalists have recognized that by deliberately selecting those plants which display desirable characteristics, for example, resistance to disease or rate of growth, then the quality and productivity of a crop can be improved.

As a result of improved understanding of genetics it has become possible to create genetic combinations which were hitherto impossible under natural conditions. Most contemporary agricultural crops and animals are the result of genetic engineering. The sunflower, now the major source for vegetable oil, has been improved by drawing on the gene pool of the wild sunflower, and the yield of cassava, a staple food crop of the tropics has been increased up to 18 times by inclusion of disease-resistant strains obtained from the wild cassava. Current plant genetic engineering research includes attempts to incorporate biological nitrogen fixation characteristics into major food grains to lessen dependence on chemical FERTILIZERS, and to develop more nutritional cereal grains, pest- and HERBICIDE-resistant strains, and crops better adapted to the needs of food processing industries.

In animal breeding, innovations in genetics have made possible almost complete control of animals' reproductive processes through, for example, embryo transfer, superovulation, oestrus detection and oestrus synchronization. In embryo transfer the developing embryo is removed from a superior female and transferred to the uterus of a recipient of otherwise inferior stock, allowing the recipient to develop the fetus without genetically influencing it, while the donor can be induced by hormonal stimulation to superovulate, that is, to produce more fertilized ova, and thus, in the case of cows artificially inseminated produce perhaps sixty genetically superior calves in a year instead of one. Developments in embryo transfer have led to laboratory fertilization, cloning, twinning and cell fusion such that one embryo may produce several genetically identical progeny.

The principal advantages in genetic engineering lie not only in the promise of raising greatly improved plants and animals but also in reducing the time taken to develop new varieties by conventional selective breeding employing the normal sexual process of fertilization. Recent innovations in genetic engineering have tended to further the influence and controlling interests of AGRIBUSINESSES which have increasingly sought to appropriate the process of agricultural production.

However, genetic engineering of agricultural species can result in the simplification of the GENE POOL through the repeated concentration upon relatively small species population. For this reason, wild gene pools in which natural variation is allowed to occur, should be retained.

**genotype,** *n.* **1.** the genetic constitution of an organism. Compare PHENOTYPE.

**2.** a group of organisms with the same genetic constitution.

**3.** the type species of a GENUS.

**gentrification,** *n.* the process whereby one social group of higher status and wealth displaces another by buying into run-down areas of old but sound houses, in attractive locations in the inner parts of cities. The in-comers individually refurbish the houses, often with the aid of IMPROVEMENT GRANTS, until the whole area takes on a renewed appearance and property values as a whole rise. It thus becomes impossible for people on low incomes to buy houses in the locality and the social composition of the local resident population changes permanently in the direction of higher socio-economic groups. See also INNER CITY, RESIDENTIAL SEGREGATION.

**genus,** *n.* any of the taxonomic groups into which a FAMILY is divided and which contains one or more SPECIES. For example, *Vulpes* (foxes) is a genus of the dog family, Canidae. See CLASSIFICATION HIERARCHY, LINNEAN CLASSIFICATION.

**geochemical prospecting,** *n.* the chemical analysis of surface water, alluvium, soil and plants to identify high local concentrations of metallic minerals which may indicate the presence of an underlying ore body.

**geographical information system (GIS),** *n.* a set of integrated techniques for collecting, storing, retrieving, transforming and displaying spatial data collected from the real world for analysis. Because of the large quantities of data involved and the repetitive calculations needed to manipulate the data, most GIS are computerized. The main components of a GIS are outlined in Fig. 37. Central to every GIS is the data base, the storehouse of information upon which the eventual quality of output depends. However, the database alone is of little value; its application to real-world problems is dependent upon the capability of the data transformation facility which is a function of the software designed to manipulate the data. Through the transformation stage the data base can be accessed, transformed and manipulated, thus enabling the data to serve as a test bed for examining environmental issues, studying trend patterns or simulating the results of specific actions made upon the physical and biological world.

Most geographical information systems have been established to study specific problems; for example, a soils information system for use by farmers and foresters. Recently, there is evidence that major government agencies are setting up comprehensive, countrywide systems, such as the Canada Land Project in which a major attempt has been made to evaluate the properties of land throughout Canada and to establish a planning procedure which links land use with land potential, thereby optimizing the use of land. Information systems are increasingly being used by physical planners, land use managers, conservationists and government departments involved with strategic land use planning. See ENVIRONMENTAL INFORMATION SYSTEM.

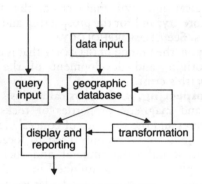

Fig. 37. **Geographical information system.**

**geological column,** see EARTH HISTORY.

**geological cycling,** *n.* the slow movement of materials from the land by means of EROSION and TRANSPORTATION to the oceans where they are deposited as sediments and eventually form SEDIMENTARY ROCKS. Eventually, OROGENESIS will result in the uplift of the sediments to form dry land upon which the cycle of erosion will begin anew. The entire process of geological cycling takes many millions of years to complete and because of this, stable geological materials such as granite have been chosen as suitable for the deposition of NUCLEAR WASTES in deep bore holes.

**geological map,** *n.* a map which details the nature of an area's underlying geology by indicating rock types, structural features and the age relationship of the various rocks. Geological maps may be of direct and indirect use in land use planning, soil surveying, hydrological and civil engineering projects, and the prospecting and exploitation of mineral resources.

**geological timescale,** see EARTH HISTORY.

**geology,** *n.* **1.** the scientific study of the origin, composition and structure of the earth.

    **2.** the study of the geological features of an area, in particular its landforms and the processes which have operated to produce the present day physical landscape. See also GEOMORPHOLOGY.

**geomorphological map** or **morphological map,** *n.* a map which details the nature and properties of landscape features and the type and scale of past and present geomorphological processes. Geomorphological maps provide a comprehensive and integrated account of the landscape and are used in LAND USE PLANNING, soil survey, hydrological and civil engineering, the development of agriculture and forestry, and for the prospecting and exploitation of mineral resources. See GEOMORPHOLOGY.

**geomorphology,** *n.* the branch of GEOLOGY that is concerned with the structure, origin, and development of the topographical features of the earth's crust.

**geophysical prospecting,** *n.* the physical analysis of the gravitational, seismic and magnetic properties of rocks to determine geological structures and the presence of concealed mineral bodies. Geophysical methods are commonly used to locate oil deposits, metallic ore bodies, industrial mineral and rock deposits, and groundwater supplies. These techniques may also be used to investigate geological structures prior to large civil engineering projects.

**geophyte,** see RAUNKIAER'S LIFEFORM CLASSIFICATION.

**geostrophic wind,** *n.* a theoretical wind which flows parallel to the ISOBARS and represents a balance between the opposing effects of the CORIOLIS FORCE and PRESSURE GRADIENT. Upper atmospheric winds may sometimes approach perfect geostrophic flow but near the earth's surface, friction causes winds to flow across the isobars at an oblique angle towards an area of low atmospheric pressure.

**geosyncline,** *n.* a large SYNCLINE in which vast thicknesses of sediment accumulate over extended periods of time. To allow such accumulations, it is thought that geosynclines subside at the same rate as sedimentation occurs. In the geological past, compressional tectonic forces have resulted in the folding of geosynclinal sediments to form FOLD MOUNTAINS, such as the Caledonian mountain chain which extends from Ireland through Wales, northern England and Scotland into Norway.

**geothermal energy,** *n.* energy derived from the heat of the earth's interior. Two main sources of geothermal energy have been recognized:

(a) hot JUVENILE WATER and steam can be harnessed in areas of volcanic and tectonic activity. This steam and hot water may be used directly for heating buildings and hothouse cultivation, as happens in Iceland. The hot water and steam may also be used to generate electricity, such as at Larderello in Italy and The Geysers in California, USA;

(b) in certain parts of the world where hot, dry, intrusive igneous rock bodies lie close to the surface, geothermal energy can be tapped by drilling holes to the hot rocks and injecting water to create steam which can then be harnessed to generate electricity. Such a water circulating system has been successfully established in the Jemerez Mountains in New Mexico, USA and further investigations into the geothermal energy potential of hot rock regions are being carried out in a number of areas throughout the world including the Piedmont Province in eastern USA and the Wessex Basin in England.

The water and steam that are used to generate geothermal energy has several useful by-products. The water is often highly mineralized and is a valuable source for a variety of minerals. The water can be demineralized for domestic and industrial use, such as at El Tatio in Chile.

Geothermal energy is relatively pollution-free compared with other energy sources although the disposal of hot mineral waters not utilized for its by-products poses a problem. To prevent the thermal and chemical pollution of surface drainage systems, this water is reinjected into deep wells. The emission of toxic and noxious gases from geothermal plants must also be carefully controlled.

**geothermal power,** *n.* energy obtained from heat sources contained within the earth. Geothermal power is obtained by tapping sources of superheated water or steam. The steam is used to drive turbines which produce electricity, while the hot water is used to heat greenhouses (as in Iceland) or for domestic central heating (as at Rotorua in New Zealand).

Geothermal energy is a 'clean' energy source in that it does not lead to a build-up of atmospheric PARTICULATE MATTER. Some gaseous release does occur, mainly of hydrogen sulphide, but in non-hazardous concentrations. Also, if the large quantities of very hot water produced using geothermal techniques are not utilized, as in the examples above, they must be cooled before being released into rivers or the sea. See HOT DRY ROCK TECHNIQUE.

**geyser,** *n.* a HOT SPRING which violently discharges water and steam, usually with great regularity, as with Old Faithful in the Yellowstone National Park, USA which rises to a height of 55 m approximately every 65 minutes. Geysers result when some of the superheated groundwater in the underground channel feeding the spring intermittently changes into steam causing the explosive upward movement of all overlying water to form a temporary fountain. In countries such as the USA, New Zealand, Iceland and Japan, geysers are a major tourist attraction. The largest geysers can sometimes be harnessed to generate electricity.

**ghetto,** *n.* any sector of a city containing a high concentration of people from one racial, religious or cultural group. Characteristic features of ghettos include close-knit internal social connections that are impenetrable by outsiders, a strong community loyalty in the face of outside attack, and a sense of being different from, and outside the institutions of, the surrounding polity. These areas are often characterized by a high density of population, together with housing and other social problems. See RESIDENTIAL SEGREGATION.

**gilgai,** *n.* an undulating surface of shallow depressions interspersed with ridges (up to 1 m high) which occurs on clay-rich soils. It results from alterations in the soil moisture level, which causes the clays to expand and contract.

**GIS,** see GEOGRAPHICAL INFORMATION SYSTEM.

**glacial deposition,** *n.* the dropping of rock and other debris by a glacier. All deposits are known collectively as *glacial drift*, of which two main types are recognized:

(a) *glacial till* is an unsorted and unstratified heterogenous mixture of rock fragments, ranging in size from CLAY particles to boulders. Glacial till is sometimes known as *boulder clay* and is associated with various landforms including MORAINES, DRUMLINS and ERRATICS.

(b) *fluvioglacial deposits* or *outwash deposits* consist of debris often crudely sorted and stratified by meltwater streams. The resulting landscape features include OUTWASH PLAINS, ESKERS, and KAME TERRACES.

**glacial drift,** see GLACIAL DEPOSITION.

**glacial erosion,** *n.* the progressive removal of bedrock, soil and other material by GLACIERS. Glacial erosion can be accomplished in two ways:

(a) *plucking* or *quarrying* entails the removal of blocks of bedrock dislodged by water freezing in BEDDING PLANES.

(b) *abrasion* is the scraping and grinding of the underlying bedrock by rock fragments carried by the glacier (see GLACIAL TRANSPORTATION).

The rate of glacial erosion is highly variable and is largely dependent upon the size and velocity of the glacier and the nature of the underlying topography. Valley glaciers are highly effective agents of erosion and in formerly glaciated regions have left a very distinctive landscape with features such as U-SHAPED VALLEYS, HANGING VALLEYS, ARÊTES, TRUNCATED SPURS and TARNS. Erosion by ice sheets often smooths out underlying relief and can lead to dramatic alterations of the previous landscape. Compare WIND EROSION, FLUVIAL EROSION

**glacial striae,** see ERRATIC.

**glacial till,** see GLACIAL DEPOSITION.

**glacial transportation,** *n.* the removal of rock and other debris by a GLACIER away from its point of origin. The LOAD of glaciers can range in size from large boulders to finely powdered rock fragments known as *rock flour*. Three methods of glacial transportation are recognized:

(a) *supraglacial transportation* involves the carrying on the glacier surface of rock debris that falls from valley sides;

(b) *englacial transportation* involves the carrying within the glacier of material that has fallen down a crevasse or has been buried by snowfalls;

(c) *subglacial transportation* involves the carrying at the base of a glacier of material that is both accumulated by, and largely responsible for, GLACIAL EROSION.

Compare FLUVIAL TRANSPORTATION.

**glacial watershed breaching,** *n.* the breaching of a pre-glacial WATERSHED by the movement of a glacier. This may lead to the permanent alteration of former river DRAINAGE PATTERNS. Watershed breaches may be utilized for the construction of easy transport routes in otherwise difficult terrain.

**glaciation,** *n.* the process by which the earth's surface is covered and altered by GLACIERS and ice sheets during an ICE AGE. The causes of periods of glaciation are unclear but may result from:

(a) variations in the level of heat received from the sun;

(b) variations in the carbon dioxide and water vapour content of the atmosphere;

(c) the drift of continents towards the polar regions.

# GLACIER

See GLACIAL EROSION, GLACIAL DEPOSITION.

**glacier,** *n.* a mass of ice that flows outward from a SNOWFIELD under the influence of gravity and the pressure of accumulating ice. During the last ICE AGE, glaciers occupied 30% of the earth's surface but presently account for around 10%. Glaciers are found on all continents except Australia.

The rate of glacier movement can vary from a few centimetres to tens of metres per day and is influenced by the size of glacier, air temperature, the amount of unfrozen water in and under the glacier, and the slope and topography of the underlying land. Glacial movement terminates when the rate of melting ice equals or exceeds the forward advance of the glacier.

Three main types of glacier have been recognized:

(a) *Valley glaciers* or *mountain glaciers* form in the snowfields at the heads of mountain valleys. Their downward movement follows pre-existing river valleys. Valley glaciers are common in the American Rockies, Scandinavia, the European Alps and New Zealand's South Island;

(b) *piedmont glaciers* result when several valley glaciers coalesce on the plains below mountain ranges. Examples include the Frederikshaab Glacier in Greenland and the Malaspina Glacier in Alaska;

(c) *ice sheets* or *continental glaciers* are very large ice masses that spread over their plateau or continental regions of supply, Greenland and Antarctica being the only two current examples. *Ice caps* are a smaller and more localized form of icesheet that exist in several areas including Spitzbergen and Iceland.

Valley glaciers in many parts of the world are being developed for all year round winter sports whilst the ice sheets are presently being explored for economic resources such as oil and gas. See GLACIATION, GLACIAL TRANSPORTATION, GLACIAL DEPOSITION, GLACIAL EROSION.

**glasshouse cultivation,** *n.* a specific form of HORTICULTURE in which glasshouses (or *greenhouses*) are used to grow crops of fruit, vegetatbles or flowers under controlled or protected conditions. The greenhouses provide a sheltered environment which permits the production of high quality crops either earlier than would be possible in the open (if at all) or out of season. Artificial heating is used and in recent years, there has been an increase in the use of environmental control systems that regulate ventilation, moisture and carbon dioxide enrichment.

Glasshouse cultivation is both labour- and capital intensive, and represents, per hectare, the highest initial investment costs in agriculture. Production, however, can be switched more easily to meet the vagaries of the market than, for example, tree crop holdings which have similarly high start-up costs. See MARKET GARDENING. See also HYDROPONICS.

**glazed frost,** *n.* a type of FROST which results when rain or dew comes into contact with a very cold ground surface and freezes. Glazed ice on roads, also known as *black ice*, is a serious hazard to traffic due to its clarity and hence its invisibility to road users.

**glei soil,** see GLEY SOIL.

**gley soil, glei soil, gleysol** or **meadow soil,** *n.* a compact and usually structureless soil in which one or more horizons have a grey, blue or olive colour due to intermittent waterlogging which causes anaerobic conditions. Gley soils may be caused by a high WATER TABLE or by a HARDPAN or SOIL HORIZON which prevents the free passage of water downwards through the profile. Given adequate drainage and the addition of lime to overcome the soil acidity brought about by the anaerobic conditions, gley soils can be utilized for agriculture. See GLEYING. See also HYDROMORPHIC SOIL.

**gleying,** *n.* the process by which iron compounds in a soil are depleted of oxygen (*reduced*) by soil bacteria under anaerobic conditions to give a blue, grey or olive colouration. Gleying is caused by the permanent or temporary waterlogging of at least part of a soil and may take two forms:

(a) *groundwater gleying* which occurs in low-lying areas where a high WATER TABLE produces anaerobic conditions,

(b) *surface water gleying* or *stagnogleying* which is due to the limited permeability of one or more SOIL HORIZONS and leads to imperfect drainage.

**gleysol,** *n.* the term used in the FAO and Canadian soil classification systems for a GLEY SOIL.

**global warming,** see GREENHOUSE EFFECT.

**Gondwanaland,** *n.* the southern protocontinent believed to have resulted from the fragmentation of the PANGAEA supercontinent in the late Palaeozoic or early Mesozoic eras (see Fig. 19). Gondwanaland is thought to have comprised Australia, Africa, India, Antarctica and areas of South America. See LAURASIA, CONTINENTAL DRIFT. See also PLATE TECTONICS.

**gorge,** *n.* a deep, narrow, steep-sided river valley. Often the walls of a gorge are near vertical. The term is synonymous with a *canyon* although the latter is generally considered to be greater in size.

Examples include the Rhine gorge in West Germany and the Grand Canyon in USA.

Gorges can occur for a variety of reasons, such as river downcutting in a FAULT or after REJUVENATION or in the early stages of gradation when a stream's capacity exceeds its load and rapid vertical erosion is experienced (see FLUVIAL TRANSPORTATION, FLUVIAL EROSION).

**graben,** see RIFT VALLEY.

**graded river,** see RIVER PROFILE.

**graffiti,** *n.* drawings, messages, etc. often obscene, scribbled on walls or other surfaces. This is a phenomenon of great antiquity, seemingly related to a basic need to leave a memorial of some kind or to express defiance towards a person or towards society in general. Graffiti is most commonly associated with run-down urban areas and with the large, faceless housing estates of the suburbs. Some sociologists have argued that urban grafitti is a quasi-acceptable art form expressed by deprived urban dwellers; a manifestation of this was seen in the decoration of New York subway trains. See VANDALISM.

**grain alcohol,** see ETHANOL.

**grain crop,** *n.* any member of the grass family (Gramineae) in which the characteristic fruit is a grain or *caryopsis* in which the seed coat is united with the ovary wall. The grains consist mainly of starch and some protein. The term grains is used collectively for wheat and all allied food-grasses, such as rye, oats, barley, maize, rice and the millets.

Grain crops or cereals are the most important sources of organic food for humans and FODDER CROPS for livestock. By 1983-84 half of the world's grain production was being fed to animals. For human consumption, grains may be cooked unground (rice), ground as flour for bread (wheat and rye) or pasta (durum wheat), malted (barley and sorghum), pounded or rolled and made into a meal or porridge (oats, maize and millet).

World average annual production of cereals for 1984-86 was 1.8 billion tonnes, 18% of which came from USA. Wheat (28%), rice (26%) and maize (26%) were the most important grain crops. In the EC in 1986, cereals occupied almost 36 million ha of land, 28% of the utilized agricultural area (excluding woodland) and 53% of all arable land.

**granite,** *n.* a coarse-grained, acidic INTRUSIVE ROCK formed by the slow cooling of MAGMA deep within the earth's crust. Granite is always associated with plutonic bodies such as BATHOLITHS and

LACCOLITHS. In temperate climates, granite is very resistant to weathering and often forms upland areas such as the Sierra Nevada in California, USA, and Bodmin Moor in England. In humid tropical climates, granite is highly susceptible to CHEMICAL WEATHERING which results in the formation of deeply weathered soils. Granite is often used for construction purposes and Aberdeen in Scotland is known as the 'Granite City' due to the extensive use of the rock in 19th and early 20th buildings.

**granitization,** *n.* the metamorphic process by which pre-existing COUNTRY ROCK is changed to granite by the action of magmatic fluids rising from deep in the earth's crust. The exact mechanism by which this metamorphism is brought about is not fully understood. See BATHOLITH.

**GRAS,** *vb. acronym for* generally recognized as safe, the term applied in the USA to any food ADDITIVE which has been extensively tested and found to be safe when consumed in small quantities by humans.

**grat,** see ARÊTE.

**grazing food chain,** *n.* a structured feeding hierarchy in which the transfer of energy is brought about by the consumption of organic matter (plants) by HERBIVORES which in turn are consumed by various levels of CARNIVORES. Compare DETRITAL FOOD CHAIN. See FOOD CHAIN.

**grazing management,** *n.* the adoption of practices to control and regulate the use and quality of grazing land to enable livestock to fulfill their productive potential while conserving the land resource. Grazing management consists of two basic activities directed towards:

(a) the encouragement of sward growth. The application of nitrogenous FERTILIZER to PASTURES and in LEY FARMING has increased considerably in recent years. On upland pastures where LEACHING is active, lime may be applied to raise the soil pH, encourage soil organisms, speed up decomposition, aid the release of nutrients, and thus improve the quality of grazing. The use of FIRE to control weeds, restrict vegetation succession and maintain a young, nutritious plant cover is a common practice in areas of extensively grazed pasture;

(b) the control of animal behaviour through the choice of grazing system. There are innumerable such regimes, each modified by local physical conditions, the type of livestock and the farmer's preference. *Free-range* systems are those in which livestock have unrestricted access to pasture, while *deferred grazing* occurs

201

when livestock are excluded from a green pasture for an extended period to enable vegetation to recover. On extensively grazed pastures (see RANCHING), the distribution and movement of livestock may be further controlled by the provision of water supplies and the placement of salt and supplemental feed. In *paddock* or *rotational grazing*, livestock are systematically rotated from one pasture to another over a period of days, or from one season to another. The reserve paddocks may be cropped for fodder when grass growth is rapid, or held back for exclusive use in the dry season. *Strip grazing* is a refinement in which strips of pasture are fenced off for short periods of time, generally by electric fencing, to give maximum control over the grazing regime. *Zero-grazing* is the practice of stall feeding livestock with freshly cut grass. It is a system often used on intensively stocked farms, providing a method of using grass from outlying and unfenced pastures.

**Great Plains,** see PRAIRIE.

**green belt,** *n.* any zone of countryside immediately surrounding a town, defined for the purpose of restricting indiscriminate outward extension of the urban area. The original purpose of green belts, first devised in the regional plans for Greater London in the earlier part of the 20th century, was to secure a permanent recreation area for people living in the then polluted and unpleasant urban environment of the capital, as well as controlling urban sprawl. The planning aim, in the case of London and the other cities which subsequently adopted the green belt principle, was to encourage new development to jump beyond the green belt into totally NEW TOWN locations or into smaller existing towns. It was felt that such sites would provide attractive locations independent of the parent city, where a new sense of COMMUNITY could develop. Such a move would benefit both those migrating from the city and those remaining who could be rehoused in less pressured and redesigned redevelopment areas. The Greater London green belt was made statutory by the Green Belt (London and Home Counties) Act of 1938 which imposed particular planning protection on areas so zoned. Around other cities in the UK, such as Glasgow, the green belt principle was incorporated piecemeal into the several component development plans covering the suburban areas. See also GARDEN CITY, GREENFIELD SITE, BROWNFIELD SITE.

**green fallow,** *n.* a type of FALLOW in which land being rested from main crop cultivation is occupied for part of the fallow period by a quickly maturing LEAFY CROP, particularly turnips or vetches, which may then be grazed in the spring.

**greenfield site** or **virgin site,** *n.* a site selected for possible development for industrial or housing use, which is presently in non-urban, usually agricultural, use. Such sites have been the principal locations of new urban development until recently, and have been preferred by developers because of the relative ease and economy with which large commercial structures and housing estates can be laid out and built. Greenfield sites, by definition, are at the edges of existing towns which presents problems of access to urban services especially, shops, schools and health care facilities for residents of new houses built on these sites. Recently, city authorities have encouraged development of the alternative locations known as BROWNFIELD SITES, which are within the body of the city. These sites are seen as important elements in the drive to renew the deprived areas of the INNER CITIES. See also GREEN BELT, INFILL SITE.

**greenhouse effect,** *n.* the gradual raising of the temperature of air in the LOWER ATMOSPHERE believed to be the result of the accumulation of gases such as carbon dioxide, METHANE, nitrous oxide, CHLOROFLUOROCARBONS and OZONE. These gases, known as *greenhouse gases*, are natural components of the lower atmosphere and act somewhat like a pane of glass in a greenhouse, absorbing the harmful short wavelength, high-energy solar rays, letting in the visible light and preventing the loss of some of the long wavelength, outgoing energy. By this process the temperature of the earth remains balanced. If there was no greenhouse effect the average temperature of the Earth would be $-18°C$.

As a result of the burning of FOSSIL FUELS such as wood, coal, oil by industrial, domestic and transportation sources, the amount of carbon dioxide in the atmosphere has increased by 26% between 1860 and 1986. If this trend continues then by about 2040 the $CO_2$ level could be double its pre-industrial level. The heat balance of the lower atmosphere would be so altered that average atmospheric temperature would rise by 4°C with polar temperatures rising by twice this amount. Melting of the polar ice caps would result in a rise in SEA LEVEL and a flooding of cities, industry and agricultural land. The distribution of world CLIMATIC REGIONS and OCEAN CURRENTS would change leading to an alteration of agricultural regions.

More recent than the build-up of $CO_2$ has been the discovery that chlorofluorocarbons destroy the ozone layer allowing ultraviolet radiation to penetrate the lower atmosphere with as yet unpredictable consequences for all life forms. This radiation is known to cause mutations in the genetic material of plants and animals.

**greenhouse gas,** see GREENHOUSE EFFECT.

**green manure,** see MANURE.

**Greenpeace,** *n.* a non-governmental organization (NGO) dedicated to world conservation issues. Of the approximately 4000 NGOs concerned with BIOSPHERE issues and operating throughout the world, Greenpeace is one of the longest established and best known. Founded in 1971, it has continuously made world headlines through the activity of its members who have often placed themselves direct confrontation with governments and with industrial corporations intent on economic development at the expense of the biosphere. Greenpeace has, for example, constantly argued for a ban on sealing and whaling, for the prevention of above-ground testing of nuclear devices, for a ban on dumping nuclear waste into the oceans and has worked for the coordination of conservation policies on a world-wide basis. See FRIENDS OF THE EARTH.

**green politics,** *n.* any political activity which gives prominence to issues involving the conservation of the world's natural resources, and the improvement and protection of the environment. Central to green political agendas are the twin concepts of the human race as custodian of the environment, and the sustainable exploitation of biospheric resources (see SUSTAINED YIELD); radical elements in the green movement believe that the application of such concepts could only follow the decentralization of political and economic power.

The most successful exponent of green politics has been the West German ecology party, the Green Party. It first emerged as a significant political force in local elections in 1979 and has since gone on to take seats in the federal parliament, the Bundestag. Similar parties have also enjoyed political success in other West European countries. In the UK, the British Ecology Party changed its name to the Greens in 1985 and made its first significant electoral successes in the 1989 European parliamentary elections.

While environmental issues have never been the exclusive concern of ecology parties, in the late 1980s green politics has moved in from the political fringes to assume a greater prominence in the agendas of the main political parties.

**green revolution,** *n.* the development of HIGH-YIELDING CEREAL varieties, and their adoption and diffusion in the THIRD WORLD as part of a package of measures including the increased use of FERTILIZERS and chemical protection, and controlled water supply. First coined in 1968, the term gained popularity following Norman E. Borlaug's award of the 1970 Nobel Peace Prize for breeding crop varieties thought likely to close the gap between population growth and food output.

Characteristic features of the high-yielding varieties (HYV) include: dwarfing genes that produce shorter, stiffer straw and yield a heavier head of grain without lodging; a much greater responsiveness to fertilizers and regulated IRRIGATION than traditional varieties; and photo-insensitivity to daylight length which reduces maturation times and permits MULTIPLE CROPPING. Elements of specific disease and pest resistance may be inbred but HYVs require increased chemical protection for full benefits to be realised. The SELECTIVE BREEDING of HYVs dates back at least to the turn of the century (see also HYBRID SEED). However, the green revolution can be said to have begun in 1943 with the Rockefeller Foundation's involvement in Mexico in the improvement of local wheat and maize varieties. Notable successes by Borlaug in the 1950s were followed by the rapid diffusion of high-yielding wheat to India and Pakistan after 1965. These advances encouraged workers at the International Rice Research Institute (IRRI) in the Philippines, and in 1966 they released IR-8, the first of the so-called 'miracle rice' strains.

Reactions to the green revolution have varied over the past two decades. Optimism in the 1960s gave way to a more radical critique in the early 1970s as output gains levelled off, and both technological and socio-economic problems began to emerge, although with the benefit of a longer time-perspective and more complete data, some of these criticisms have proved premature. Some early problems that were encountered included widening income disparities; growing regional inequalities; greater susceptibility to diseases and pests; dependence on an imported, fossil-fuel based technology; a reduction in the area planted to non-cereal crops; and the poorer taste, cooking quality and nutritional status of many of the new rice varieties.

In the West, the response to the greater susceptibility of HYVs to disease and pest attacks has been continuous plant breeding and heavier applications of chemical protection. This same strategy must be employed in Third World countries less able to pay for the imported technology, and by farmers ever more locked into dependence on international AGRIBUSINESS. Internationally, the growing grain surplus both from traditional exporters such as the USA and formerly import-dependent countries such as India has unsettled commodity markets, with as yet uncertain consequences for grain prices and the ability to sustain high levels of agro-chemical inputs in the production process. Further, there is increasing concern over the ecological effects of these inputs, in particular the use in the Third World of toxic INSECTICIDES often banned in the West.

The successes at IRRI and the International Centre for Maize and Wheat Improvement (CIMMYT) in Mexico have stimulated interest in improving other tropical FOOD CROPS, and led to the setting up of the Consultative Group on International Agricultural Research (CGIAR) in 1971, a multinational consortia supported by western governments, multilateral agencies and philanthropic foundations. CGIAR mobilizes funds to finance international institutions each with a specific research focus or a particular regional orientation, for example, the West African Rice Development Association in Liberia, and the International Crops Research Institute for the Semi-Arid Tropics in India.

**groin,** see GROYNE.

**gross primary productivity (GPP),** see PRIMARY PRODUCTIVITY.

**gross residential density,** see RESIDENTIAL DENSITY.

**groundmass,** *n.* the MATRIX of IGNEOUS ROCKS in which larger crystals are embedded. See PORPHYRY.

**ground moraine,** see MORAINE.

**groundwater,** *n.* the body of water which occupies the earth's mantle and which forms the sub-surface section of the HYDROLOGICAL CYCLE. Groundwater occupies the PORE SPACES, BEDDING PLANES and joints of rocks, and originates from two main sources: as hot mineral water of magmatic origin rising from deep within the earth (JUVENILE WATER), or resulting from the percolation of precipitation and meltwater (*meteoric water*).

Groundwater may return to the surface by seepage or through springs. Water may also be artificially withdrawn through the use of wells (see ARTESIAN WELL). The tapping of groundwater is long established and modern technology allows wells to be bored progressively deeper. As a result, in many parts of the world much of the domestic, agricultural and industrial water supply comes from groundwater sources. In semi-arid and arid areas, agriculture is heavily dependent on ground water for IRRIGATION, as surface supplies of water may run dry or are fully utilized.

However, the increasing use of groundwater has had serious environmental implications. When the rate of removal exceeds that of recharge then the level of the WATER TABLE may drop. In coastal areas such drops may result in the intrusion of salt water into groundwater supplies, as has occurred on the Mediterranean island of Majorca. Excessive groundwater withdrawal may also lead to subsidence; for example, in Houston, Texas, subsidence of up to 1 m has been recorded in the metropolitan area. Another serious problem is the pollution of groundwater from various sources. These include contaminants leaking from LANDFILL sites, rubbish dumps, rivers carrying industrial EFFLUENT, surface RUNOFF from roads and NUCLEAR WASTE storage sites.

**groundwater gleying,** see GLEYING.

**growing season,** *n.* the period of the year when climatic conditions, especially temperature, sustain plant growth. Sufficient daylight hours and adequate moisture provision are important, but in temperate mid-latitudes the critical factors are generally the attainment of mean daily temperatures of at least 6°C and the lack of damaging frosts. The growing season for many crops is often expressed in terms of the number of continuously frost-free days necessary to ensure full maturity between planting and harvesting; for example, cotton requires 180-200 frost-free days. Extremely high temperatures, especially when accompanied by a lack of rainfall, also inhibit plant growth. In semi-arid and MEDITERRA-NEAN areas the growing season may be restricted to a relatively mild winter period. The length of the growing season normally decreases with rising latitude and altitude. In the UK the limit for ARABLE FARMING is generally reached at a height of about 225 m where the conditions favourable for crop growth exist for at least one month less than at sea level.

**groyne** or **groin,** *n.* an artificial wall or embankment, usually constructed at right angles to the shore and designed to limit the erosive effects of LONGSHORE DRIFT on beaches. Groynes can be constructed from wooden pilings, concrete or rocks and are intended to act as a temporary sediment trap until a more acceptable rate of longshore drift has been established.

**gully erosion,** *n.* a form of SOIL EROSION that usually occurs near the bottom of slopes and which results from the removal of soil and soft rock by a concentrated RUNOFF to form a deep channel or gully. Minor gullying can occur in almost any part of the world and is commonly found on compacted dirt tracks or on industrial spoil heaps. Large-scale gullying usually occurs in tropical and arid regions where high-intensity rainstorms cause severe erosion wherever the soil surface has been disturbed by cultivation or the removal of vegetation. Gully erosion can cause the loss of large tracts of formerly cultivable land, as in the BADLANDS of the USA.

**gusher,** see OIL.

**Gutenberg discontinuity,** *n.* the lower boundary between the earth's mantle and core. The Gutenberg discontinuity is found at a depth of approximately 2900 km. See MOHOROVIČIĆ DISCONTINUITY.

**guyot,** *n.* a steep-sided, flat-topped underwater mountain. Guyots occur in association with SEAMOUNTS to form underwater ranges, and are common in the Pacific Ocean. The summits of guyots are usually at least 1500 m below sea level and result from the subsidence of wave-planed volcanoes. In tropical waters, guyots often provide the foundation for CORAL REEFS and atolls.

**gymnosperm,** *n.* any seed-bearing plant of the division Gymnospermae in which the seed is poorly protected (naked) in a cone. As the cones ripen and dry the seeds are allowed to fly free. The dominant members of the gymnosperms are the conifers. These form extensive forests in northern latitudes (see BOREAL FOREST). Three orders can be distinguished: Coniferales (conifers); Taxales (yew) and Ginkoales (maidenhair tree). All conifers are much branched and can attain a height of 100 m. Reproductive organs are borne on separate cones. From fossil evidence the gymnosperms appear to have been unchanged since the end of Mesozoic times.

Conifers now form an important biotic resource providing valuable SOFTWOOD for conversion to PAPER PULP.

**gypsum,** *n.* a SEDIMENTARY ROCK composed of a calcium sulphate mineral precipitated by the evaporation of seawater. Gypsum is used in the manufacture of paint, glass, cement, porcelain, plasterboard and plaster of Paris.

# H

**habitat,** *n*. **1.** the natural home of an animal or plant.

**2.** the sum of the environmental conditions which determine the existence of a COMMUNITY in a specific place. The habitat is the result of interaction of edaphic (soil) factors, climatic factors, ANTHROPOGENIC FACTORS and BIOTIC FACTORS.

**habitat corridor,** *n*. narrow strips of undeveloped or disused land which link larger areas of conserved, open and/or waste ground and along which animals and plants can migrate. Typical habitat corridors are canals, river banks, railway embankments, and motorway verges. See MINIMAL AREA.

**Hadley cell,** *n*. a low-latitude ATMOSPHERIC CELL, which along with a mid-latitude and high-latitude cell, operates in each hemisphere to offset the horizontal temperature gradient between the equator and the poles. Hadley cells consist of an equatorward low-level air flow and a high-level poleward compensating flow and are located between the equator and the sub-tropical anticyclone belts around the latitudes of 30°N and S (see Fig. 10). These air flows are initiated by the intense heating of the earth's surface in equatorial regions and the resulting vertical and horizontal movement of large masses of warmed air. The CORIOLIS FORCE causes the deflection of the surface air flows within the Hadley cells to form the TRADE WINDS (see Fig. 70).

**hail,** *n*. PRECIPITATION in the form of small pellets of ice with diameters usually ranging from between 5 and 50 mm. Showers of these pellets (*hailstorms*) are associated with atmospheric INSTABILITY which causes the rapid ascent of air masses and the CONDENSATION of water vapour. The resulting raindrops continue to be carried by updraughts until they freeze and grow heavy enough to fall from the clouds.

**half-life,** *n*. **1.** the time taken for half of the ATOMS in a given amount of radioactive material to decay. The amount of time necessary for the half-life stage to be reached varies considerably: for example, xenon-135 has a half-life of only 9.2 hours while uranium-238 requires $4.5 \times 10^9$ years to reach its half-life.

**2. biological half-life,** the time required for half of a quantity of radioactive material absorbed by a living tissue or organism to be naturally eliminated.

**halite,** see ROCK SALT.

**halomorphic soil,** *n.* any one of a range of soil types which develop under saline GROUNDWATER conditions. SOLONCHAK and SOLONETZ soils are typical of this group.

**hamlet,** *n.* a small settlement consisting of a seldom more than 10 houses in number and which does not possess a parish church. It is a descriptive term only and has no current statistical or functional definition.

**hanging valley,** *n.* a tributary valley whose floor lies above that of the main valley. Hanging valleys are most common in formerly glaciated regions and are due to the greater erosive power of the main valley's GLACIER (see Fig. 66). In other areas hanging valleys may result from:

(a) the truncation of a valley as a result of the MARINE EROSION of cliffs;

(b) a new river valley interrupted by a fault;

(c) the differential erosion of a major river and its tributaries.

**hard chain,** see DETERGENT.

**hardpan,** *n.* a hard, compact cemented layer usually found in the B-HORIZONS of soils and formed by the ILLUVIATION of chemical compounds from the A-HORIZON. Three main types of hardpan are recognized:

(a) *claypans* which result from the accumulation of clay-sized particles in the B-horizon;

(b) *ironpans* which comprise thin crusts of mainly ferric oxide and are characteristic of PODZOLS;

(c) *moorpans* which may occasionally result from the deposition of humic materials.

Hardpans restrict root formation and lead to instability amongst trees. They can impede soil drainage and often result in water-logging in the SOIL HORIZONS immediately above the hardpan. If the hardpan becomes exposed at the surface by soil erosion then it can severely limit agricultural activities particularly in the low technology agricultural systems common in DEVELOPING COUNTRIES. See also DURICRUST, CEMENTATION.

**hard shoulder,** see MOTORWAY.

**hardwood,** *n.* **1.** the wood of any broad-leaved dicotyledonous trees such as oak, beech or ash. Hardwood timber is much used in the construction of good quality furniture.

**2.** the name given to the broadleaved (Dicotyledoneae) forests.

Vast hardwoods still exists within the tropics despite recent DEFORESTATION. Hardwoods were also once extensive throughout North America, western and central Europe and in China. These hardwoods have been largely cleared to make way for permanent agriculture.

Compare SOFTWOODS.

**harvest index,** *n.* the ratio of grain weight to total plant weight in a cereal crop. The development of HIGH YIELDING CEREAL varieties has markedly lowered the harvest index by selectively breeding heavier heads of grain and shorter stalks.

**hay,** *n.* any dried grasses or legumes used as a FODDER CROP, in which the moisture content has been reduced to 22% or less to prevent spoilage in storage.

**hazardous waste,** SEE TOXIC WASTE.

**haze,** *n.* an atmospheric condition marked by a slight reduction in atmospheric visibility. Hazes may result from the formation of PHOTOCHEMICAL SMOG, the radiation of heat from the ground surface on hot days, or the development of a very thin MIST. Visibility in a haze is usually greater than 2 km and rarely affects traffic flow or aircraft operations.

**headward erosion,** *n.* the action of a river in lengthening its upstream valley by eroding back from its original source. Headward erosion is particularly prominent in areas of weak, underlying bedrock or in areas affected by faulting. Common causes include the upstream migration of waterfalls, the localized erosion of areas adjacent to springs (*spring sapping*), along with GULLY EROSION and SHEET EROSION. Headward erosion may ultimately lead to RIVER CAPTURE.

**headwater,** SEE WATERSHED.

**heathland,** *n.* any tract of land which is typically the habitat of many of the ericaceous shrubs, mostly of the genus *Erica*. Heathland is commonly associated with sandy acidic or PODZOL-type soil and as such is of very low agricultural value. See MOORLAND.

**heat island,** *n.* any zone (usually a large urban area) that frequently records higher air temperatures than those of the surrounding countryside. Heat islands are created by the release of waste heat from a variety of domestic, commercial and industrial sources and can be up to several degrees centigrade warmer than adjacent areas.

**heavy industry,** *n.* any industry such as steel-making, ship-building and engineering. These industries were formerly charac-terized by large (almost exclusively male) workforces, by the

211

movement of large quantities of bulky and heavy raw materials and equally bulky finished products, by the complex differentiation of craft skills and by a high dependency on team organization.

These kinds of industry have been contracting for half a century in the old industrialized countries of the northern hemisphere and those which remain have undergone dramatic changes, mainly as a result of the automation of production processes which has greatly reduced labour forces per plant. The greatest proportional decline in manpower has occurred in the manual trades.

**heavy metal,** *n.* any metal with a high atomic weight (usually, although not exclusively, greater than 100), for example mercury, lead, cadmium, chromium and plutonium. Heavy metals have a widespread industrial use, and many are released into the BIOSPHERE via air, water and solids pollution. Heavy metals are poisonous and tend to persist in living organisms once consumed. Typically, they enter the FOOD CHAIN and accumulate in selective organs, particularly in the brain, liver and kidneys of animals located towards the upper TROPHIC LEVELS, gradually poisoning their hosts. Quite small quantities of heavy metals ($1$-$2 \times 10^{-6}$ g) can prove lethal. Substantial disturbance of energy and material movements in an ecosystem can occur through the elimination of whole animal populations brought about by accumulation of heavy metals, for example in estuaries in Japan, mercury accumulation has resulted in the disappearance of many CRUSTACEANS.

Some scientists consider the threat of heavy metal pollution to be the single most important threat to the immediate stability of the biosphere.

**helophyte,** SEE RAUNKIAER'S LIFEFORM CLASSIFICATION.

**hemicryptophyte,** SEE RAUNKIAER'S LIFEFORM CLASSIFICATION.

**hemicryptophyte climate,** SEE BIOCHORE.

**herbicides,** *n.* any chemical substance, usually synthetic, that injures or kills plant life by disrupting normal hormonal functions. Early herbicides were cheap by-products from the embryonic chemical industry of the 1900s. Arsenic and sodium chloride were examples of early, and highly dangerous, herbicides. From 1939 onwards the 'selective' herbicides were developed, for example DNOC, 2,4-D, and 2,4,5-T. Herbicides have made possible a sustained increase in agricultural productivity, thereby allowing a much improved quality and quantity of food. See also PESTICIDE, INSECTICIDE, FUNGICIDE.

**herbivore,** *n.* any animal which obtains most of its food from plants. Herbivores are the primary consumer organisms and form the second TROPHIC LEVEL in a FOOD CHAIN. They are inefficient convertors of plant protoplasm to animal tissue and between 80-90% of energy input is used to digest the large masses of vegetable protoplasm, to search for more food and, in mammals, to generate heat. Compare CARNIVORE, OMNIVORE.

**heterotroph,** *n.* any organism (usually animal) incapable of generating its own basic food supply. Instead, heterotrophs feed on AUTOTROPH organisms, from which they derive a supply of manufactured sugars and starches. Through its own metabolism, the heterotroph then rearranges these organic substances into more complex forms such as animal fats, proteins and carbohydrates. Heterotrophs form the second, third and fourth TROPHIC LEVELS of a FOOD CHAIN. Humans are heterotrophic feeders.

**H-horizon** or **O-horizon,** *n.* the uppermost layer in a mineral soil which contains organic matter in various stages of decay (see Fig. 55). The H-horizon is sometimes regarded as part of the A-HORIZON.

**high,** see ANTICYCLONE.

**high forest,** *n.* any forest which approximates to an undisturbed COMMUNITY. In the UK, the term is often associated with the ancient and protected royal forests such as Epping Forest.

**high-rise development,** *n.* a group of buildings in which many or all of the structures are over 10 storeys high. The term is commonly applied in the UK to multistorey housing built during the 1960-75 period, when massive urban renewal projects were undertaken to rehouse people displaced from areas of overcrowded housing which had been demolished.

The high-rise style was adopted for several reasons. The style was then fashionable in architectural circles as an adaptation of the ideas advocated by the French architect and town planner Le Corbusier (1887-1965). It also made possible the speedy provision of many housing units through the adoption of *system building techniques* by which modular design and construction enabled large buildings to be erected quickly; this was politically popular at a time when the demand for new housing was great. Furthermore, it was mistakenly believed that it was possible to achieve high densities of population in such developments and thus retain the population within the administrative limits of the city, diminishing the need for OVERSPILL.

However, high-rise housing proved unsuitable for housing the elderly and young families and many problems arose because of the forbidding scale of the structures and the problems of access to the higher floors.

High-rise office development occurred in the centre of major European cities at the same time and still continues to be used in the business cores of capital cities where administrative and financial activities interact. In the UK, planning controls attempt, with varying degrees of success, to restrict the scale of such development in the interests of preserving the character of city centres.

**highway,** see MOTORWAY.

**high-yielding cereal,** *n.* any variety of cereal GRAIN CROP selectively bred to produce a higher yield than the indigenous seed stock. The most important advances in SELECTIVE BREEDING have been with hybrid maize and sorghum (see HYBRID SEED) and with true-breeding varieties of wheat and rice. Traditional varieties of wheat and rice in both tropical and temperate zones respond poorly to FERTILIZER application, tending to grow too tall and to *lodge*, that is, to be flattened by wind and rain. Plant breeding, therefore, has concentrated on developing short-strawed varieties that will carry a much heavier head of grain produced in response to markedly increased dressings of nitrogen fertilizers. In the UK, new varieties of genetically improved wheat out-perform those sown in the 1940s by more than 60%. Extensive work in Taiwan during the Japanese colonial administration saw the emergence of improved rice strains suited to sub-tropical environments, but it was not until the 1960s with the GREEN REVOLUTION that high-yielding cereal varieties became widely available in the tropics.

The continual process of cross-breeding to counter particular diseases and pests and to produce varieties better suited to location-specific conditions has increased the dependence of farmers throughout the world on the AGRIBUSINESSES that control the development, marketing and distribution of new seeds and the chemical inputs necessary for their successful cultivation.

**hill farming,** *n.* any agricultural activity in hill areas where the harshness of the physical environment imposes severe constraints on agricultural production. In the UK it has been customary to distinguish between hill farming areas in which the environmental conditions have led to the rough grazing of cattle and sheep at a low stocking density, producing breeding and store stock rather than finished animals, and UPLAND FARMING at lower altitudes where the general environment tends to be less severe, allowing a wider range of production alternatives than hill farming. Prior to the UK joining the EC, hill farming areas were more precisely defined as those eligible for government SUBSIDIES and grants intended to promote a continued agricultural presence in the mountains and hills. The subsidies were provided to improve the social and economic fabric of hill farming areas.

With the adoption by the EC in 1975 (revised in 1986) of a *Directive on Less Favoured Areas*, livestock subsidies have been replaced by compensatory allowances of roughly similar effect. Almost 50% of the total utilized agricultural area (UAA) of the EC comprises less favoured areas, much of which is in France, Italy and Greece. The hill farming areas of the UK are classed as areas of infertile land, below average PRODUCTIVITY and with a low or dwindling population dependent on agricultural activity. See also ALPINE FARMING.

**hinterland,** *n.* the area of land immediately inland from a port and from within which trade and commerce originated. See also CATCHMENT AREA.

**historic building,** *n.* any building which by reason of age and/or architectural merit is especially valued locally, nationally or internationally. Such buildings together with their immediate surroundings are especially protected by government legislation. Buildings are 'listed' in categories of excellence, and equivalent levels of stringency protection are afforded to those threatened by new development proposals which might involve their damage or demolition. Some historic buildings are publicly owned and are preserved, together with their contents, for the purpose of public education and recreation. Those in private hands may be retained in good order with some financial assistance in the form of government grants. Archaeological sites and monuments are similarly protected because of their historical, scientific or traditional interest.

Historic buildings and monuments can become liabilities to their owners because of ongoing structural repairs. Abandoned churches are a special problem, especially those which have no national significance. See ANCIENT MONUMENT, CONSERVATION AREA, AMENITY.

**hoar frost,** *n.* a type of FROST which results from the rapid nocturnal cooling of air above the ground. This air may freeze immediately or condense first and then gradually freeze to deposit a layer of white, needle-like ice particles on the ground and other objects such as trees and cars.

**hobby farming,** see PART-TIME FARMING.

**Holocene epoch,** see RECENT EPOCH.

**homeostasis,** *n.* the attainment of a *steady state* or complete balance between the inputs and outputs of a system. Until recently it was thought that natural ecosystems existed in a climax condition which was equivalent to homeostasis (see CLIMAX COMMUNITY). However, because of the operation of TIME LAGS which exist between system inputs responding to the system outputs it is now consisdered very unlikely that ecosystems can attain complete homeostasis. It also appears that natural changes in climate and soils occurs much faster and more extensively than was previously considered possible and accordingly the ability of ecosystems to attain homeostasis is now regarded as being abnormal and due to extremely favourable circumstances.

**homestead,** *n.* an area of land, usually less than 65 ha, granted to settlers in the USA under the Homestead Act of 1852.

**homesteading,** *n.* a housing tenure arrangement in the UK whereby a public authority repairs vacant rented houses in a declining area to make them weatherproof, and then sells them at very low prices to buyers willing to put their own time and resources into upgrading them. The system brings vacant houses back into use, thereby helping the environmental regeneration of semi-derelict areas within PUBLIC SECTOR HOUSING estates. It also stabilizes population numbers in these areas and so aids the continuance and improvement of local shopping and service facilities.

**horse latitudes,** *n.* the subtropical belts of high air pressure which occur at latitudes of 30° north and south. The upper atmospheric convergence of AIR MASSES and the resulting descent and surface divergence of these masses polewards and equatorwards, causes calm weather conditions with light variable winds.

**horticulture,** *n.* a form of INTENSIVE AGRICULTURE for the commercial production of crops, especially fruit, vegetables and flowers, in which the produce is generally sold for direct human consumption rather than for processing. Crops may be grown in fields to provide in-season produce or advanced by means of GLASSSHOUSE CULTIVATION. See MARKET GARDENING.

**host rock,** see COUNTRY ROCK.

**hot desert,** see DESERT.

**hot dry rock technique,** *n.* a method for tapping GEOTHERMAL POWER. A hole is drilled several kilometres into the parent rock and the base rock cracked. This may be done hydraulically, using conventional oil field technology, or with explosives. Water is then injected into the fractured rock, whereupon it becomes superheated due to the great pressure and temperature existing at depth in the earth's crust. The steam is extracted, usually via a separate vent, to be used as an energy source.

The means of producing geothermal power in this way are currently still at an experimental stage. Geologists are actively involved in the location of *hot spots* near the surface of the earth (where the ground temperature is significantly higher than that of the surrounding land). Promising sites are usually located near large INTRUSIVE ROCK masses, such as the granite masses of Devon and Cornwall.

**hot spring** or **thermal spring,** *n.* a SPRING of continuously flowing hot water, ranging in temperature from 20°C to boiling point. Hot springs are usually found in dormant volcanic areas or where MAGMA lying near the earth's surface warms GROUNDWATER. In several countries, including Iceland and New Zealand, the water from hot springs is used for domestic, commercial and industrial heating and for hothouse cultivation. The water from many hot springs is rich in minerals, which has led to the development of spa resorts such as at Bath in England. Certain minerals such as sulphur can sometimes be commercially exploited from hot springs. Compare GEYSER.

**house condition survey,** *n.* an official assessment of the quality of a dwelling that enables decisions to be made about its suitability for habitation. The condition of a structure is judged upon a combination of several features relating to the characteristics of the building itself. The most important of these features are as follows:

(a) *structural condition*: this entails the assessment and recording of items relating to the safety and weatherproofing of the building; the soundness of the roof and other structural timbers, in respect of age, wet or dry rot damage or woodworm infestation; potentially dangerous structural faults such as cracks in stairwell or floors; subsidence damage arising from ground instability;

(b) *sanitary condition*: this entails the assessment of the quality of basic amenities available, that is, the access by each household to bathroom, toilet and clean piped water supply, and adequacy of arrangements for waste disposal; the adequacy of kitchen facilities are also judged;

(c) *Other features* include the adequacy of access to outdoors from all rooms in the event of fire; the condition of electrical wiring and/or gas-pipe connections; and the presence of dangerous materials such as asbestos used in the construction of the house. See also IMPROVEMENT GRANT, REPLACEMENT RATE, REHABILITATION.

**housing layout,** *n.* the design for the construction of a residential area, showing the proposed size, location and orientation of the houses in relation to connecting roadways and paths. The layout will be designed to meet any prescriptions of RESIDENTIAL DENSITY (the number of houses per unit area) which the planning authority may have attached to the granting of consent to develop the site.

**housing plan,** *n.* a statement of intent by a housing authority (or district in the UK) regarding its future provision of new PUBLIC SECTOR HOUSING, its ideas for the rehabilitation of run-down stock (both private and public) and its anticipated processes of housing management. Accompanying this will be estimates of costs of implementing the proposals and a programme of resulting expenditure for which the approval of the elected council will be required.

**housing tenure,** *n.* the conditions under which a house or dwelling is occupied; conditions of tenure define the relationship between the owner of a house and the occupier. In the private sector, the owner and the occupier are usually the same person. The private rented sector is more complex in that one landlord may own one or several houses which he rents to tenants, the management of the maintenance of the fabric being in the hands of an agent. The landlord may be an individual or a private company. Within the private rented sector there is a further distinction between furnished and unfurnished rental.

In PUBLIC SECTOR HOUSING, the owner is the district council or an agency of central government. The management of the letting and the maintenance of the stock are both in the hands of the professional officers of the local authority.

**humic acids,** *n.* a group of complex organic acids which originate from the roots of plants such as heather (*Calluna vulgaris*), and from rotting HUMUS material. Humic acid can cause the extensive CHEMICAL WEATHERING of normally resistant rocks, such as granite.

**humidity,** *n.* the amount of WATER VAPOUR in the atmosphere. At a specific temperature there is a maximum limit to the quantity of moisture that can be held by a body of air. When this state has been reached the air is said to be *saturated*. The proportion of water vapour present relative to the maximum quantity possible is the RELATIVE HUMIDITY value, and is expressed as a percentage. Relative humidity can change due to a gradual diffusion of water into an air mass, or from a change in the air temperature. The warmer the air mass then the greater its moisture-holding capacity.

The actual quantity of moisture held in the air is the ABSOLUTE HUMIDITY level and is the weight of water vapour contained in a given volume of air measured in grams per cubic metre. Due to the constant changes in air temperature then the absolute humidity of an air mass is liable to rapid fluctuation.

**humification,** *n.* the decomposition of organic material at the top of a SOIL PROFILE and the subsequent mixing of HUMUS with mineral soil. This results from physical processes such as ELUVIATION, chemical processes such as LEACHING and biological processes such as the activity of burrowing animals.

**humus,** *n.* a brown or black amorphous mass of decayed organic material found in soils. Humus is derived from the natural biological and chemical decomposition of organic matter, such as leaf litter, dead animals and plants and animal faeces; these may be supplemented in agricultural areas through the application of MANURE. Humus is an important source of many soil nutrients and combines in the A-HORIZON with clay minerals in a complex chemical relationship to form the CLAY-HUMUS COMPLEX. See HUMIFICATION, MULL, MOR, MODER.

**hunter-gatherer,** *n.* any individual who maintains a subsistence way of life based on the hunting of wild animals, fishing, and gathering of wild plants, roots, seeds, insects, grubs, etc. Although many pastoralists and farmers, especially in tropical regions, supplement their diet by hunting and gathering, true hunter-gathers depend solely on this mode of existence, moving in small, isolated bands over a well-defined territory to utilize known sources of water and food plants, and to follow migrating game. Peoples such as the Australian aborigines, the Kalahari bushmen and the Innuit (Eskimo) of Canada survive without AGRICULTURE and without domestic animals other than dogs. Such an existence requires a detailed knowledge of the environment and a close adaptation to the seasonal rhythm.

**hurricane, typhoon** or **tropical cyclone,** *n.* **1.** an intense tropical NON-FRONTAL DEPRESSION which usually originates in the western part of the Atlantic, Pacific and Indian Oceans between latitudes 5°-10°N and S.

The development of hurricanes is closely associated with the movements of equatorial zones of low pressure and most occur during the summer and early autumn. Hurricanes are generated and driven by the atmospheric INSTABILITY which results from the vast amount of energy released by the large-scale condensation of warm moist air. They may be up to 650 km in diameter and are characterized by a calm, cloudless central area of very low pressure, known as the *eye* which may have a diameter up to 11 km , and a surrounding vortex of massive cloud development, heavy precipitation, thunderstorms and very strong winds. Sustained wind speeds of 160 km per hour are common but gusting of up to 360 km per hour may occur occasionally. Hurricanes advance at around 20 km per hour and initially track westwards but then gradually curve towards the poles.

Hurricanes can cause considerable loss of life and damage to property, as happened with Hurricane Gilbert in 1988 which devastated large areas of the Caribbean and Central America.

Hurricanes degenerate as they cross cooler water or land surfaces and usually have a lifespan of only 9 days, although they may be revitalized and modified in the mid-latitude circulation systems and move eastwards to affect countries such as the UK.

Attempts are being made to reduce the effects of hurricanes in a variety of ways, including the construction of hurricane-proof buildings and seawalls to prevent the flooding of low lying areas. CLOUD SEEDING techniques have proved successful in reducing the intensity of some hurricanes. Satellites are increasingly being used to detect the initial formation and movement of hurricanes and to provide early warning for shipping and populations in coastal areas in the likely path of the hurricane.

**2.** a Force 12 wind of the BEAUFORT SCALE.

**hybrid,** *n.* **1.** (Evolutionary biology) a sexually produced cross between two closely related species.

**2.** (Genetics) a cross between two similar genetic types (see GENETIC ENGINEERING).

**hybrid seed,** *n.* a type of seed developed by cross-breeding two inbred lines in order to achieve heterosis or hybrid vigour, that is, qualities such as hardiness, growth rate or YIELD that out perform either parent.

Hybridization techniques were developed in the USA from the 1870s onwards, particularly on varieties of maize (corn). Double-crossing, that is, the crossing of two single crosses each having been derived from crossing two inbred lines, produced commercial quantities of hybrid seed corn by the 1920s. Subsequent developments in the American mid-western corn belt focused on location-specific varieties suited to sub-regional ECOSYSTEMS, and helped to account for a yield increase of 25% by the end of the 1930s. Since the early 1960s, high-yielding single-cross hybrids with greater disease and pest resistance have been produced, contributing to a three-fold increase in corn yields in the USA this century. In the 1950s, plant breeders in Mexico were successful in developing hybrid seeds suited to tropical environments, and their subsequent diffusion to the THIRD WORLD heralded the arrival of the GREEN REVOLUTION.

**hydration,** see CHEMICAL WEATHERING.

**hydraulic action,** see FLUVIAL EROSION.

**hydrocarbon,** *n.* a gaseous liquid or solid organic compound consisting of carbon and hydrogen. Hydrocarbons are the most important constituents of OIL and NATURAL GAS.

**hydrography,** *n.* the description, surveying and charting of rivers, lakes, seas and oceans, especially for navigation purposes.

**hydrological cycle,** *n.* the continuous and complex transfer of water through its gaseous, liquid and solid states from the oceans to

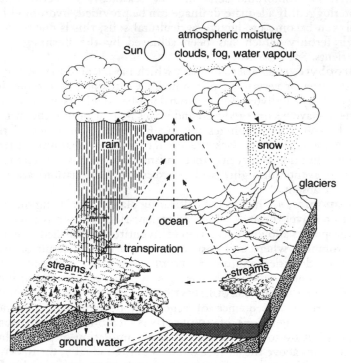

Fig. 38. **Hydrological cycle.** See main entry.

land and back again. Water is evaporated from the OCEANS into the ATMOSPHERE and transported by WINDS to the land masses where the air condenses and falls as PRECIPITATION. Some precipitation is evaporated as it falls and is returned to the oceans by winds moving in the opposite direction, where it will then fall as precipitation. Once on the ground, the remaining water finds its way back to the sea via EVAPORATION from the land surface, EVAPOTRANSPIRATION from vegetation, RUNOFF from rivers and GROUNDWATER movement (see Fig. 38).

**hydrologic sequence,** see CATENA.

**hydrology,** *n.* the study of the characteristics, occurence, movement and utilization of water on and below the earth's surface and within its atmosphere.

**hydrolysis,** see CHEMICAL WEATHERING.

**hydromorphic soil,** *n.* any one of a range of soil types whose pedogenic processes are dominated by the presence of abundant water. Hydromorphic soils may be seasonally or permanently waterlogged. If adequate drainage can be provided, hydromorphic soils can become productive agricultural soils; this is due to their high fertility which has been achieved by the flushing in of nutrients. See GLEYING.

**hydrophyte,** *n.* a variety of plant which has adapted to existing in fresh water or very wet environments. Consequently, the physiology of hydrophytes differs from land-based plants in that root systems have become reduced and leaves float at or near the surface of the water. Hydrophytes rely mainly on the buoyancy of the water for support and lack the strengthening mechanisms of land plants (lignin). Protective mechanisms against adverse environmental events (drought, cold or excessive illumination) are also notably absent.

**hydroponics,** *n.* a method of cultivating plants by growing them in peat or gravel through which a nutrient-rich solution is pumped. Hydroponics allows the creation of an entirely controlled growing environment where unit costs can be reduced as a result of high planting densities, shortened GROWING SEASONS, and continuous and automated production methods. It is usually associated with small-scale GLASSHOUSE CULTIVATION.

**hydrosere,** *n.* a sequence of vegetation development in which successive generations of HYDROPHYTES colonize a fresh water habitat but by so doing, result in an accumulation of plant debris which rises above the surface of the water. Gradually, the HABITAT becomes drier and conditions for growth become favourable for terrestrial plants so leading to the disappearance of true HYDRO-PHYTES. A complete hydrosere sequence would involve the following stages: submerged aquatic plants, shallow water aquatic plants, mixed aquatic and aerial community, marsh, moist woods, dry woodland.

**hydrosphere,** *n.* the total mass of free water in solid or liquid state on the earth's surface. It consists mainly of the oceans but includes lakes, rivers, glaciers and ice sheets. The movement of moisture between the hydrosphere, ATMOSPHERE and LITHOSPHERE is known as the HYDROLOGICAL CYCLE.

**hygroscopic nuclei,** *n.* minute particles in the ATMOSPHERE, such as salt particles, which actively attract condensing water vapour. Without hygroscopic nuclei for water droplets to form around, the scale of atmospheric condensation and resulting meteorological phenomena such as PRECIPITATION, clouds and FOG, would be greatly reduced.

**hygroscopic water,** see WILTING POINT.

**hypabyssal rock,** see INTRUSIVE ROCK.

**hypermarket,** *n.* a very large retail establishment, usually single-storey, built on a site large enough to include extensive car parks. A hypermarket caters for car-borne shoppers and keeps prices to customers down through the economies achieved by its large scale of purchasing. They are usually found at the edges of towns or in GREENFIELD SITES, since by virtue of their size it is difficult to accommodate them within the boundary of urban areas. Hyper-markets are usually established at points of maximum accessibility on regional road networks. Their success, especially when part of a multistore shopping centre at the edge of town, can have serious effects on the viability of existing shopping areas. This, in turn, has social consequences if the established shops which are accessible to citizens without cars are forced to close or if their range of products or standard of service is reduced. See OUT-OF-TOWN SHOPPING CENTRE, SHOPPING MALL.

**hypolimnion,** see THERMOCLINE.

# I

**IAEA,** see INTERNATIONAL ATOMIC ENERGY AGENCY.

**ice age,** *n.* a phase of large-scale GLACIATION which extends over a usually ice-free region, such as occurred in the PLEISTOCENE EPOCH. Ice ages have occurred at numerous times in the geological history of our planet and geologists can provide evidence that each of the main geologic periods have experienced major glaciation. This evidence is based mainly upon the lithology of the rocks and also upon the fossils preserved in the strata. It is generally thought that a temperature decline of between 5 and 9°C would be sufficient to bring about a permanent accumulation of snow and ice in mid- and high-latitudes. Such a temperature decline could be attributed to a variety of factors including a fluctuation in the solar energy output, a rise in volcanic activity thereby increasing the amount of atmospheric dust and reducing INSOLATION, a change in the earth's orbit, and to CONTINENTAL DRIFT causing disruptions to the planetary circulation system.

Unravelling the history of ice ages is a complex process and is accompanied by many controversies. Ice accumulation and recession are not contemporary over parts of even one continent which leads to difficulties in establishing chronologies of events. Also, because of the great erosive power of moving ice, the more recent ice ages have obliterated the evidence of earlier ice ages.

**iceberg,** *n.* a large, floating mass of ice that has broken off from an icesheet or a GLACIER flowing into the sea. Most icebergs originate from the icesheets of Antarctica and Greenland and the Arctic Ocean and may be transported many thousands of kilometres by wind and OCEAN CURRENTS. Icebergs vary greatly in shape and size, but 90% of their mass always remains below water. Various schemes have been unsuccessfully proposed to transport icebergs into arid areas where they would be utilized as major sources of fresh water for human use and for IRRIGATION purposes.

**ice cap,** see GLACIER.

**ice fog,** see STEAM FOG.

**ice sheet,** see GLACIER.

**IDC,** see INDUSTRIAL DEVELOPMENT CERTIFICATE.

**identified mineral resources,** *n.* specific mineral deposits of known quality and quantity.

**igneous rock,** *n.* a rock formed by the cooling and solidification of MAGMA. Igneous rocks may be either intrusive or extrusive in origin and may be classified in several ways:

(a) by silica content, which determines if igneous rocks are acid, intermediate, basic or ultrabasic in nature;

(b) by grain size of the groundmass, which is dependent upon the rate of cooling of the magma. Coarse-grained GRANITES cool slowly at great depths whilst BASALTS cool rapidly on the earth's surface and are fine-grained. An arbitrary scale may be used to indicate grain size:

(i)   Very coarse: larger than 3 cm
(ii)  Coarse      : 5 mm - 3 cm
(iii) Medium      : 1-5 mm
(iv)  Fine        : smaller than 1 mm
(v)   Glassy      : no apparent crystalline structure

(c) the texture of the rock, which results from its mineralogical and chemical characteristics;

(d) the level of dark-coloured minerals in the rock.

Major deposits of economic resources are often associated with igneous rocks including gold, silver, tin, tungsten, iron, uranium, lithium, lead, zinc, arsenic, sulphur and diamond. Hard igneous rocks, such as granite, may be used as a building material. See EXTRUSIVE ROCK, INTRUSIVE ROCK.

**illuviation,** *n.* the precipitation and deposition of organic materials and soluble salts, in particular iron and aluminium compounds, within the B-HORIZON of a soil. These materials have been washed down from the A-HORIZON by the process of ELUVIATION Illuviation is one of the processes of TRANSLOCATION and may result in the formation of a HARDPAN. The occurrence and rate of illuviation is usually determined by the character of the soil (especially the texture) and climatic conditions. See PODZOL.

**immature soil,** *n.* any soil with poorly developed horizons due to the relatively recent commencement of PEDOGENESIS. Compare MATURE SOIL.

**improvement grant,** *n.* a grant made available by a local authority in the UK to a private householder for the specific purpose of upgrading property to modern standards. Improvement grants were first made available in 1949.

**inceptisol,** *n.* a term used in the American soil classification system for a BROWN EARTH soil.

**incised meander,** *n.* a MEANDER lying below the level of the surrounding landscape and formed by the downcutting of a river undergoing REJUVENATION. ENTRENCHED MEANDERS and INGROWN MEANDERS are subtypes of incised meander.

**indeterminate species,** see RED DATA BOOK.

**indicator species,** see ECOLOGICAL EVALUATION.

**indirect recycling,** SEE RECYCLING.

**industrial development certificate (IDC),** *n.* a certificate required in the UK, of any industrial building or extension of such stating that it conforms to the national location policy. Introduced by the Town and Country Planning Act of 1947, IDC were a means for central government to control the national distribution of industry and specifically to channel industrial development into areas of high unemployment.

IDC were used extensively in the period before 1980 in the interests of encouraging industrial expansion in areas of the UK outside the prosperous south east of England. This was a negative control operating to inhibit development in the more prosperous areas where the issue of a certificate could be denied. The positive counterpart of this instrument was the system of grants to induce industry to locate in the less prosperous areas.

**industrial estate,** *n.* a planned grouping of workplace units, laid out to maximize vehicular access for raw materials or finished product handling. Industrial estates were first used in the UK in the late 1930s, in an attempt to create new kinds of jobs in old industrial areas of high unemployment. The idea has continued in use to the present day, with variations in the size of estates and of location in relation to the centre of cities. Certain economies were expected originally from the grouping together of several enterprises, for example the possibility of sharing common services such as recreational or shopping facilities and special transport for workers.

Early industrial estates were very large and could be built only on GREENFIELD SITES at the then edge of towns and cities. Since major redevelopment of INNER CITIES began in the 1960s, smaller industrial estates have been fitted into the abandoned sites formerly occupied by heavy industrial or dockland users. These INFILL SITE developments may also include training workshops and in recent years have played an important role in attempts to regenerate the economy of deprived areas within major cities.

**infield,** see INFIELD-OUTFIELD SYSTEM.

**infield-outfield system,** *n.* a system of farming and an arrangement of fields commonly found in Scotland and Ireland. The land closest to the farmstead or village (the *infield*) was intensively cultivated and kept in continuous arable cultivation by the application of all available sources of MANURE, including kelp and household waste, while land further afield (the *outfield*) was cultivated on a temporary basis, without benefit of manure, or else utilized as rough grazing. Although originally associated with farming in areas with unrewarding soils in northern latitudes, many tropical agricultural systems also incorporate an element of infield-outfield differentiation.

**infill site,** *n.* any site abandoned by its former user for which a new use has been found. Infill sites can be found in two locales:

(a) within the fabric of an existing city, where former housing, factory or transport sites have become vacant and available for renewal; the replacement developments may be of any kind and even new adaptations of the previous site uses;

(b) in unbuilt areas, where former extractive industries such as quarrying have left large gaps in the surface of the landscape. These sites may be infilled by the CONTROLLED TIPPING of waste and eventual topped off by soil to return the site to agricultural use, or in some cases, to selected kinds of urban development such as recreation.

**infiltration,** *n.* the movement of meltwater or rainfall into a soil through PORE SPACES or small openings. The rate of infiltration is determined by the soil's initial moisture content, its surface permeability, its physical and chemical characteristics, its temperature, and the duration and intensity of precipitation.

*Infiltration capacity* is the maximum rate at which water can enter a soil. Initially, infiltration capacity may be high but it declines over time as soil particles become swollen and the surface soil structure breaks down, causing pore spaces to fill with water or fine washed-in material. Infiltration capacities are usually highest on soils covered by dense vegetation whereas on unvegetated soils the surface soil layers become compacted by rain splash thus reducing infiltration capacity. If precipitation exceeds the infiltration capacity then OVERLAND FLOW will result. See also RAINDROP EROSION, SOIL WATER.

**infiltration capacity,** see INFILTRATION.

**infrastructure,** *n.* **1.** the network of facilities and installations that transport water, energy, information or vehicles, for example, water pipes, power lines, gas or oil pipes, telecommunication cables, roads, or railways.

**2. Social infrastructure**. buildings such as hospitals or schools which accommodate important social services. The term refers both to the buildings and to the functional networks of the services they accomodate.

**ingrown meander,** *n.* an INCISED MEANDER formed by both lateral and vertical FLUVIAL EROSION. River valleys containing ingrown meanders usually have one steep valley side and another gentler slope, resulting in an asymetrical CROSS-SECTION. Examples of ingrown meanders can often be found in rejuvenated river systems such as the River Rheidol in Wales and the lower reaches of the Seine in France (see REJUVENATION).

**inner city, 1.** *adj.* (of a policy) directed towards regenerating areas of DEPRIVATION. The use of the term became common in the late 1960s and early 1970s as public policy in the UK turned its attention to urban social and economic deprivation. This was observed as being both a problem which affected individuals living in many locales, and a problem of certain areas within major cities, where deprived households were grouped together and where physical decay of the environment reinforced the problem. Because the physical decay was more visible, it came to be used as a marker for the definition of areas within which measures to alleviate deprivation could be focused. Many of these areas were in older and more derelict parts of cities and were therefore 'inner city' in location. However, it was observed early on that deprivation affected other areas also and was most marked in large PUBLIC SECTOR HOUSING estates built on the edges of cities in the 1950s and 1960s. These areas were neither old in structural terms nor 'inner city' in location. However, the term 'inner city' came to be applied to all areas of deprivation and to the policies directed towards improving the quality of life within those areas. See also PRIORITY TREATMENT AREA.

**2.** *n.* that part of a major urban area which adjoins the city-centre business area.

**inorganic (mineral) fertilizer,** see FERTILIZER.

**insecticide,** *n.* any chemical substance, usually a synthetically manufactured organochloride or organophosphate, used to prevent, destroy or repel insects. Many insecticides work on the systemic principle, that is, they are absorbed by plant surfaces and translocated to various parts of the plant in amounts lethal to insects feeding thereon. Other insecticides can be sprayed directly onto the creature, in which case the insecticide is absorbed through the skin. Yet another alternative is to incorporate the insecticide into foodstock, then feed to a domestic animal which may be parasitized by grubs, worms, larvae, etc. See also PESTICIDES, FUNGICIDE, HERBICIDE.

**inselberg,** *n.* an isolated, steep-sided, round-topped hill formed either by the differential deep WEATHERING of bedrock to leave an upstanding feature of resistant rock, or by the SLOPE RETREAT of an escarpment to leave a residual hill.

**insolation,** *n.* the quantity of solar radiation falling upon the earth's surface per unit area. The proportion of solar energy that reaches the outer atmosphere of the earth is called the SOLAR CONSTANT. Of this, slightly less than 50% penetrates the atmosphere and arrives at the earth's surface. The remainder is reflected back into space by dust and clouds or is scattered by air molecules, water vapour and dust or is absorbed by the atmosphere.

The energy which reaches the surface of the earth may take the following pathways:

(a) it may be reflected back into the atmosphere and eventually into space (see ALBEDO);

(b) it may be absorbed by the land masses and oceans but is subsequently lost to the atmosphere by long wavelength infrared radiation convection and conduction.

The amount of insolation received by an area is dependent upon the solar constant, latitudinal position, season of year and the transparency of the atmosphere.

Insolation patterns are an important factor in the development of natural environments and also influence human activities such as agriculture.

**instability,** *n.* the atmospheric condition existing when a parcel of air which is warmer and less dense than the surrounding air moves upwards; in doing so, the air cools and if it is moist, water vapour condenses and PRECIPITATION often results. Such vertical movement in humid air often results in the formation of CLOUD and the generation of associated precipitation and atmospheric disturbances such as THUNDERSTORMS. See also LAPSE RATE

**institutional tenure,** see TENURE.

**integrated rural development planning (IRDP),** *n.* a form of RURAL DEVELOPMENT where projects to promote all aspects of the local economy and improve infrastructural provision such as roads, water supply and schools are designed to be interdependent. IRDP is characterized by a BOTTOM-UP APPROACH to development through the deliberate setting up of planning machinery at the local level, involving cooperation betweeen local government departments and agencies. Key elements of IRDP include multidisciplinary inputs in plan formulation, multi-agency involvement in implementation, local community participation, and self-help.

**intensive agriculture,** *n.* a method of farming in which there are high inputs of capital and labour leading to high output per unit area. The most intensive farming is generally confined to relatively small areas of land, as with FACTORY FARMING and FEEDLOTS where the STOCKING RATES are high, and in MARKET GARDENING involving GLASSHOUSE CULTIVATION. However, agriculture can be described as intensive whenever inputs of MANURE and AGROCHEMICALS are high, when more specialized forms of IRRIGATION and MECHANIZATION are required, and when labour demands are considerable, as with WET-RICE CULTIVATION and tropical farming systems based on MULTIPLE CROPPING. The intensity of agricultural activity is very often regulated by the effort of moving capital inputs and labour to the fields, and the product to market. There comes a point at which the return is no longer worth the cost of application; EXTENSIVE AGRICULTURE will then yield higher profits. Thus, intensive farming is generally found close to the farmstead or to markets.

**interchange,** see FLYOVER.

**intercropping,** see MIXED CROPPING.

**interculture,** see MEDITERRANEAN AGRICULTURE.

**interglacial period,** *n.* a period of large scale but temporary retreat of the continental ice sheets during the PLEISTOCENE EPOCH. Some interglacial periods are known to have extended over 10 000 years of time and were characterized by a relatively mild climate.

**interlocking spurs,** *n.* alternating extensions of a valley side (*spurs*) which project onto a valley floor containing a meandering river.

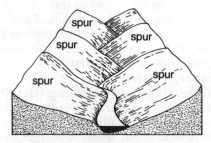

Fig. 39. **Interlocking spurs.**

When viewed up- or downstream these features appear to overlap or interlock (see Fig. 39). See also TRUNCATED SPUR.

**intermediate rock,** *n.* any IGNEOUS ROCK with a chemical composition between ACID ROCK and BASIC ROCK.

**intermediate technology,** see APPROPRIATE TECHNOLOGY.

**International Atomic Energy Agency (IAEA),** *n.* A United Nations agency responsible for all aspects af atomic energy, particularly the scientific and commercial applicationsof RADIO-ISOTOPES.

**International Bank for the Reconstruction and Development (IBRD),** see WORLD BANK.

**International Biological Programme (IBP),** *n.* a major international biological research programme organized by the International Council for Scientific Union and which operated from 1964 to 1974. Over 40 countries participated in the programme the main aims of which were to examine biological productivity and human adaptations to different BIOMES, and also to develop predictive mathematical models of ECOSYSTEM structure and function which would be of help when attempting to resynthesize degraded ecosystems. Although the IBP failed to meet all its objectives, it did make major advances in explaining the mechanisms responsible for the PRIMARY PRODUCTIVITY of ecosystems.

**International Development Association,** see WORLD BANK.

**International Union for the Conservation of Nature and Natural Resources (IUCN),** *n.* an independent, international organization founded in 1948 whose main purpose is to promote and initiate the conservation of wildlife, habitats and natural resources throughout the world. Its members include over 400 government agencies and conservation organizations from more than 100 countries; it also has close links with a variety of UN agencies.

Its chief concern is to ameliorate the worst exesses of human modification of the natural environment resulting from the rapid spread of urban and industrial development and the excessive exploitation of the earth's natural resources.

The IUCN works through six Commissions: ecology, education, landscape planning, legislation, national parks and survival service. The last of these commissions is responsible for the collection of data on all plant and animal species in danger of extinction and for initiating action to prevent it. The IUCN RED

DATA BOOKS provide information on the status of all threatened species.

**interstadial period,** *n.* a minor retreat of continental icesheets during a phase of GLACIATION. Interstadial periods were shorter than INTERGLACIAL PERIODS, lasting for between 50 and 500 years.

**intertidal zone,** *n.* the area between the positions of low and high tides. See LITTORAL.

**Intertropical Convergence Zone (ITCZ),** *n.* the zone of converging TRADE WINDS along the equator which cause rising air currents and low atmospheric pressure. The ITCZ, which is most sharply defined over the continental landmasses, migrates according to the season. The convergence of tropical air masses, usually of different humidities rather than temperatures, results in the formation of a distinct FRONT, and the associated INSTABILITY causes the development of clouds and convection rainfall. Weak DEPRESSIONS sometimes form in the ITCZ which can deepen and form embryo hurricanes if they move polewards.

**intrazonal soil,** *n.* a type of soil for which local features such as drainage, slope and PARENT MATERIAL are the major features of the soil-forming process, rather than climate and vegetation. The concept of the intrazonal soil is too simplistic to explain the multitude of soil types to be found in a region and is considered outdated. See AZONAL SOIL, ZONAL SOIL.

**intrenched meander,** see ENTRENCHED MEANDER.

**intrusive rock,** *n.* a type of IGNEOUS ROCK that forms within the existing rocks of the earth's crust following the injection of MAGMA along lines of weakness such as FAULTS and BEDDING PLANES. Two types of intrusive rock are recognized:

(a) *plutonic* or *abyssal rock*, which results when magma cools at great depths within the crust. Due to the slow rate of cooling, these rocks are characterized by the presence of large crystals and a coarse texture; GRANITE is an example of a plutonic rock.

(b) *hypabyssal rock*, which occurs closer to the surface. The faster rate at which magma cools at these shallower depths leads to the formation of medium-grained rocks.

There are several types of intrusive rock bodies including BATHOLITHS, LACCOLITHS, DYKES and SILLS. Intrusive rocks are found at the earth's surface only after long periods of erosion. Major deposits of economic resources are often associated with intrusive rocks, such as diamonds, gold, silver, iron, chromite, uranium, tin tungsten, lead, zinc, and sulphur. Intrusive rocks such as granite, may be used for buildings and other construction projects. Compare EXTRUSIVE ROCK. See COUNTRY ROCK.

**inversion layer,** *n.* **1.** a layer of air in which temperature increases with altitude (see Fig. 9). This is opposite from the usual tropospheric temperature gradient in which temperature decreases with altitude according to the LAPSE RATE.

Low-level inversions may be due to rapid nocturnal cooling of the ground surface which chills the overlying layer of air to temperatures below those of subsequently higher layers. This type of inversion is common in dissected terrain when cold air collects in the bottoms of valleys (see FROST HOLLOW) and causes the formation of RADIATION FOG and HOAR FROST.

High-altitude inversions are caused by the undercutting of a warm air mass by a cold air mass, as happens along COLD FRONTS or when a warm air mass overrides a cold air mass such as at a WARM FRONT.

The gradual subsidence and adiabatic warming of a subsiding high-altitude air mass in an ANTICYCLONE may also cause an inversion.

The accumulation of industrially produced pollutants is made worse by the presence of inversion layers. PHOTOCHEMICAL SMOG in the Los Angeles basin is trapped when cold air from above the cold California Current moves inland and undercuts warm air from the Mojave Desert to form an inversion.

**2.** a layer of water in a reservoir, lake or sea within which temperature increases with depth. It may be due to the presence of a warm water current (in oceans), from thermal pollution released from coastal power stations or from natural hot water vents occurring on the beds of oceans or lakes and usually associated with tectonic plate boundaries. See THERMOCLINE.

**invertebrate,** *n.* any animal that does not possess a backbone, loosely used to include all animals except the chordates. Compare VERTEBRATE.

**inverted tree line,** see TREE LINE.

**ion,** *n.* an electrically charged atom or group of atoms formed by the loss or gain of one or more electrons. An *anion* has an overall positive charge, while a *cation* has an overall negative charge. See IONIZATION.

**ionization,** *n.* the formation of IONS as a result of a chemical reaction, electrical discharge or radiation. See IONIZING RADIATION.

**ionizing radiation,** *n.* a form of RADIATION capable of causing IONIZATION in the medium in which it is travelling. Examples of ionizing radiation include X-rays, gamma rays, or flows of high-energy particles such as electrons, ALPHA PARTICLES or protons. Ionizing radiation can be of natural origin (solar or cosmic), or may be the result of human activity (nuclear explosions or accidents). Apprehension over the effects of such radiation is one of the greatest concerns of modern times.

The exact nature of the biological repercussions of exposure to ionizing radiation is not known. In contact with organic material, ionization produces alterations to its molecular structure and can lead to the the development of cancerous tissue in animal cells. It may also cause random genetic changes in the DNA material and may produce unfavourable mutations in future generations. In many cases, the result of exposure is not seen until many years later, often as much as a quarter of a century.

Any RADIATION DOSE, no matter how small or how slowly received, is now regarded as potentially harmful to protoplasm. A maximum whole body exposure value for the general public of 50 rems per year has been set by the International Commission on Radiological Protection. Maximum average dose levels are set at one-third of the maximum value (17 rems per year). See ION, ALPHA RADIATION, BETA PARTICLE.

**IRDP,** see INTEGRATED RURAL DEVELOPMENT PLANNING.

**ironpan,** see HARDPAN.

**irrigation,** *n.* the practice of applying water to the soil artificially and in a controlled manner to promote plant growth and, through the creation of a cooler and more humid MICRO-ENVIRONMENT, to enhance the prospects of successful crop production. Irrigation is used to supply any deficit existing between PRECIPITATION and potential EVAPOTRANSPIRATION during part or all of the year.

Including land flooded by river water for crop production or PASTURE improvement, 220 million ha of land were under irrigation in 1984, a 17% increase over the 1974-76 figure. 62% of the world's irrigated area is in Asia, with China (21%), India (18%) and the USA (9%) the leading countries. Almost 18% of the world's CROPLAND is now irrigated, producing about 30% of the world's food.

**island arc,** *n.* an extended, curvilinear group of islands that lie adjacent to a DEEP-SEA TRENCH on their convex side. Island arcs, such as the West Indies, the Kuril Islands and the Aleutian Islands, occur close to the boundaries of LITHOSPHERIC PLATES and consequently, are often subject to earthquakes and intense seismic activity.

Island arcs are thought to form in several ways:

(a) by the deposition of rock material scraped off the lithospheric plate descending into the ZONE OF SUBDUCTION below the deep-sea trench;

(b) the deformation and uplift of the overriding plate;

(c) by the extrusion of MAGMA, formed by the melting of the descending plate in the zone of subduction, onto the overriding plate.

See also PLATE TECTONICS.

**island biogeography,** *n.* the specialized study of plants and animals which inhabit remote islands. Because of their isolation, and in some examples the lack of human exploitation, islands can provide ideal natural laboratories in which to study evolutionary change, the colonization rates of successive immigrant species, and species extinction rates.

The number of different species found on an island depends upon the size, range and diversity of HABITAT types, on the island's accessibility from the source of its colonists, on the richness of that source, and upon the equilibrium between the rate of colonization by new species and the rate of extinction of existing species.

It is notable that both Charles Darwin and A.R. Wallace made major studies of the flora and fauna of islands before contributing to the modern theories of evolution. See WALLACE LINE.

**isobar,** *n.* a line on a weather chart which joins all points of the same air pressure at a particular time.

**isostasy,** *n.* a theory that propounds equilibrium within the earth's crust between the lighter SIAL of the continental landmasses and the heavier SIMA of the oceanic crust. The sialic continents are considered to float on the denser sima. Isostatic compensation dictates that equal mass at depth must underlie equal aerial mass with the result that the roots of mountain ranges must penetrate more deeply into the sima than the roots of areas of lesser relief.

The accumulation of continentally derived sediment causes the oceanic crust to subside. To compensate and maintain the equilibrium, isostatic adjustment leads to the uplift of the continental landmasses. It is thought that there is also a compensating subcrustal movement of rock material from below the area of subsidence to below the area of uplift.

**isotherm,** *n.* a line on a weather chart which joins all points of the same temperature at a particular time.

**isotope,** *n.* one of two or more forms of the same element whose atoms contain the same number of protons and electrons but different numbers of neutrons. Such substances share common chemical properties but have differing physical properties. There are over 300 naturally occurring isotopes, some of which are radioactive. Some radioisotopes are only slightly unstable and these take many tens of thousands of years to decay. For example, uranium-238 requires 4.5 × 10⁹ years for half of a given amount of that material to decay. Isotopes are frequently produced in nuclear reactions. See HALF-LIFE, ALPHA PARTICLE, BETA PARTICLE.

**ITCZ,** see INTERTROPICAL CONVERGENCE ZONE.

**IUCN,** see INTERNATIONAL UNION FOR CONSERVATION OF NATURE AND NATURAL RESOURCES.

# J

**jet stream,** *n.* a strong, narrow, thermally driven high-altitude wind. Jet streams are located in various latitudes and usually occur in the vicinity of a steep gradient or fracture in the TROPOPAUSE, at altitudes of between 9000 and 15 000 m. The velocity of most jet streams ranges from 160-320 km per hour.

Two major jet streams are recognized:

(a) the *polar jet stream* is a strong, discontinuous westerly airflow which is associated with the POLAR FRONT and its seasonal movements. The occurrence of the polar jet stream is closely linked to the origin and movements of mid-latitude DEPRESSIONS.

(b) the *subtropical jet stream* is a strong and continuous westerly airflow which occurs in latitudes around 30°N and S. The subtropical jet stream has been linked with the formation of HURRICANES and MONSOONS.

**journey to work,** *n.* any journey between a domestic residence and a place of work. Because most COMMUTERS travel at similar times of the day (0700-0900 and 1630-1900) travel conditions are often characterized by overcrowding and journey times can be double that of other times of day. Some workers are able to practice *flexitime* in which starting and ending times of the working day can be arranged outside the peak travel times.

The pressures exerted by many tens of thousands of people entering and leaving a city create many environmental problems. For example, AIR POLLUTION levels from motor vehicles show a diurnal trend similar to the flow rate of traffic, while some cities in North America have attempted to solve the overcrowding of roads by commuter vehicles by building yet more highways thus using more valuable building land. In Europe, by tradition, greater use is made of public transport for travelling to work than in the USA.

**Jurassic period,** *n.* the geological PERIOD that followed the TRIASSIC PERIOD about 195 million years ago and that ended 136 million years ago when the CRETACEOUS PERIOD began. Conditions in Europe and Africa were characterized by OROGENESIS, and volcanic and igneous activity. In the Jurassic period dinosaurs attained their maximum development, and lizards and crocodiles first appeared. Flying reptiles (pterosaurs) were abundant although modern birds who together with mammals appeared for the first time, evolved from a different group of reptiles (archaeopteryx). Economic resources found in Jurassic rocks include CLAYS. See EARTH HISTORY.

**juvenile water** or **magmatic water,** *n.* the hot and often mineralized water that supplies HOT SPRINGS and originates from magmatic sources deep within the earth's interior. Juvenile water is an important source of GROUNDWATER and is sometimes utilized in the harnessing of GEOTHERMAL POWER.

# K

**Kainozoic era,** see CENOZOIC ERA.

**kame,** *n.* a mound of stratified GLACIAL TILL deposited by meltwater streams. Kames vary considerably in nature and origin; some developed as DELTAS on the margins of a stagnating glacier whilst others resulted from the accumulation of glacial till in large crevasses. Kames are sometimes considered synonomous with ESKERS. Sands and gravel in kames form an exploitable economic resource and are often extracted. See KAME TERRACE, GLACIAL DEPOSITION.

**kaolin,** see KAOLINITE.

**kaolinite,** *n.* a white clay mineral formed by hydrothermal activity at the time of rock formation or by the CHEMICAL WEATHERING of rocks with a high feldspar content. It is usually associated with intrusive granitic rocks. *Kaolin*, a fine white CLAY whose main constituent is kaolinite, is used in the manufacture of pottery and ceramics, paints, paper and rubber.

**Karoo,** *n.* **1.** any of several high, arid plateaus in South Africa, especially the Great Karoo and the Little Karoo of Cape province. An extensive and distinctive form of VELD, the Karoos are of great antiquity and appear to have evolved out of FYNBOS, tropical forest and SAVANNA vegetation types. They typically include grasses, large shrubs and succulent plant species; trees are poorly represented. These areas are extensively mismanaged, primarily as a result of overgrazing. This has led to SOIL EROSION and a pronounced lowering of species diversity. Partly as a result of the mismanagement and partly due to a succession of very dry years in the 1980s, the Karoo vegetation has shown a marked increase in its distribution. As it extends, it leaves behind an almost useless desert area.

**2.** a PERIOD or rock system in southern Africa equivalent to the period or system extending from the Upper Carboniferous to the Lower Jurassic. See also GEOLOGICAL COLUMN.

**karst,** *n.* a rugged, barren limestone or dolomite region with little or no surface drainage. The main characteristics of karst areas include extensive UNDERGROUND DRAINAGE and cave systems. Surface features associated with these areas include DOLINES, UVALAS and POLJE, all resulting from solution and carbonation weathering (see CHEMICAL WEATHERING). Dry valleys are also common as rainfall rapidly disappears underground through SWALLOW HOLES. Ground collapse is a prominent feature of karst landscapes leading to the destruction of the foundations of roads, railways and buildings.

The term 'karst' refers specifically to a limestone area near the northern Adriatic Coast in Yugoslavia, but the use of the term has been extended to many areas which display some or all karstic features.

**katabatic wind** or **mountain wind,** *n.* a localized wind which flows down valley slopes, usually at night. Katabatic winds are caused by the rapid nocturnal cooling of valley slopes and their overlying layers of air and the subsequent drainage of this cold air under gravity onto the valley floor. Compare ANABATIC WIND.

**key diagram,** see STRUCTURE PLAN.

**kibbutz,** *n.* a form of co-operative rural settlement in Israel based on communal ownership of resources, collective organization, and a policy of pooling labour and income. Kibbutzim differ from the COLLECTIVE FARMS of the Soviet model in being voluntary organizations, free from state control, and having a democratically elected leadership. Kibbutzim have been characterized by an idealism among members towards the communal way of life, with the nuclear family de-emphasized in favour of collective social organization. Economically, the kibbutz controls purchasing, production, marketing and consumption, paying no wages but ensuring that members' basic needs are met. There are some 250 kibbutzim in Israel varying in size from less than 100 to more than 2000 individuals, with a total population of 100 000, responsible for producing half of the country's food. Since the 1930s, but mostly in recent years, some industry has been established on kibbutzim. See also AGRICULTURAL COOPERATIVE.

**knickpoint,** *n.* a marked change of slope in the long profile of a river caused by REJUVENATION or the outcropping of a resistant rock formation such as a DYKE.

**kolkhoz,** see COLLECTIVE FARM.

**krill,** *n.* any small marine CRUSTACEAN of the order Euphausiacea. Krill is a collective Norwegian term for several dozen different species of shrimp-like creatures of about 4-6 cm in length which live exclusively in the cold waters around the poles. Krill feed on PLANKTON and form the second TROPHIC LEVEL of the marine FOOD CHAIN. They in turn constitute the main food source for whales, penguins, seals and squid. Whales used to consume an estimated 190 m tonnes of krill per annum although the overfishing of whales has caused a reduction in the feeding pressure to about 60 m tonnes p.a. As a consequence, krill have become an overabundant food source for sea birds and seals.

Krill exist in vast shoals covering several square kilometres with a density of up to 24 kg/m$^3$. Because of this density and their propensity to 'shoal', trawling for krill by Far Eastern fishing fleets has become highly organized. Approximately 450 000 tonnes of krill are harvested each year by the Russian and Japanese fishing fleets, 45% of the total harvest. The potential harvest may be as high as 50 m tonnes of krill per year. Much of the catch is currently used for animal feedstock although krill could also form a nutritious food source for humans.

Krill, as a biotic resource, require careful management if they are not to become over-exploited. POLLUTION of the marine environment poses a particular threat as it could destroy the plankton on which the krill feed.

**krummholz,** see TREE LINE.

**k-species,** *n.* any species of plant or animal characterized by a large body size, a slow rate of development, considerable competitive ability, a reproductive phase occuring mid-way or late in the life history and which can be undertaken several or many times over. Such species often live for many years and are characterized by terrestrial vertebrates. Compare R-SPECIES.

# L

**laccolith,** *n.* a large dome-shaped mass of INTRUSIVE ROCK formed by the injection and solidification of MAGMA along BEDDING PLANES and the resulting upward arching of overlying COUNTRY ROCKS. The erosion of the overlying rock results in the formation of a domed landscape, such as the Henry Mountains in Utah, USA.

**lagoon,** *n.* **1.** a quiet coastal backwater separated from the ocean by a physical barrier such as a baymouth bar or CORAL REEF (see ESTUARY).

**2.** a calm area of sea water within an ATOLL.

Lagoons are often linked to the open sea by passes and the reduced effect of waves and currents makes them good natural harbours.

**lahar,** *n.* a volcanic MUDFLOW formed by the mixing of water and unconsolidated volcanic ash and other PYROCLASTIC MATERIAL. Lahars can be triggered by precipitation, rapidly melting snow and the explosive expulsion of water from a caldera LAKE due to a volcanic eruption. Lahars can travel considerable distances from their point of origin and some have recorded speeds of over 90 km/h. Great loss of life and damage to property can result from lahars, such as the lahar from the Kelut volcano in Java which in 1919 caused 5000 deaths and the loss of over 200 km² of agricultural land. In areas susceptible to lahars, such as Mount Ranier in Washington, USA, their destructive effects can be minimized by careful LAND USE PLANNING.

**LAI,** see LEAF AREA INDEX.

**lake,** *n.* a body of standing water that occupies a depression on the earth's surface and is completely surrounded by land. Ranging greatly is size and depth, lakes form an important part of the HYDROLOGICAL CYCLE and account for 75% of all the water (apart from that contained in the OCEANS) on the earth's surface.

Lake water may be fresh or saline depending on whether the lake has an outlet such as a surface stream. If no outlet is present then the lake acts as an inland drainage area. This type of lake is often associated with dry climates and high evaporation rates, giving rise to highly saline lake waters. Saline lakes may form in a variety of ways; for example, the Caspian Sea, situated between South East Europe and Asia, was formerly an arm of the sea whereas the Great Salt Lake in Utah and the Dead Sea on the borders of Israel and Jordan were formerly fresh water lakes. These lakes all have a high salinity, ranging between 170 and 250 parts per thousand.

Lakes can be the result of:

(a) depressions caused by earth movements. These form some of the largest lakes, for example, the Caspian Sea and the Dead Sea;

(b) glacial erosion and/or deposition, for example the Great Lakes of North America and Dove Lake in Tasmania;

(c) the work of rivers and the formation of OX-BOW LAKES;

(d) natural dams due to MASS MOVEMENT, for example Slide Lake in Montana;

(e) volcanic activity. Lava flows can dam valleys, as occurred for example, to create the Sea of Galilee. Collapse of a volcanic crater can produce a *crater lake* or *caldera*, for example Crater Lake in south west Oregon;

(f) LAGOONS becoming land locked behind bars and barrier beaches;

(g) groundwater movements leading to sink holes and caverns in areas of soluble underlying bedrock. If these become choked with debris then a lake can form;

(h) river damming by animals or by the accumulation of organic material. Beaver dams have led to the formation of Beaver Lake in Yellowstone National Park, USA, while swamp vegetation and log jams can also cause ponding;

(i) human activity. Reservoirs have been deliberately created to store water whereas other lakes have formed through the abandonment of mines, quarries and peat diggings.

Lakes may become filled with sediment or organic matter and become increasingly swampy. Alternatively, the downcutting action of an outlet stream can cause the lake to become drained. Increased evaporation due to climatic changes can also lead to the disappearance of a lake. Some lakes, especially the larger ones, are of great importance to humans. They are used for the storage and supply of water for both domestic and industrial purposes as well as being important transportation routes. Lakes also have a value as a recreational facility.

**land,** *n.* **1.** the solid part of the earth's surface.

**2.** a factor of production comprising all natural and man-made resources including all of the earth's surface, plants grown in it, structures built upon it, mineral deposits and water resources.

**land attribute,** *n.* any of the physical, chemical and biological conditions of the land which collectively influence the LAND USE pattern. See LAND USE CLASSIFICATION, LAND USE SURVEY.

**land breeze,** *n.* any light wind which occurs at night in coastal areas and blows offshore due to the nocturnal cooling of the land surface and the resulting slight increase in air pressure.

**land colonization,** *n.* the act of settling on and farming land following an influx of a migrant population. With European settlement in the Americas and Australia, pioneering communities pushed into the continental interiors, colonizing VIRGIN LAND or territory so sparsely populated that indigenous groups such as the Plains Indians of North America and the Australian aborigines could, with relative ease, be absorbed or more usually destroyed or displaced to ever more marginal locations. In more recent times land colonization has generally depended on state support, especially for land clearance, infrastructural provision and financial aid to settlers. In many Latin American countries, for example, land colonization is regarded as a component of AGRARIAN REFORM, relieving population pressures in settled areas and providing the landless with means of access to land. However, many of the contemporary land colonization programmes have a high failure rate in terms of the permanence of settlement and agricultural PRODUCTIVITY. In part, this is due to misplaced optimism based on inadequate knowledge of environmental conditions and responses, and to settlers continuing with farming practices unsuited to their new environment. In many cases, especially in the Amazonian states of Rondonia and Acre in Brazil, the livelihoods of communities employing sustainable methods of forest exploitation are threatened by the short-term gains from wholesale forest clearance for land colonists.

**land devil,** see WHIRLWIND.

**land ethic,** *n.* the principle developed in 1949 by the American ecologist, Aldo Leopold advocating cooperation between humans and other biospheric components. Leopold used the land ethic principle to explain the attitude of humans towards the BIOSPHERE. He argued that human instinct is to compete for supremacy with other species in pursuit of dominance, whereas a land ethic approach promotes cooperation in order to retain maximum ecological diversity while at the same time permitting the desired level of human land management. Traditional ecological theory has suggested that cooperation operates at an intra-species level while Leopold envisaged cooperation as an inter-species relationship between humans and many other species. More recently, the land ethic has been redefined in anthropocentric and utilitarian terms in that it advocates the maintenance of natural resources and environments which, in turn, ensures the well-being of the human species. It is the acceptance of this new interpretation which has given the conservation movement a greater realism.

**landfill,** *n.* the disposal of domestic and/or industrial waste in sites such as former quarries, disused mine workings and gravel-or clay-pits. Unlike traditional disposal methods for domestic refuse in which material was burnt in 'open' sites, landfill does not utilize burning and so avoids the problems of smoke and odour. Wastes are spread in layers, between which soil or building rubble is laid in order to minimize the generation of spontaneous heat through the decomposition of organic wastes. This decomposition produces METHANE gas and in large landfills a pipe system is installed during dumping to collect the gas which can then be used as an energy source. In Staten Island, New York, such a system provided sufficient gas to heat 10 000 homes a year. However, sites are often selected with little regard to possible pollution of groundwater and surface water due to run-off and LEACHING. When the site is full, a layer of soil is placed on top and the surface countours regraded to suit the local topography.

Old landfill sites can sometimes be returned to agriculture although care must be taken that concentrations of pollutants in the soil and vegetation do not exceed safety levels. Housing or industry can sometimes be eventually constructed on old landfill sites, although problems of settling (*subsidence*) can create major problems for subsequent land users. See CONTROLLED TIPPING.

**land evaluation,** *n.* an assessment of the performance or potential of land with respect to a particular purpose, designed to assist in LAND USE planning and management. Land evaluation involves the interpretation of data concerning the physical environment (for example, soil, climate, and vegetational patterns), and past and present land use in terms of its resource potential. It is thus concerned with seeking solutions to problems such as the possible long-term degradation of land quality as a result of current usage, the viability or likely return of alternative land uses, the extent to which the management of existing land uses can be improved, and the impact of inputs on PRODUCTIVITY and land quality.

Most land evaluation projects adopt a three-phase approach: description, appraisal and development. Natural resource surveys that are chiefly quantitative are followed by appraisal that combines environmental data with technological information, for example, agricultural methods and crop needs. When expressed in terms of resource potential this information can be used in the design of economically based development options for any area of land under survey.

To counter the increasing variability in land evaluation procedures the FAO published a *Framework for Land Evaluation* in 1976, intended to be flexible enough to be applicable to any land evaluation problem. With an emphasis on a multidisciplinary approach to achieve an evaluation relevant to the physical, economic, and social context of the area, the system involves an assessment and classification of land suitability with respect to specified kinds of sustainable use.

Due to the complexity and magnitude of land evaluation most projects now resort to computer storage of data (see ENVIRONMENTAL DATA BASE) and rely on computer modelling and GEOGRAPHIC INFORMATION SYSTEMS for analysis of the data.

**land inventory,** see LAND USE SURVEY.

**landnam,** *n.* a primitive agricultural practice formerly in common use throughout northern Europe whereby small clearings were made in the virgin forest and in which crops were grown until soil fertility declined or the adjacent forest over-ran the clearing. See SHIFTING CULTIVATION, SLASH-AND-BURN.

**land reclamation,** *n.* any of a number of processes that brings an area of land back into productive use. The types of land most subject to land reclamation include:

(a) waterlogged areas which may be made suitable for agriculture following drainage;

(b) coastal and lakeside margins which may be enclosed and subsequently filled-in and drained, as with POLDERS;

(c) arid areas made productive through IRRIGATION;

(d) areas where over-irrigation has led to serious problems of SALINIZATION requiring marked reductions in the intensity of land use and costly investment in systems of pumped drainage;

(e) land with undesirable or unproductive vegetation cover such as HEATHLANDS and MOORLAND where it may be possible to establish agriculturally preferred plant species by clearing the natural vegetation, drainage, deep ploughing, and repeated seasonal burning (see FIRE);

(f) areas rendered unfit for cultivation by industrial POLLUTION which may require chemical treatment;

(g) land blighted by extractive and manufacturing industries, necessitating wholesale land REHABILITATION, including the filling-in of mines and quarries, and the stabilization and landscaping of waste tips.

The deliberate planting of vegetation often together with the construction of TERRACES and embankments in areas prone to SOIL EROSION not only acts as a form of SOIL CONSERVATION but also helps to reclaim degraded land otherwise unfit for cultivation.

**land reform,** *n.* any STRUCTURAL REFORM of an agrarian system involving a change in the form and distribution of LAND TENURE. Land reform is generally tackled by land redistribution and the reform of tenancies, and to be successful, must be tailored to the particular circumstances of the country or region where it will take effect. It must take specific account of the nature of the existing tenurial arrangements and extent of land concentration, the level of urbanization, and the role of agriculture in the economic structure. A common feature of redistributive reform is the expropriation of land on estates above a specified size, with or without compensation or exemption. This land is then allocated to SMALLHOLDERS, landless labourers, COLLECTIVE FARMS or STATE FARM organizations according to the political motivation of those instituting the reforms. Land reforms concerned with changes in tenancy systems aim to establish security of tenure, to regulate payments to landlords, particularly under SHARECROPPING arrangements, and to legislate for the transfer of ownership of land to its cultivator, especially where absentee landlords are common.

Land reform has been proposed or undertaken as a means of achieving greater social equity, increased agricultural PRODUCTIVITY, national control over productive forces, and to create rural structures compatible with the prevailing national ideology. However, it is often ineffective without accompanying AGRARIAN REFORMS.

**Landsat,** *n.* any of a series of orbiting earth-survey satellites used to monitor civilian land use, as well as forest, mineral, energy and water resources. The first satellite was launched in 1972, as part of the EARTH RESOURCES TECHNOLOGY SATELLITE (ERTS) project, and two ERTS satellites were launched before the programme was renamed. The first named Landsat satellite (Landsat 3) was launched in the spring of 1978 followed by Landsat 4 and by Landsat 5 in 1984. Orbital characteristics are similar to those described for ERTS.

Landsat scans a section of the earth's surface every 33 milliseconds, providing an assigned field of view of 56 x 79 m. An image of ground conditions is made by a multispectral scanner in the satellite and converted into FALSE-COLOUR IMAGERY to reveal information not normally visible on true-colour photographs. MSS information is coordinated by computer and produced in standard 'Landsat scenes' covering an area 185 x 178 km. (For comparison, a similar area covered by conventional air photography at a scale of 1:15000 would require 5000 photographs).

Landsat imagery has proved of great use for agriculturalists, foresters, resource use planners, conservationists, geologists, pollution monitoring and meteorological use. Famine-struck areas in northern Africa and cereal production in the USSR are just two themes studied by means of Landsat.

All Landsat data has been collected in accordance with an 'open skies' principle, that is, that there will be non-discriminatary access to data collected from anywhere in the world. Landsat data is on offer to all nations but is currently underused due to its expense, the need for specialized manpower and equipment to analyze the imagery, the lack of efficient, accurate ways of analyzing the data and problems in verifying the imagery with real world ground conditions.

Up to the end of 1988, the US Department of Commerce funded Landsat's estimated annual running costs of $20 million (£11.6 million). However, the future of the programme is in danger because of threats by the US government to abandon the project. See SPOT.

**landslide** or **landslip,** *n.* the rapid and sudden movement of soil and rock debris under the influence of gravity. Landslides are common in areas of undercut cliffs or weak underlying bedrock and are usually triggered by heavy rainfall or melting snow increasing the weight and reducing the cohesion of slope material. Major EARTHQUAKES may also cause landslides. In 1989 an earthquake triggered a landslide in the Soviet Republic of Tadzhikistan killing 1400 people in a densely peopled region 30 km south west of the capital, Dushanbe. See also SLUMP, MUDFLOW, EARTHFLOW.

**land subsidence,** *n.* the gradual sinking of the ground surface to a lower level. Subsidence may occur for natural reasons such as the collapse of caves in limestone areas or large-scale tectonic movements. Artificial subsidence may follow the extraction of solid material or fluids from underlying bedrock. The settling of ground due to the collapse of former mine workings can cause considerable damage to buildings, roads, railways and other civil engineering structures. To prevent further subsidence, it is often possible to inject water back into AQUIFERS or to pump concrete into the ground to stabilize underground workings.

**land-systems mapping,** *n.* an integrated environmental assessment of the potential usefulness of a body of land which takes into account topography, geomorphological processes, soils, geology, climate and vegetation. Land-systems mapping is used mainly for agricultural, engineering and planning purposes. Maps are prepared from information gathered by REMOTE SENSING and from fieldwork. Numerous countries make use of land-systems mapping, especially Canada, Holland, Australia and the USA. Land-systems mapping can be particularly useful in providing an environmental reconnaissance survey of poorly mapped or unmapped regions, as well as for those countries in which land-use pressure is intense.

**land tenure,** *n.* the system of rights regulating the ownership or use of land, and the arrangement governing the relationship between landlords and tenants. There is great variety in the form of land tenure systems. The following represent only the major types:

(a) *usufruct* or *use right*, in which individuals, or family groups have the right to use land and consume its produce without claiming formal ownership of the land itself. This tenurial system is common amongst tribes in Africa;

(b) *owner-occupation*, in which the rights to land may be inherited or purchased. Common forms of this type of tenurial holding include PEASANT farms, FAMILY FARMS, and LATIFUNDIA. Hereditary land ownership ensures the stability and continuity of property and encourages efficient use of the land;

(c) *tenancy*, in which the user of the land repays the landowner for that right in money, in kind, or in the provision of services, notably of labour. Well-regulated tenancies enable farmers with limited capital resources to concentrate on purchases of agricultural implements and livestock while the landlord provides the items of fixed capital. Short leases, however, act as disincentives to investment by tenants;

(d) *institutional tenure with wage labour*, in which the land is owned by a private company or institution and is farmed by paid employees. The tropical PLANTATION and the modern AGRIBUSINESS are the most familiar and widespread examples of this type of tenure;

(e) *collectivist tenure*, in which there is a common ownership of resources and a pooling of labour and income. Ownership of the land may rest with the state, as with the COLLECTIVE FARMS of the

Soviet Union, or else with the occupying community, as in the Israeli KIBBUTZIM and Tanzania's Ujamaa villages.

(f) *state ownership with wage labour,* that is, the STATE FARMS of the Soviet Union and other communist countries.

**land use,** *n.* **1.** *functional land use.* The use of an area of land to meets the requirements of the inhabitants of the area.

**2.** *current ground cover.* The present use of land, for example, for agriculture, industry or housing.

In practice, there is ambiguity in the use of this term. In the strictest sense, the term should be confined to the specific uses placed upon the land. See LAND ATTRIBUTE, LAND USE CLASSIFICATION, LAND USE SURVEY, LAND USE PLANNING.

**land use classification,** *n.* the categorizing of land into a series of types of use. Most countries have produced an inventory of the proportion of land devoted to forestry, agriculture, urban and industrial land. Alternatively, some countries such as Holland have produced highly detailed land use maps in which individual fields have been classified along with every industrial type. The maps are sometimes accompanied by a descriptive memoir. The *Second Land Use Classification of Britain* undertaken from 1960 recognizes 13 land use groups: settlement, industry, transport, derelict land, open spaces, grass, arable, market gardening, orchards, woodland, heath and rough grazing, water and marsh, and unvegetated land.

Traditionally, the surveying of land use patterns necessitated the manual mapping of every land area. Nowadays, such surveys are undertaken from aerial surveys or make use of satellite-derived information (see also LANDSAT, SPOT, COSMOS-1870).

**land use competition,** *n.* the principle by which land is allocated to a particular use rather than any alternative use because it will yield the highest and best return. Various uses are said to 'compete' for land in that each will provide different average returns over a given period of time. A land owner or decision maker will evaluate these uses and adopt that which will maximize returns.

**land use planning** or **physical planning,** *n.* the process of zoning land parcels for a particular designated use, under powers given by specific statutes or acts passed by elected governments at national, regional or local level. The zoning reflects the expected future land needs for housing, employment, recreation, social services, and transport networks. This is based on calculations of expected changes in population numbers and likely social and economic changes which will effect the demand for particular kinds of space at particular kinds of location. See also DEVELOP-

MENT, DEVELOPMENT CONTROL, DEVELOPMENT PLAN, LOCAL PLAN, STRUCTURE PLAN.

**land use survey** or **land inventory,** *n.* an assessment of the LAND USE of an area presented in tabular and/or mapped form. Most developed nations require its farmers, foresters and land use planners to provide regular information concerning, for example, how much land on a farm is devoted to a specific crop, or on the quantity of land that is forested or devoted to urban land use. Traditionally, a land inventory has been made by means of a labour-intensive field surveys although in recent years aerial photography and earth-survey satellites such as LANDSAT and SPOT have allowed inventories to be made for remote and sparsely populated areas in developing countries. See LAND USE, LAND USE COMPETITION, LAND USE PLANNING, LAND-SYSTEM MAPPING. See also CROP COMBINATION ANALYSIS, FARM ENTERPRISE COMBINATION ANALYSIS.

**land value,** *n.* the contribution of land as a factor of production in the process of producing goods and services. The *present value* or *capitalized value* of land to the producer is the sum of all future incomes which the land will yield at current values and subject to a rate of interest. The capitalized value is unlikely to be the same as the price that might be realized on sale, that is, the market value of the land. Also in the case of rural land, the possibility of urban development may raise prices beyond the agricultural value of the land. Rural land prices in developed countries are generally highest where there are concentrations of good quality land, and in the RURAL-URBAN FRINGE where LAND USE COMPETITION is high.

**lapse rate** or **vertical temperature gradient,** *n.* the rate of change of atmospheric termperature with altitude. The lapse rate records drops in temperature ranging from 0.2°-1.0°C per 100 m of ascent, depending on whether the ENVIRONMENTAL LAPSE RATE, the DRY ADIABATIC LAPSE RATE or the SATURATED ADIABATIC LAPSE RATE is being recorded. The lapse rate is usually continuous to the TROPOPAUSE unless an INVERSION LAYER occurs.

**lateral moraine,** see MORAINE.

**laterite,** *n.* **1.** a porous, reddish DURICRUST found in humid tropical regions. Laterites are formed through the concentration of iron and aluminium oxides in the soil profile. resulting from the heavy LEACHING of other minerals, such as silica, and the deposition of iron and aluminium from groundwater which had been drawn upwards by CAPILLARITY. Laterites can occur both as subsurface and surface features. When wet they are plastic like but when exposed to the drying action of sun and wind they set to a rock-hard consistency. The aluminium content of some laterites allows its commercial exploitation as BAUXITE.

**2.** the soil formed from laterite. See LATOSOL.

**laterization, ferralization** or **latosolization,** *n.* the accumulation of clays, sand, iron oxides and aluminium oxides in soils as a result of the WEATHERING of rocks in humid tropical regions. The conditions of high rainfall and temperature experienced in these regions causes intense LEACHING of the soils and, in particular, leads to the process of *desilication*. By this process all of the silica (quartz and mica) is dissolved from the soil, leaving behind a clay-rich medium impregnated with ferric iron compounds (giving a predominantly red colouration) or aluminium (giving yellow colouration). Laterization is responsible for the formation of LATOSOLS.

**latifundia,** *n.* large estates most commonly found in Central and South America on which much or all of the land is worked by means of SHARECROPPING and day labourers. Latifundia are symptomatic of the gross inequalities in land distribution found in many Latin American countries. In many cases the land owners' influence is all-pervasive and extends beyond the farming activities of their employees to include effective social control over the lives of the mass of the rural population. Compare MINIFUNDIA.

**latosol, ferrisol, sol ferralatique** or **oxisol,** *n.* a red, ochre or yellow soil with a high clay, aluminium oxide, and iron oxide content which occurs in humid tropical regions. The formation of LATERITES particularly within the B-HORIZON of a latosol is a common feature of this soil type. Latosols were formerly covered by extensive TROPICAL RAIN FOREST, the luxuriance of which suggested that latosols were fertile soils. This, however, is untrue. Latosols are inherently infertile, the luxuriant forest being supported by the rapid cycling of minerals between living organisms and the soil. High soil temperatures (often in excess of 21°C) support an active soil fauna which is responsible for the rapid decomposition of HUMUS material. Despite the abundance of clay-sized particles (see SOIL TEXTURE), a CLAY-HUMUS COMPLEX onto which soil nutrients can be attached does not usually exist.

The maintenance of agricultural crops on latosols can rarely be sustained for more than two or three years. After this time, soil nutrients become depleted and crop failure occurs. The most successful agricultural practice occurring on latosols involves PLANTATION crops. See LATERIZATION.

**latosolization,** see LATERIZATION.

**Laurasia,** *n.* the northern protocontinent believed to have resulted from the fragmentation of the PANGAEA supercontinent in the late Palaeozoic or early Mesozoic eras (see Fig. 19). Laurasia is thought to have comprised North America, Greenland and all of Eurasia (excluding India). See GONDWANALAND, CONTINENTAL DRIFT. See also PLATE TECTONICS.

**lava,** *n.* the MAGMA extruded onto the earth's surface from a VOLCANO or FISSURE. The velocity of lava flows depends upon the nature of the underlying topography, and the viscosity of the lava itself, which in turn is dependent upon its temperature and chemical composition. The temperature of recently erupted lava ranges from around 900°C to 1200°c, which decreases rapidly as the lava flows away from the source of eruption and begins to solidify. Lavas that are basic in composition (containing less than 50% silica) are able to flow further and faster than more viscous acid lavas (containing more than 65% silica). Some basic lavas can travel up to 300 m per day.

**law of limiting factors,** see LIEBIG'S LAW OF THE MINIMUM.

**law of the minimum,** see LIEBIG'S LAW OF THE MINIMUM.

**Law of the Sea,** *n.* a convention agreed in 1982 after nine years of negotiations at the Third United Nations Conference on the Law of the Sea (*UNCLOS III*) concerning jurisdiction over ocean waters and the use of ocean resources. This followed earlier conferences in 1958 and 1960. Although only a minority of the signatories have, as yet, ratified the convention, most of its provisions regarding the regulation of maritime activities within defined zones and boundaries have been widely accepted as customary international law.

The convention proposes a universal right for coastal states to exercise total sovereignty over a territorial sea extending 12 nautical miles offshore. With regard to the CONTINENTAL SHELF and the already widespread practice of coastal states unilaterally declaring more extensive zones from which foreign fishing vessels would be excluded, UNCLOS III established the principle of an Exclusive Economic Zone granting coastal states exclusive rights to the fisheries and sea-bed resources up to 200 miles (322 km) from the base line of their coasts, and rights over adjacent continental shelf up to 350 miles from the shore under specified circumstances. On the high seas not otherwise separately defined, there should be no restrictions on passage or on FISHING, although sea bed resources would be controlled by the United Nations through an appointed

International Sea-Bed Authority (ISA). States are further bound by the convention to cooperate in forming preventative rules to control POLLUTION.

The convention was hailed by many THIRD WORLD members as a triumph for the NEW INTERNATIONAL ECONOMIC ORDER in declaring sea-bed mineral resources as the common heritage of mankind. However, the USA and several other DEVELOPED COUNTRIES with vested interests in sea-bed mining voted against the convention, or abstained. Their activities, which have been declared illegal by the Preparatory Commission of the ISA, threaten to destabilize the delicate balance achieved by UNCLOS III.

**Lawyers' Ecology Group,** see ENVIRONMENTAL DEFENSE FUND

**layering,** *n.* the existance of horizontal strata within undisturbed vegetation assemblages. In mature vegetation up to 4 layers can be distinguished:
- (a) a tree layer;
- (b) a shrub or bush layer;
- (c) a field or herb layer;
- (d) a ground or moss layer.

Such layers develop in response to the decreasing amounts of light which penetrate to successively greater depths into the vegetation. The above ground layering is sometimes reflected in below ground layers; thus there is a ground root layer, field root layer, shrub root layer and tree root layer. Wherever the vegetation has been disturbed by burning, deforestation or grazing then one or more layers may disappear. Where the disturbance is considerable, only one layer of vegetation may exist.

**lazy-bed,** see MOUND CULTIVATION.

**leaching,** *n.* the process by which soluble salts are removed from the A-HORIZON of a soil and washed downwards to the B-HORIZON through the action of percolating rainwater (see ELUVIATION). The subsequent deposition of most of these substances in the B-horizon is known as ILLUVIATION. Leaching is an important process in most well-drained soils throughout humid regions and may lead to the formation of soils with chemically deficient A-horizons, and in extreme cases lead to the development of PODZOL soils. See also TRANSLOCATION.

**lead pollution,** *n.* the accumulation of lead in organic tissue, primarily in plants, animals and humans. Lead is released from a wide variety of industrial processes involving combustion and in particular from car exhausts. It has been estimated that 450 000 tonnes of lead are released into the environment each year and over half of this is produced by motor vehicles. Lead has been traditionally added to petrol (gas) in order to improve the combustion process in high compression engines.

Once deposited in the soil and on vegetation surfaces, lead moves quickly through the FOOD CHAIN. Humans may ingest lead directly through inhalation or indirectly through contaminated food. Small doses of lead produce behavioural changes while larger doses produce blindness and ultimately, death. The maximum permitted adult dose is 6 $\mu$g/kg body weight/day; for children (who absorb lead more quickly than adults) the maximum permitted dose is 1.2 $\mu$g/kg/day.

The use of UNLEADED PETROL is now becoming widespread in Western Europe and North America having been available in California for the past decade. See also HEAVY METAL.

**leaf area index (LAI),** *n.* the ratio of the leaf surface area of a plant to 1 m$^2$ of the ground surface beneath it. For example, if the LAI is 4, the plants have developed a leaf surface four times larger than the soil surface beneath, that is, the leaf surface covers an area of 4 m$^2$. Ratios of greater than 1:1 occur because plants can produce many layers of leaves, or grow in dense bunches. High LAI values are usually found in undisturbed plant COMMUNITIES while low LAI values are typical of agricultural crops and PIONEER COMMUNITIES.

**leafy crop,** *n.* any crop in a CROP ROTATION, other than a GRAIN CROP. Leafy crops are important both as FOOD CROPS, consumed either raw or in cooked form, and as FODDER CROPS. They may also be consumed after processing (for example tobacco).

The most important leafy crops include sugar beet and cane, oil-producing plants other than those grown as PERMANENT CROPS, legumes and pulses, salad plants, brassicas and other leaf vegetables, onions and root crops. According to type, the useful part of the plant may be the root, tuber or storage stem, the stem itself, the bud or bulb, the leaf stalk, the immature flower, the seed, the immature fruit or the mature fruit.

**lean-burn engine,** *n.* a type of internal combustion engine in which the fuel-to-air ratio is carefully controlled by a microprocessor in a fuel injection unit. This 'lean' mixture of fuel and air reduces petrol consumption and exhaust emissions. See also CATALYTIC CONVERTOR, AUTOMOBILE EMISSIONS.

**lee depression,** *n.* a type of NON-FRONTAL DEPRESSION comprising an area of low pressure on the leeward side of a mountain chain. The mountains force the air flowing over them to rise and, consequently, to contract. As the air descends on the leeward side of the range, it expands and establishes an area of low pressure. Examples of lee depressions are usually found in winter and include the areas south of the Alps in northern Italy, to the east of the Southern Alps in New Zealand and east of the Rockies in the USA.

**leeward,** *n.* the side or direction sheltered from the wind. The leeward side of a range of hills is often in a RAIN SHADOW while the temperature regime may be several degrees warmer due to the effects of FÖHN-like winds. Often, vegetation shows pronounced differences in luxuriance between the leeward and WINDWARD sides of a mountain.

**lethal zone,** see TOLERANCE.

**levee,** *n.* a raised natural river bank formed by the deposition of sediment during periods of flood. Levees are formed when a river regularly breaches its banks thereby spilling onto its FLOODPLAIN. As the speed of flow is checked sediments are deposited in bands parallel to the river bank (see Fig. 40). Over time the deposited material increases to form linear banks which confine the river and help prevent further flooding. Under extreme conditions the river will overflow its levees and on these occasions very serious flooding can occur with the resultant loss of life and damage to crops and property.

Levees can sometimes be used to provide routeways for roads and railways. In densely populated areas, levees are often artificially strengthened and built up to reduce the threat of flooding. There are more than 4000 km of artificial levees along the Mississippi in the USA, some of which are up to 10 m in height.

**ley farming,** *n.* a method of ARABLE FARMING in which artificial PASTURES of sown grasses and legumes, occupying the field for a year or more, are grown in rotation with annually cultivated crops.

**lichen,** *n.* any plant of the division Lichens which is formed by the symbiotic association of a fungus and an alga and occurs as crusty patches or bushy growths on tree trunks, stone walls, roofs, etc. Lichens form the first stage of PLANT SUCCESSION. They have no true roots and gain their sustenance directly from the atmosphere and rainwater. Lichens are highly sensitive to atmospheric pollution with different species disappearing when pollution reaches specific levels. As such, lichens can be used as *indicator species* to plot the pattern of air pollution distribution. See LICHEN DESERT,

**lichen desert,** *n.* any geographical area from which LICHENS are absent as a result of ATMOSPHERIC POLLUTION. See ECOLOGICAL EVALUATION.

**Liebig's law of the minimum,** *n.* a principle established in 1840 which states that the rate of growth of a plant is dependent upon the minimum amount of essential nutrients that is available to it. Liebig's law is now usually combined with Blackman's *law of limiting factors* which states that the rate of PHOTOSYNTHESIS is governed by the level of the environmental factor that is operating at a limiting intensity. Named after the German biochemist, Baron Justus von Liebig (1803-73). See also SHELFORD'S LAW OF TOLERANCE, FACTOR INTERACTION.

**life table,** see SURVIVORSHIP CURVE.

**life zone,** *n.* **1.** on a world scale, the major bioclimatic regions characterized by distinctive plant types, for example the desert regions or TUNDRA.

**2.** any local variation in the environmental conditions and which result in regional or small-scale changes in distribution of plants and animals, for example the altitudinal zonation of communities in mountainous regions. Compare DEATH ZONE.

**lightning,** *n.* a visible flash of electrical discharge within the clouds of a THUNDERSTORM or between the clouds and the ground. The exact processes in the generation of electrical charges in thunderstorms are unclear but it is thought that lightning operates to neutralize differing electrical charges.

On occasion, lightning may start fires, melt metal, cause trees to explode, damage electrical and telephone installations, cause disruption to communication systems as well as cause injury and death to people and animals through direct strikes. In the USA between 1953 and 1963, 160 persons were killed by lightning while in 1963 alone, it caused $100 million damage. However, these are comparatively rare incidents and must be put into perspective given the vast numbers of thunderstorms occurring throughout the world each year. At any particular moment, there are 1800 thunderstorms in progress across the world. See also THUNDER.

**light rapid transit system (LRT),** *n.* an urban railway, UNDERGROUND or tram system built to lighter construction standards than conventional rail systems. As such they are cheaper to build and require less energy to operate. Normally, they use electric propulsion thus making them pollution-free and relatively quiet. LRT systems are most common in North America although some examples can now be seen in Europe where they are often integral to the REDEVELOPMENT of INNER CITIES, for example the Tyneside Metro and the Docklands railway in east London.

**lignite,** *n.* a low-grade, brown-black COAL containing around 70% carbon and 20% oxygen. Of all the types of coal, lignite gives off the most smoke and least heat. Lignite is most commonly used for the generation of electricity. Compare BITUMINOUS COAL, ANTHRACITE.

**limestone,** *n.* a SEDIMENTARY ROCK largely composed of calcium carbonate. Limestones are formed from the calcareous remains of animals such as molluscs and corals, the precipitation of calcium carbonate from seawater, or the eroded fragments of pre-existing limestones and can be classified according to their chemical composition and texture. Limestone is often used as a building material and in the manufacture of cement and fertilizer.

**limiting factor,** *n.* any environmental factor (such as light, temperature or moisture) which defines an organism's minimum and maximum limits of TOLERANCE and restricts its distribution and/or activity.

**limits to growth,** *n.* the limits imposed on industrialization and population growth by the finite nature of the earth's resources. Once these limits are reached, it is argued, rapid decline will set in, leading to the inability of the BIOSPHERE to support its population at existing levels of industrialization and social welfare.

The term was introduced in 1972 in the CLUB OF ROME's report, *The Limits to Growth*, which employed computer models to project population forecasts, estimates of resource depletion, agricultural output, capital investment and levels of POLLUTION. This predicted that if present trends continued the limits to economic growth could be reached within 100 years, resulting in over-development and collapse. It resurrected the pessimistic spectre of the MALTHUSIAN MODEL and generated considerable critical debate. Although received favourably by many ecologists, the report was widely condemned by technologists, economists and other social scientists.

**line transect,** see TRANSECT.

**line fishing,** *n.* a method of fishing whereby a single line is laid on the sea bed with up to 100 shorter, baited lines (or *snoods*) attached to it. Line fishing is used in sites where the sea bed is too rough to permit TRAWLING. Line fishing should not be confused with rod-and-line fishing or angling. See FISHING. Compare SEINE NETTING, TRAWLING.

**Linnean classification,** *n.* the modern zoological classification devised by the Swedish naturalist Carl von Linné or Linnaeus (1707-1778). The Linnean classification places every species into a category based upon its evolutionary position in relation to other species. The largest *taxonomic* (classificatory) group is the phylum followed by the class, order, family, genus and species.

A full example of a Linnean classification for the polar bear would be:

| | |
|---|---|
| Phylum: | Chordata |
| Class: | Mammalia |
| Order: | Carnivora |
| Family: | Ursidae |
| Genus and species: | *Thalarctos maritimus* |

**lithic haplumbrept,** see RANKER.

**lithification** or **diagenesis,** *n.* the COMPACTION and CEMENTATION of unconsolidated sediment to form SEDIMENTARY ROCK.

**lithosequence,** *n.* a sequence of related soils which differ in certain characteristics due to variations in the PARENT MATERIAL from which they are formed. See also CHRONOSEQUENCE, CLIMOSEQUENCE, TOPOSEQUENCE.

**lithosphere,** *n.* the rigid outer layers of the EARTH'S CRUST and MANTLE. The lithosphere consists of the continental SIAL, the SIMA, and the upper part of the MANTLE and is bounded by the ASTHENOSPHERE (see Fig. 28).

**lithospheric plate,** *n.* a large independent segment of the LITHOSPHERE that is able to move on the partially molten ASTHENOSPHERE. Seven major plates and twelve or more smaller plates have been identified (see Fig. 47) and are composed of continental or oceanic crust, or a combination of both SIMA and SIAL. New sections of plates are formed as a result of SEA-FLOOR SPREADING at MID-OCEANIC RIDGES whilst the margins of plates are destroyed along ZONES OF SUBDUCTION. The movement of these plates is believed to have contributed to CONTINENTAL DRIFT, a theory supported by PLATE TECTONICS

**Little Ice Age,** see CLIMATIC CHANGE.

**littoral,** *n.* **1.** the area between the low and high SPRING TIDE levels (see BENTHIC ZONE).

**2.** a coastal area including the SHORELINE.

**load,** *n.* the material transported by rivers, glaciers and wind. See FLUVIAL TRANSPORTATION, GLACIAL TRANSPORTATION, WIND TRANSPORTATION.

**loam,** *n.* an aerated, well-drained and often fertile soil with a medium texture composed of roughly equal proportions of CLAY, SILT and SAND. Loams usually possess only the favourable qualities of their sand and clay components and form a productive, easily cultivated agricultural soil. Various sub-types of loams are classified according to the specific levels of the constituents. See LOESS.

**local plan,** *n.* a statutory document produced by local planning authorities in the UK, consisting of a statement setting out the land use ZONING policy for a sub-area within a local authority territory, and the programming of the sequence of sites to be developed over a given timescale. A large-scale map accompanies the policy statement, and shows the precise boundaries and expected zoning sites which are scheduled for development or rehabilitation within the plan period.

Public opinion is consulted before a local plan is finalized and, once it is officially adopted by the local authority, it becomes the reference document against which DEVELOPMENT CONTROL decisions are made in respect of applications to develop any site within the plan's area. Local plans may be amended to meet changed local circumstances and are, in any case, subject periodically to statutory overall review.

In some instances, the term is used to refer to plans without statutory backing which may or may not have local authority approval. Such 'bottom drawer' plans are sometimes used to guide development in the absence of a statutory local plan.

**lock,** *n.* a device for raising or lowering vessels between different levels of waterway by means of altering the water level in an enclosed basin containing the boat. Movable lock gates are closed behind the entering vessel on, for example, the lower side of the river or canal and water is released into the basin until it rises to the same level as that above the gate on the higher side. The upper gate is then opened and the boat moves forward on this higher level of the waterway. Locks may vary in size from small, one-man operated structures to massive hydraulic locks such as those found on the Welland Canal joining Lakes Erie and Ontario in North America.

**lode,** see VEIN.

**loess,** *n.* a yellow-brown, fine-grained, wind- deposited LOAM found in parts of central Europe, Asia and North America. Loess deposits are composed chiefly of angular particles of quartz, feldspar and calcite in a clay MATRIX, and may be extensive and of considerable depth (up to 50 m). They are largely derived from the wind erosion of fine surface material on glacial OUTWASH PLAINS, DELTAS, river FLOODPLAINS and deserts. Soil derived from loess forms some of the world's most productive farmland such as the southern plains of the USSR, the Argentinian PAMPAS, and parts of the American Midwest. However, roads and houses built on loess deposits often suffer from subsidence.

**logistic curve,** see S-SHAPED CURVE.

**Lomé Agreements,** *n.* a series of agreements initiated in 1975 at Lomé, Togo, and renewed in 1979 and 1984, that give a group of 66 African, Caribbean, and Pacific (ACP) countries preferential access to the markets of the EC. The Lomé Agreements provide ACP countries with better terms for their exports to EC countries than can be expected under the generalized system of preferences operating between other developing countries and the industrialized nations. All ACP industrial exports and most agricultural exports to the EC enter free of duty, and are less constrained by non-tarriff barriers and trade regulations than non-ACP exports.

The Agreements have also been responsible for channelling loans and grants to finance development projects, especially those oriented towards product diversification, and for establishing *STABEX*, a fund designed to stabilize export earnings from certain COMMODITIES by compensating ACP countries when adverse climatic or market conditions result in a loss of expected revenue. In practice, the stringent qualifying conditions, and persistent underfunding have limited the effectiveness of STABEX.

**longitudinal dune,** see DUNE

**Lotka-Volterra equations,** *n.* mathematical MODELS describing the competition between organisms occupying the same area for a finite supply of resources; they describe the predator-prey situation between organisms and the non-predatory situation involving competition between organisms for food or space. Although derived from separate research, the two equations are now used together and are much used in the mathematical prediction of growth rates of simple organisms such as bacteria, insects and annual plants.

**low,** see DEPRESSION.

**lower atmosphere,** *n.* the section of the ATMOSPHERE which lies between the ground and the STRATOPAUSE and incorporates the troposphere and stratosphere (see Fig. 9).

**LRT,** see LIGHT RAPID TRANSIT SYSTEM.

**LUC,** see LAND USE CLASSIFICATION.

# M

**MAB,** see MAN AND THE BIOSPHERE PROGRAMME.

**macchia,** see MAQUIS.

**macroclimate,** *n.* the general climatic conditions which prevail over large geographic areas, for example, the Mediterranean climate and Monsoon climate. Compare MICROCLIMATE, MESOCLIMATE.

**macro-environment,** see MICRO-ENVIRONMENT.

**macronutrient,** see SOIL NUTRIENT.

**magma,** *n.* the high-temperature mobile fluid comprising molten rock, water and other VOLATILES found within the earth's crust Magma is formed under great pressure by the localized melting of solid rock deep within the MANTLE; the heat is believed to emanate from decaying radioactive elements contained within the rock itself.

Magma is the source of IGNEOUS ROCK. Magma that consolidates within the crust solidifies to form INTRUSIVE ROCK, whilst that emerging onto the earth's surface cools to form EXTRUSIVE ROCK. Magma gathers in *magmatic chambers* within the crust which act as LAVA reservoirs for volcanoes.

**magmatic chamber,** see MAGMA.

**magmatic water,** see JUVENILE WATER.

**magnetosphere,** *n.* the comet-shaped region surrounding the earth at an altitude of between 2000 km and 65 000 km on the side of the sun and in which the behaviour of charged particles is controlled by the earth's magnetic field.

**Magnox reactor,** *n.* a type of gas-cooled NUCLEAR REACTOR in which the MODERATOR is graphite and the core coolant gas is carbon dioxide; they operate at the relatively low temperature of 300-330°C. The enclosure of the uranium fuel in a magnesium alloy called magnox gives the reactor type its name. Magnox reactors were first built in the UK (1956) and a total of nine commercial electricity stations were constructed around this type of reactor (see Fig. 44); a further seven were built in France. They are now reaching the end of their commercial life and will be progressively decommissioned from about 1990 onwards. Magnox reactors are inefficient in terms of their useful generating capacity per tonne of fuel used but have an excellent safety record. See ADVANCED GAS-COOLED REACTOR, WATER-COOLED REACTOR, DECOMMISSIONING.

**MAI,** see MEAN ANNUAL INCREMENT.

**malnutrition,** *n.* a compromised state of human health which results from an inadequate intake of food or an inability to assimilate and metabolize nutrients properly. The most common form of malnutrition is associated with insufficient intake of food or nutrients essential for a balanced development of the body.

The World Health Organization has estimated that between 50 and 66% of the world population may be malnourished. In the DEVELOPING COUNTRIES, malnutrition is a chronic problem and has greatest impact upon children, pregnant women and adolescents. Common symptoms of prolonged malnutrition include complete growth arrest, a marked wasting of body tissues and profound apathy. In addition, the body's immune system is destroyed, leading to a range of secondary disorders.

**Malthusian model,** *n.* a representation of the theories of Thomas Malthus on the interaction of population growth and food supply, as outlined in 1798 in *An Essay on the Principle of Population.* Malthus's basic proposition was that while population numbers tend to increase exponentially (that is, 1, 2, 4, 8, 16, etc.) the means of feeding them grows only arithmetically (1, 2, 3, 4, 5, etc.); unchecked increases in population thus tend to outrun increases in the food supply. People would be condemned to live in poverty and hunger unless their numbers could be controlled, either as a result of wars, disease or famine, or because of measures of 'moral restraint' to reduce the BIRTH RATE.

The Malthusian model has been strongly attacked by Marxists who argue that the reasoning behind it ignores the fact that it is only the poor that go hungry, and claim that poverty is due to the poor distribution of resources, not the physical limits on production. Furthermore, the notion of food supply increasing only arithmetically is not supported by the evidence of the past two centuries. The Malthusian model has been revived in a modern form with the extension of the ecological concept of CARRYING CAPACITY to human populations, and by the debate over the existence of finite LIMITS TO GROWTH.

**mammal,** *n.* any mammal of the class Mammalia possessing a characteristic rigid back bone (vertebrate) and a complex vascular circulation system linked to lungs for breathing air. Other distinctive features include extensive body hair, mammary glands for suckling young, and highly developed sensory perceptors (eyes, ears, olfactory organ) linked to an elaborate nervous system and controlled by an enlarged brain. Members of the class often display complex social behaviour, a fact well illustrated by humans, the most successful of all mammals.

**Man and the Biosphere Programme (MAB),** *n.* a UNESCO project of the 1970s designed to promote long-term conservation of all the most typical world ECOSYSTEMS. Unlike other attempts to conserve ecosystems, MAB emphasized the need for conservation based on sound scientific grounds rather than for purely subjective reasons. The effect of MAB was to greatly increase the scientific understanding of individual ecosystems and the inter-relationships between different ecosystems. This was particularly true for tropical ecosystems, both terrestrial and aquatic, both of which had suffered from broad generalizations until the detailed research associated with MAB projects. Many totally new species were discovered, rain forest climates were found to contain dry spells, while of greatest importance for humans, many forest plant species were shown to possess considerable medicinal potential. Indeed, by the 1980s some 25% of western medicines contain at least one extract from rain forest species.

**man-day,** *n.* a measure of the labour requirements of an enterprise, normally in agriculture. In 1963 the UK Ministry of Agriculture, Fisheries and Food adopted *standard man-days* (smd) as a means of classifying FARM SIZE: 1 smd is equivalent to 8 hours of adult male manual work under average conditions. Farm sizes can be measured in aggregate smd on the basis of their annual manual labour requirements needed to produce crops and livestock, and carry out maintenance and other necessary tasks. Each item of farm production can be converted into smd units. At the time of its introduction, a dairy cow required 12 smd, while the standard for barley was 2 smd per acre.

In tropical agriculture, the notion of an 8-hour working day is less practical, where between 5 and 7 hours work per day is more usual. Labour inputs may thus be quoted in terms of man-hours per crop. Alternatively, labour units may be expressed in man-years, equivalent to a full-time worker working a prescribed number of man-hours annually. The EC employs the measure of *annual work units* (awu) where one awu is taken to be 2200 hours per annum, or the minimum number of hours laid down in national collective agreements if different.

**man-environment relationship,** *n.* the inter-relationship and interaction between humans and the environment in which they live. The man-environment, or man-land relationship has been seen by some in purely deterministic terms, holding that human settlement and economic activity are controlled by environmental conditions, to the extent that the role of human DECISION-MAKING can be largely disregarded (see ENVIRONMENTAL DETERMINISM). To others, the relationship demonstrates the true worth of the role of humans in deciding between alternative responses to environmental conditions. A danger in too narrow a search for the causal links between the environment and human actions is the failure to recognize cultural and socio-economic factors that may, in practice, be at least if not more significant influences. The notion of man-environment relationships and the response of, for example, farmers to variations in land quality is not at odds with economic principles of LAND USE COMPETITION. Making allowance for nature, that is, being sensitive to variations in land quality and climate, is entirely rational in seeking to maximize returns or to satisfy some other set of goals. The decisions of income optimizers depend upon production functions, the relationship between the quantity of output of a good and quantities of inputs needed to make it, and the costs and prices of inputs and outputs. Where regional variations in production functions due to environmental factors are greater than variations in costs and prices, land use is likely to bear a strong relationship to land quality.

The relationship between humans and the environment is most strikingly borne out where overpopulation and misuse of resouces, such as the overgrazing of land beyond its CARRYING CAPACITY, leads to a breakdown of the ECOSYSTEM, and problems of SOIL EROSION and DESERTIFICATION. While this may represent a failure to observe the physical limitations to the SUSTAINED YIELD, it can equally be stated in economic terms as the failure to maximize the land's use over a long period in the pursuit of short-term gains.

**mangrove,** *n.* a halophyllous (salt-adapted) evergreen forest of the intertidal zone of the tropical and subtropical latitudes. Typically mangroves are found along shore lines sheltered from strong currents or wave action and where alluvium can accumulate. Mangrove forests show appreciable species variation between their South American, African, South East Asian and Australasian distributions.

Mangrove species have adapted to anaerobic conditions by producing frequent stilt roots (*rhizophores*) which project above the mud thereby allowing the breathing pores (*pneumatophores*) to absorb and transfer oxygen into the roots. Other adaptations include heavily cuticulized, fleshy leaves and seeds which can float in water and begin germination before they become embedded in the mud.

Mangrove communities are amongst the most productive ecosystems on this planet with an average net PRIMARY PRODUCTIVITY (NPP) of 2000 g/m²/year and a maximum NPP of 4500 g/m² year. Despite their prolific biologic production they are hazardous areas for humans mainly due to the large number of disease-carrying insects which breed in the hot, damp environment.

**mantle,** *n.* the section of the earth's interior lying between the EARTH'S CRUST and the CORE. The upper and lower boundaries of the mantle are marked by the MOHOROVIČIĆ DISCONTINUITY and the GUTENBERG DISCONTINUITY respectively. The mantle is largely composed of silicate-rich ULTRABASIC ROCKS which increase in density with depth. The upper part of the mantle is solid to depths of 100 km and along with the crust forms the LITHOSPHERE. Between 100 and 200 km below the surface, the mantle is partially molten and is called the ASTHENOSPHERE.

**mantle rock,** see REGOLITH.

**manure,** *n.* any plant or animal residue containing animal excreta. Farmyard manure most often contains plant residue such as straw used as litter to absorb urine, and is stored in heaps to allow degradation. Once its structure has deteriorated it can be spread on the land to improve soil structure and act as a FERTILIZER in the form of its main constituents, compounds of nitrogen, potassium and phosphorus. Manure requires greater care in preparation and application than most artificial fertilizers, often needs fertilizer supplements to release the full value of the organic substances contained in it, and is needed in greater quantities to achieve the same level of field input as chemical fertilizers. However, manure remains a key feature of MIXED FARMING.

The term manure has been applied to any substance placed on the land to improve fertility, including seaweed, fish waste, bones, ashes mixed with lime, sooty thatch from houses, and NIGHT SOIL. A *green manure* is a crop that is ploughed under while still green and growing to improve soil structure.

**maquis, mattoral** or **macchia,** *n.* a mixed, SCLEROPHYLLOUS vegetation of the Mediterranean basin in which pines, oak, wild olive, carob and lentisk tree grow to a height of about 5 m. Beneath the tree layer is found a wide variety of aromatic herbaceous plants, the luxuriance and diversity of which is dependent upon the annual rainfall and the degree of human interference brought about by grazing pressure and the frequency of burning. Vegetation is interspersed with extensive areas of bare ground. The maquis is drab and desolate in the long dry summer season although in the cooler, moister spring time it is transformed into a short lasting blaze of colour when broom, heathers, gorse and asphodel along with members of the Labiatae and Thymus families complete their life cycles. See GARRIGUE, CHAPARRAL.

**marble,** *n.* **1.** a fine- to coarse-grained METAMORPHIC ROCK formed by contact and regional METAMORPHISM of limestone. Marble is often used as a building and ornamental stone.

**2.** any ornamental stone which can be polished.

**marginal land,** *n.* land which yields a very low economic return for a given use, or on which there is a high probability of crop failure. Common usage of the term does not always accord with the strict economic definition. In physical terms, land may be marginal for any particular type of farming or indeed all agricultural activities on account of biological and environmental limitations beyond the normal tolerance of a crop or livestock. Thus, uplands in temperate zones are frequently described as marginal land because physical deficiencies of temperature, poor drainage, thin soils and difficult relief preclude all or most forms of cultivation and permit only extensive and low productivity grazing of livestock. In this sense marginal land is equated with poor quality land. Economically, however, land is only marginal in relation to some form of use. It is marginal if its use fluctuates, if agricultural activity is abandoned altogether because of its poverty, and if it goes in and out of production in response to changing economic circumstances, including the availability or otherwise of grants and SUBSIDIES.

Every form of agricultural activity has its OPTIMUM GROWTH conditions and optimum areas of production. Away from these areas, productivity will decline until at the margins of production the revenue from that activity will equal the cost of production, that is, where the ECONOMIC RENT is zero. The crop may continue physically to tolerate conditions beyond this point until extreme limits are reached but can only be produced at a loss. The crop is likely to come out of production, however, before the margins are reached as alternative land uses become more profitable. It is this area that may be defined as marginal land for the crop or farming activity in question. There have been many attempts to extend the sown acreage into areas of low and unreliable rainfall formerly given over to extensive livestock rearing, for example, in Australia, south-west USA, and during the VIRGIN LANDS campaign in the USSR. This land is physically marginal for cereal cultivation, it being close to the biological and environmental limits of crop tolerance, and economically in that returns are low, almost to the point where production is unprofitable, leading to abandonment and alternative agricultural activities.

**mariculture,** see FISH FARMING.

**marine deposition,** *n.* the dropping of rock fragments and other debris carried by waves and currents as their velocity reduces. Localized deposition can lead to the silting up of harbours and navigation channels which then have to be artifically dredged to allow the free movement of shipping. Various features are associated with marine deposition such as BEACHES, SPITS and OFFSHORE BARS.

**marine erosion,** *n.* the progressive removal of material from a COAST by the sea. Marine erosion is caused by four processes:

(a) *hydraulic action*, which is the physical impact of waves and currents on the rocks and cliffs that form a coastline;

(b) *corrasion*, which is the grinding of rock fragments carried in the waves and currents against coastlines;

(c) *attrition*, which is the wearing down of transported rock fragments through impact and friction;

(d) *corrosion*, which is the chemical action of sea water on rocks such as limestone.

The rate of erosion is dependent upon the intensity and FETCH of the waves and currents and the nature of the material that comprises the coastline. On coasts of unconsolidated materials, such as glacial till (see GLACIAL DEPOSITION), erosion can be a major problem. There are numerous examples of marine erosion causing the loss of houses, roads and farmland into the sea. Some of the most common features of coastal scenery, such as cliffs and WAVE CUT PLATFORMS, are the result of marine erosion.

**marine pollution,** *n.* any noxious and toxic material released intentionally or by accident into seas, oceans and estuaries. Marine pollutants can originate from a number of sources and include SEWAGE and industrial EFFLUENT which may be piped directly into the sea or transported further out to sea by ship and dumped. Domestic and industrial refuse, and until recently NUCLEAR WASTE, is also dumped at sea by many countries

Oil is another major marine pollutant which may result from

tanker collisions, the accidental release of oil from terminals and the illegal cleaning of ships' bilges at sea. Between 2 and 5 million tonnes of crude and processed oil are spilled into the seas each year although about two-thirds of the oil of this comes not from ships, but from land-based sources, in particular from the illegal disposal of waste industrial and vehicle oil directly into rivers and drains which ultimately flow into the sea (see OIL SPILL).

Public attention was directed to marine pollution by the wrecks of the oil tankers *Torrey Canyon* (1967) and *Amoco Cadiz* (1978). In both of these incidents severe pollution of coastlines resulted and massive losses of sea birds were recorded. Despite the spectacular nature of these events, the long-term impact on coastal ecosystems appears to have been over-emphasized. However, in 1989 following an accident involving the supertanker *Exxon Valdez*, 10 million gallons of oil were released into the Gulf of Alaska; the ultimate effects of this pollution on the FRAGILE ECOSYSTEMS of this region will become known in due course.

**market gardening,** *n.* a form of HORTICULTURE used for the intensive production of crops, especially fruit, vegetables and flowers, in which the produce is generally sold for direct human consumption. Crops may be grown in fields to provide in-season produce or advanced by means of GLASSHOUSE CULTIVATION. Until the dawn of the railway era, most market gardening was located within a few kilometres of urban areas due to the perishable and sometimes fragile nature of the produce; today, however, produce may travel hundreds or even thousands of kilometres to market, and across international boundaries, for example, from Mediterranean areas to northern Europe, and from Mexico and the Caribbean to the USA.

In North America, market gardening is known as *truck farming*, originally so called because of the dependence of farmers on trucks or pick-ups for transport to market. Truck farming, however, is characterized by a generally larger scale of operation, in locations further from the market, and above all by the production of a more limited range of produce, perhaps a single crop best suited to a locality.

**marketing board,** *n.* a public body established by a government with legal powers of compulsion over the producers and handlers of an agricultural product, and sanctioned to perform certain marketing operations in the interests of the producers of the commodity in question.

In practice, the functions of marketing boards have varied

considerably. In the UK, for example, producer-elected boards were established for hops, milk, pigs, bacon, and potatoes in the 1930s, each with powers to regulate all sales of the product, to negotiate prices with distributers and to specify re-sale terms. Those for hops and potatoes also attempted to control production by means of acreage quotas, while the Milk Marketing Boards have played a prominent role in dairy herd improvement, mainly by encouraging SELECTIVE BREEDING through the provision of artificial insemination services. In the 1950s, the Boards' primary objective shifted towards producer protection through the administration of price support mechanisms.

Marketing boards in THIRD WORLD countries generally have more direct government participation in management, particularly where the board is the sole buyer and seller of a COMMODITY destined for export. Developed originally in the former colonies of British West Africa, these monopoly export marketing boards have become an effective means for independent African governments to generate revenue by purchasing domestic produce at prices fixed substantially below the world market level and then re-selling on the open market. Other types of marketing board include those that operate alongside pre-existing marketing enterprises with a view to stabilizing supplies and prices, and those that implement uniform and stable prices throughout the country through a state-controlled monopoly of trading and processing activities. Marketing boards may also be restricted to essentially advisory and promotional objectives or to regulating quality standards and packing procedures for export produce.

**marsh,** *n.* an area of spongy waterlogged ground with large numbers of surface water pools. Marshes are found in most climatic regimes (apart from deserts) and usually result from:

(a) an impermeable underlying bedrock;

(b) surface deposits of glacial boulder clay;

(c) a basin-like topography from which natural drainage is poor;

(d) very heavy rainfall in conjunction with a correspondingly low evaporation rate;

(e) low-lying land, particularly at estuarine sites at or below sea level.

Under any of these situations the natural drainage of the soil may become permanently or seasonally impeded. In many temperate and tropical countries, marshes have been drained as they provided a breeding ground for disease-carrying species of insects, for example, the malaria-carrying Anopheles mosquito. If drained, the former marshy areas can often be converted to good quality agricultural land as the resulting soils are highly fertile.

**marsh gas,** see METHANE.

**mass movement** or **mass wasting,** *n.* the downslope movement of rock debris and soil under the influence of gravity. Mass movements range in scale from minor shifts to major LANDSLIDES which can cause considerable loss of life and damage to property. Some mass movements are slow and continuous, such as CREEP and SOLIFLUCTION, whilst others are rapid but occasional such as AVALANCHES, SLUMPS and MUDFLOWS.

Various factors assist gravity in the initiation and continuation of mass movement:

(a) the movement of water through rock debris and soil may trigger several types of mass movements either by increasing the weight of the material or decreasing its resistance and cohesion.

(b) PHYSICAL WEATHERING, particularly freeze-thaw action, causes the downslope displacement of material.

(c) root growth and animal movement on and under slopes can further contribute to downslope movement.

(d) river undercutting, quarrying, tunnelling or similar excavations may remove the support for overlying material and lead to certain types of mass movement.

(e) vibrations from earthquakes, heavy traffic and blasting may initiate some mass movement.

**mass wasting,** see MASS MOVEMENT.

**master factor,** see ANTHROPOGENIC FACTOR.

**matrix,** *n.* **1.** the fine-grained material of a SEDIMENTARY ROCK in which larger or coarser grains are lodged. See also GROUNDMASS.

**2.** the background material in which a fossil is embedded.

**mattoral,** see MAQUIS.

**mature soil,** *n.* any SOIL which has remained undisturbed for a sufficient length of time to permit the formation of distinct SOIL HORIZONS. HUMUS is usually abundant both as a distinct surface horizon and in the form of a CLAY-HUMUS COMPLEX. WEATHERING processes have allowed the formation of an abundance of CLAY MINERALS. Such soils are usually formed over many thousands of years during which time natural vegetation (most usually forest) established a strong cycle of nutrient movements between the soil and vegetation. Nowadays, these soils are extensivley used for ARABLE FARMING. Compare IMMATURE SOIL.

**MCI,** see MULTIPLE CROPPING.

**M discontinuity,** see MOHOROVIČIĆ DISCONTINUITY.

**meadow,** *n.* **1.** a low-lying and well-watered field suitable for grazing but too water-logged for ARABLE FARMING. Although often flooded in winter, meadows yield rich PASTURE in summer. *Watermeadows*, created deliberately in the Middle Ages by channelling water across the land to improve pasture, could be flooded and drained as needed.

In the USA the term *old field* is applied to such sites and a number have been given CONSERVATION status. In the UK, meadows are rare and where they still exist may have been designated SITES OF SPECIAL SCIENTIFIC INTEREST.

**2.** an upland pasture, grazed seasonally for short periods (see TRANSHUMANCE).

**meadow soil,** *n.* the term used in the Soviet soil classification system for a GLEY SOIL.

**mean annual increment (MAI),** *n.* the average volume increase of timber within the trunk of a tree and measured over a time span extending from the date of planting to the present. When plotted on a graph against the CURRENT ANNUAL INCREMENT, the MAI value attains its maximum value when the MAI curve crosses the CAI curve (see Fig. 21). The crossing point specifies the maximum rate of volume increment which a specific species can attain at a given location.

**meander,** *n.* a naturally occurring bend which a river may develop along its course. Over time, the river accentuates a meander as the bank on the concave side of the bend is eroded by river flow while ALLUVIUM is deposited by the slacker flow on the convex side. Meanders not only grow laterally but also move downstream in a migratory movement known as *down-valley sweep*. In a valley containing a meandering river, the area of land between the outermost limits of the meanders are described as the *meander belt* (see Fig. 40).

The precise cause of meanders remains controversial, although meanders resulting from obstacles in rivers or chance occurrences have been discredited. It is generally considered that the existence of POOLS AND RIFFLES in the river bed are the precursors of meander formation. The main characteristics of a meander, such as its *wave length*, *meander amplitude* and *radius of curvature* are related to channel size which is itself a function of the river *discharge* (the average volume of water flowing in the river). Extensive studies on many different rivers have shown that meander wave length is usually 7-11 times the width of the river channel.

The growth and movement of meanders may cause the loss of

agricultural land and result in the undermining of foundations, roads, railways and engineering structures such as bridges. The migration of meanders can be detected from the examination of air photographs and from a comparison of maps constructed at various time intervals.

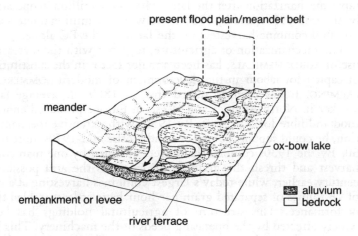

Fig. 40. **Meander.** A river meander belt with characteristic features.

**meander belt,** see MEANDER.

**mechanical weathering,** see PHYSICAL WEATHERING.

**mechanization (of agriculture),** *n.* the process of applying mechanical implements to assist in farm production. These include materials-handling and processing equipment on the farmstead, machinery to assist in livestock feeding and milking, and implements, whether powered or pulled by tractors or self-propelled, to further all stages of crop production from TILLAGE to harvesting.

Although more efficient and labour-saving farm equipment was developed in Europe in the 18th and early 19th centuries, significant advances awaited the evolution of standardized mass-produced iron and steel implements after the mid-19th century, together with the development of new energy bases. In the latter half of the 19th century steam power was harnessed to activities such as threshing that could be carried out at a central point on the farm but it had limited application to the actual process of production which was not much changed until the advent of the internal combustion engine and the spread of tractors and combine harvesters. There were only some 10 000 tractors on American farms by 1910, rising to 250 000 by 1921, 1.6 million by 1940, 3.4 million by 1950, and reaching a peak of 4.9 million in 1965. Numbers have since fallen, but the average power output and work capacity has increased steadily. In Western Europe there were fewer than 200 000 tractors before the Second World War, but rapid mechanization after the late 1940s saw 2 million in operation by the mid-1950s, and by 1983 there were 5.4 million tractors and 461 000 combine harvesters on the farms of the EC alone.

The mechanization of agriculture, together with the increasing use of AGROCHEMICALS, has been a major factor in the substitution of capital for labour in the development of modern COMMERCIAL FARMING. It has been estimated that in 1860 the average farm worker in North America and Western Europe produced enough food and fibre for 5 persons. By 1950, with extensive mechanization, he could support 25 persons and by the mid-1970s more than 50. By the 1930s combine harvesters operated by one man could harvest and thresh the grain from 75 times the area possible a century earlier, while today's largest combines harvesting at a rate of 18 tonnes of separated grain per hour have at least trebled that performance. The structure of agricultural holdings has been widely affected by the operating needs of the machinery. This has contributed to land CONSOLIDATION and the removal of hedgerows, and has been an economic force behind the rationale for farm enlargement that has seen average farm sizes and the concentration of farm production increase throughout the developed world in the post-war period. Although mechanization is generally associated with increased agricultural productivity and output, it can, in opening up MARGINAL LANDS lead to over-cultivation, the depletion of soil fertility, declining crop yields and SOIL EROSION, as occurred in parts of North America in the early 20th century.

**medial moraine,** see MORAINE.

**medicine (from plants),** *n.* any botanical extract used for the treatment of human ailments. Reliance upon plant extracts for use as medicines has a long history and although largely abandoned in DEVELOPED COUNTRIES, the practice is still followed by the native peoples of DEVELOPING COUNTRIES. For example, in South East Asia, some 6500 plants are used for treatment of diseases such as stomach ulcers, malaria and syphilis and for the preparation of astringents, disinfectants, purgatives, emetics and sedatives.

In recent years, however, pharmacologists from the developed world have re-examined the scientific value of many traditional medicinal substances while searching for new organic-based cures. Among the most important materials to be gained from plants are the complex biocompounds known as alkaloids. They are found in greatest abundance in the TROPICAL RAIN FORESTS and include such substances as cocaine, quinine, caffeine and nicotine. A plant's ability to produce alkaloids is genetically controlled and it is thought that with the application of modern GENETIC ENGINEERING techniques, plants may be specifically bred to yield useful alkaloids. Perhaps the most famous alkaloid medicines are vincristine and vinblastine, obtained from the Vinca species (rosy periwinkle) of the rain forest. These drugs have provided spectacular control over Hodgkin's disease and malignant lymphomas, breast cancer, and cervical and testicular cancers.

At present superficial studies have been made of only 10% of tropical plants as potential sources of medicines while intensive studies have been made on only 1%. However, on present evidence it is probable that the tropical rain forests represent a far more important resource for medicines than as sources of HARDWOOD timber. For example, it has been estimated that 25% of medicines in the western world contain at least one extract from tropical rain forests and the commercial value of these substances could be as much as $20 billion per annum.

**Mediterranean,** *adj.* (of an ecosystem) characterized by long, hot and dry summers with mild, damp winters. The type area is located in the Mediterranean basin but similar bioclimatic regions can be found in coastal areas of California, Chile, South Africa and the south and south-eastern parts of Australia. Intensive human activity has resulted in substantial change to the fauna, flora and soils of these regions. See FYNBOS, GARRIGUE, MAQUIS.

**Mediterranean agriculture,** *n.* a particular mix of ARABLE FARMING and ANIMAL HUSBANDRY that evolved in, and is associated with, the Mediterranean basin and areas with similar climatic characteristics elsewhere in the world. Its distinctive features ensures its place as the only geographically specific type of AGRICULTURAL REGION in traditional AGRICULTURAL TYPOLOGIES.

The climate, vegetation and landforms found in the Mediterranean basin have enabled an unusual combination of farming practices to emerge. The long dry summer limits crop cultivation unless under IRRIGATION, but the markedly wetter winter supports the growing of vegetables and cereals for harvest in early summer. Consequently, the main summer crops are tree-grown such as olives, figs and grapes (see VITICULTURE). A characteristic feature of land use is that of *interculture*, a form of MIXED CROPPING in which arable crops are grown or livestock grazed beneath PERMANENT CROPS of fruit trees. DRY FARMING is frequently resorted to in the drier areas. With the increasing use of irrigation, citrus fruits, peaches, apricots and almonds have become important in Mediterranean agriculture, and are among the major export products of the region. In recent years, a form of GLASSHOUSE CULTIVATION utilizing low-cost plastic-covered temporary structures has enabled Mediterranean farmers to exploit the 'early' market for soft fruit and salad vegetables in cooler climates, and to obtain a second crop to meet the later influx of north European tourists.

Animal husbandry has been based largely on the grazing of sheep and goats, often involving TRANSHUMANCE. Cattle are generally stall-fed with FODDER CROPS.

Until recently, agriculture across much of the Mediterranean basin was characterized by PEASANT farming structures, technologically simple irrigation systems and little MECHANIZATION, but it has become increasingly commercialized in many areas, such as coastal Spain, northern Italy and Israel. In examples of Mediterranean agriculture elsewhere in the world, such as in California, South Africa and Chile, intensive COMMERCIAL FARMING is the norm. FARM SIZES vary considerably from a few hectares in the more fertile lowlands to very large holdings in mountain regions where extensive olive cultivation predominates; the citrus groves of California and South Africa also support large farm holdings.

**Mediterranean and trans-Asiatic zone,** see EARTHQUAKE.

**medium,** *n.* **1.** either of the two surrounding substances in which organisms exist, that is, either water or air. Media provide the supply of oxygen or carbon dioxide required for respiration and thus are necessary for the survival of all plants and animals. See also SUBSTRATUM.

**2.** any substance on or in which microorganisms can be grown (*cultured*). Media can be solid or liquid, and contain the essential nutrients required for the growth of the organism.

**megalopolis,** see CITY.

**mesa,** *n.* **1.** an isolated, steep-sided, flat-topped hill that is a remnant of a PLATEAU which has undergone prolonged FLUVIAL EROSION and SLOPE RETREAT. Mesas differ from BUTTES in that they have undergone less erosion and thus tend to be more extensive.

**2.** (USA) a flat upland extending back from an ESCARPMENT.

**mesoclimate,** *n.* the distinct climate which may prevail over an area of several square kilometres, such as urban climates, forest climates and mountain climates. Compare MACROCLIMATE, MICROCLIMATE.

**mesopause,** *n.* the transitional zone of minimum temperature between the MESOSPHERE and the THERMOSPHERE, found at an altitude of between 50 and 80 km (see Fig. 9). Compare TROPOPAUSE, MESOPAUASE.

**mesosphere,** *n.* the layer of the ATMOSPHERE that exists between the STRATOSPHERE and the THERMOSPHERE. Extending between 50 and 80 km above the earth's surface, it ends at the MESOPAUSE. Air temperature decreases with altitude in the mesosphere.

**Mesozoic era** or **Secondary era,** *n.* the geological ERA that commenced after the PALEOZOIC ERA about 225 million years ago and which ended about 65 million years ago when the CENOZOIC ERA began. Lifeforms that developed during the Mesozoic included the mammals, ANGIOSPERMS and birds. The era is popularly known as the Age of the Reptiles on account of the predominance of the dinosaurs during this era. See EARTH HISTORY.

**metabolism,** see ATP.

**metamorphic rock,** *n.* a rock formed by the application of great heat and pressure to igneous and sedimentary rocks over immense lengths of time which results in alterations to the original texture, structure and composition of the rocks. See METAMOPHISM.

| Metamorphic rock | Original igneous or sedimentary rock |
| --- | --- |
| Gneiss | Granite |
| Slate | Shale |
| Pure limestone | Marble |
| Bituminous coal | Anthracitecoal/Peat/Lignite |

Fig. 41. **Metamorphic rock.** The igneous and sedimentary compositional equivalents of some metamorphic rocks.

**metamorphism,** *n.* the action of heat, pressure and chemically active magmatic fluids deep within the earth's interior which results in the alteration of the structure, texture and composition of existing igneous and sedimentary rocks to form new rock types.

Three main types of metamorphism have been recognized:

(a) *contact metamorphism* is usually associated with the intrusion of igneous rock bodies such as BATHOLITHS and DYKES. The heat from the intrusion is largely responsible for the alteration of surrounding COUNTRY ROCK to form a metamorphic aureole although pressure and chemical activity also play a part;

(b) *dynamic metamorphism* involves the alteration of existing rocks due to the intense compression associated with folding and faulting;

(c) *regional metamorphism* results from OROGENESIS and entails the alteration of rocks by heat, pressure and chemical activity. Due to the scale of orogenic processes, regional metamorphism can affect many thousands of square kilometres.

**meteorite,** *n.* any solid extraterrestial body that reaches the earth's surface, having passed through the atmosphere without being vapourized. Three types of meteorites have been identified:

(a) *siderites*, composed principally of nickel-iron;

(b) *siderolites*, composed of both metals and silicates;

(c) *aerolites*, composed principally of silicate minerals.

Meteorites may range in size from just a few centimetres in diameter to the meteorite which formed the Great Meteor Crater in Arizona which is 1 km in diameter and 180 m deep.

**methane** or **marsh gas,** *n.* an ordourless, colourless hydrocarbon gas produced either by natural or artificial anaerobic decomposition of organic material. Formula: $CH_4$. It is nearly insoluble in water, burns with a pale flame to produce water and carbon dioxide and releases no hazardous air pollutants. It is the principle consituent of NATURAL GAS and can be used as a fuel. Methane gas can be produced from a BIOGAS DIGESTER.

Some intensive agricultural systems, such as stall-fed cattle rearing, have such high densities of animals per unit area and generate such considerable amounts of animal urea (a major source of methane) that they have been cited as a major cause of the increase in carbon in the atmosphere. See GREENHOUSE EFFECT, BIOGAS ENERGY.

**methanol,** *n.* a colourless, volatile, flammable and poisonous alcohol traditionally formed by the destructive distillation of wood or, more recently, as a result of synthetic distillation in chemical plants. Formula: $CH_3OH$. Methanol is used especially as a solvent, as anti-freeze and in the synthesis of other fuels.

**micelle,** see CLAY-HUMUS COMPLEX.

**microclimate,** *n.* the climate of a small area perhaps measuring only 10 or 20 m$^3$, such as a garden or crop field. Compare MACROCLIMATE, MESOCLIMATE.

**micro-environment,** *n.* the small-scale, detailed conditions which prevail upon the species of an area. Whereas the general distribution pattern of plants and animals will be determined by the overall conditions or *macro-environment*, such as climate and main soil group, the micro-environment allows an infinite variation in local conditions. For example, variations in shelter or exposure values, in the compaction of soils by trampling and grazing pressure by HERBIVORES or between freely drained and waterlogged soils, will result in a broad range of responses by individual plants and animals and results in the seemingly patternless distribution that may be observed at the local level.

**micronutrient,** see SOIL NUTRIENT.

**microphyllous forest,** *n.* a type of forest found in semi-DESERT areas, typically in regions with less than 250 mm rainfall per annum. Microphyllous forest occurs extensively in South America, on the Indian sub-continent, and less widely in Africa and Australia. Its appearance is broadly similar in all regions comprising low- to medium-sized trees (6-8 m tall) and thorny SCRUB (3-5 m tall), all of which show adaptations to survive dry periods of up to 6 months duration. Leaves are small, spiney and thickly covered with cuticle to minimize evapotranspiration losses. A thick protective bark is also typical, as are deep roots to tap the groundwater supplies.

The most extensive microphyllous forest can be found in Brazil (see CAATINGA). The productivity value for this vegetation type reflects the severity of the climate. Net PRIMARY PRODUCTIVITY values average 200 g/m$^2$/year.

Microphyllous forest can provide some grazing cover in the damper season although if used for agriculture, great care must be taken to prevent overgrazing and consequent SOIL EROSION. This vegetation type has been severely cut over in recent years for firewood supplies.

**mid-oceanic ridge,** *n.* a high basaltic ridge on the ocean floor specifically in the north and south Atlantic. Mid-oceanic ridges usually form narrow tracts of submarine mountains that lie approximately parallel to coastlines. Occasionally these ridges may rise close to the ocean surface to form reefs, or may rise above sea level to form oceanic islands, such as Bouvet Island and Tristan da Cuhna. Mid-oceanic ridges occur close to diverging LITHOSPHERIC PLATE boundaries and are probably due to the extrusion of lava resulting from SEA-FLOOR SPREADING. A well-formed example occurs beneath both the north and south Atlantic ocean. Earthquakes are often associated with mid-oceanic ridges (see Fig. 47).

**migration,** *n.* the journey made by many animal species (especially birds and fishes) between different HABITATS at specific times of year along well-defined routes, particularly those involving a return to breeding grounds.

**mineral,** *n.* **1.** any naturally occuring solid inorganic substance of definite chemical composition and crystalline structure, such as FELDSPAR or MICA. Minerals can be classified according to chemical composition, crystal form and a range of physical properties including hardness, colour, lustre, specific gravity, tenacity and cleavage/fracture.

**2.** any natural material obtained through mining, quarrying or drilling.

**mineral resource** or **mineral deposit,** *n.* a naturally occuring concentration of economically valuable and workable MINERALS. Mineral deposits can be classified according to their origins and may be formed by a variety of igneous, metamorphic and sedimentary processes.

**minifundia,** *n.* small farms commonly found in Central and South America. Some are independently owned SMALLHOLDINGS but most are tenanted farms or squatter holdings too small to support the family unit. Compare LATIFUNDIA.

**minimal area,** *n.* **1.** the smallest area of ground required by a species in which to fulfill its entire life history. For most species minimal areas become critical during the reproductive phase when availabilty of nesting sites and guaranteed food supply becomes of paramount importance. Physically large species require more space than small species (although birds are an exception to this rule: a pair of robins may need up to 1 ha of land in order to survive).

Many species can no longer attain their minimal area due to the expansion of urban areas, development of intensive agriculture and to DEFORESTATION. Modern conservation theory now supports the linking of remaining conservation, open, waste and wilderness areas by means of HABITAT CORRIDORS.

**2.** the smallest QUADRAT size that ensures an adequate representation of the variety of vegetation species under examination.

**minimal tillage,** see TILLAGE.

**mire,** see BOG.

**Mississippian period,** see CARBONIFEROUS PERIOD.

**mist,** *n.* a less dense form of FOG, in which visibility ranges from 1-2 km.

**Mistral,** *n.* a strong, dry, cold, northwesterly WIND which blows offshore along the Mediterranean coasts of France and Spain.

**misfit river** or **underfit river,** *n.* a river that is apparently too small for its valley. Misfit rivers can occur for a variety of reasons including:

(a) the removal of a stream's headwaters by RIVER CAPTURE (see Fig. 54),

(b) climatic changes resulting in a reduced stream flow,

(c) the deepening and widening of a river valley by GLACIAL EROSION,

(d) the disappearance of a river into an UNDERGROUND DRAINAGE system in chalk and limestone areas.

**mixed cropping,** *n.* a form of MULTIPLE CROPPING in which two or more crops are grown simultaneously in the same field. Mixed cropping is most commonly achieved by one of the following means:

(a) *multistorey cropping* in which perennial and annual crops are grown together utilizing space and light at different heights above the ground. Where arable crops are grown below perennial crops this is known as *interculture*. The practice is widespread in Mediterranean areas where cereals are grown beneath fruit trees, and in the tropics under systems of SHIFTING CULTIVATION and ROTATIONAL BUSH FALLOW where it is adapted to the surface conditions in a largely informal and unstructured manner. Tree crops such as oil palms or coconuts may form an upper storey, beneath which, at intermediate height, cocoa or coffee may grow as a CASH CROP. Alternatively, a FOOD CROP such as cassava or climbing yams may take up the middle storey. Cereals, root crops or green vegetables may form a ground cover where available light permits;

(b) *intercropping* in which two or more crops are planted in separate but proximate rows in the field. Intercropping is often used synonymously with mixed cropping but it is more correctly a particular form of mixed cropping where all crops are planted in a fixed pattern of spacings and rows to facilitate individual crop care. Intercropping may be achieved by relay planting (see MULTIPLE CROPPING), or by planting the crops at the same time. In drier parts of North America prone to SOIL EROSION, an intercrop of small grains or crops such as alfalfa or clover may be grown beneath the rows of a main crop primarily to provide a ground cover to counter RUNOFF, and when cut, a protective MULCH;

(c) *relay cropping* which is a sub-type of MULTIPLE CROPPING in its own right.

Mixed cropping is a rational response to both physical and socio-economic circumstances. It makes fuller use of available light, water and nutrients, and reduces runoff, leaching and erosion. Although each crop may yield less than it would as a sole STAND the gross return per hectare and per year may be higher and have less annual variation. Due to natural weed suppression, labour demands per unit of output may be less than for sole stands in the most intensive forms of mixed cropping, and the need for HERBICIDES is much reduced. Mixed cropping, and especially relay planting, extends the planting season, staggers the harvest and generally distributes labour demands more evenly throughout the year. Above all, it is a risk-minimization strategy that provides PEASANT smallholders with a continuous and varied supply of fresh foodstuffs. On the other hand, mixed cropping regimes cannot easily be mechanized or improved through the use of selective chemical inputs and crop-specific production innovations.

In the tropics, mixed cropping is the predominant cropping principle. In West and Central Africa some 80% of all CROPLAND is managed in this way. In the forest zone especially, there is an almost infinite variety of crop combinations.

**mixed farming,** *n.* a type of agriculture based on a combination of CROP production and ANIMAL HUSBANDRY. Mixed farming has been the basis of most agricultural systems throughout recorded history, with varying degrees of interdependence of the use of arable land for the production of FODDER CROPS and the use of MANURE on CROPLAND. From the late 19th century there was a trend in North America and Western Europe towards increasing specialization at the expense of mixed farming, largely due to the growing proportion of non-farm inputs in the productive processes and the general growth of AGRICULTURAL INDUSTRIALIZATION. Recent evidence from parts of the USA, Australia and Argentina indicates a return to mixed farming after periods of all-arable MONOCULTURE.

**model,** *n.* any idealized or abstract representation of reality designed to clarify the relationships between selected variables, to provide insights on operating conditions, and to predict likely outcomes. Environmental systems are necessarily complex phenomena, further complicated by interactions with humans often behaving in irrational or unpredictable fashion. Models of these systems may, according to preference, ignore detail through generalization or be narrowly selective in specialization, but all

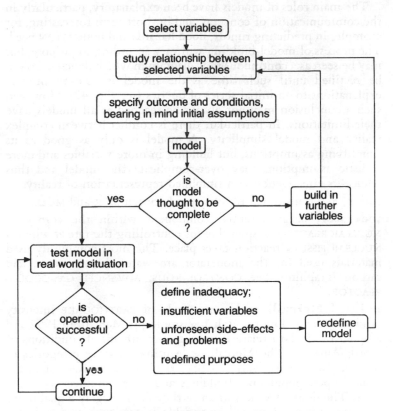

Fig. 42. **Model**. The process of model building and testing.

should be based on logical reasoning, aim to identify the most important relationships and the variables which control them, and be operationally valid in the real world in meeting a clearly stated objective. Among the many kinds of model are: *analogues* in which selected characteristics of a particular environmental phenomenon are taken to be analogous to some simpler and more familiar system, and include computer models allowing for increased data handling and greater simulation potential; *conceptual* and *theoretical models*; *hardware models*, including scale models, wave tanks and maps; and *mathematical models* in which the constituent elements are replaced by mathematical variables, parameters and constants expressed in abstract symbols.

The main roles of models have been explanatory, particularly in the communication of concepts, and in short-term forecasting, for example, in predicting runoff, flood hazards and IRRIGATION need. The process of model-building will vary according to purpose, but may be seen as a continuous flow diagram in which deficiencies can be rectified until such time as the model is able to provide explanations to an initial series of queries (see Fig. 42). However, such a conclusion may not always be reached as all models have their limitations. In particular, there is conflict between complex reality and model simplicity. A model is only as good as its simplifying assumptions, but building in more variables and more realistic assumptions may over-complicate the model and thus reduce its effectiveness as a simplified representation of reality.

**moder,** *n.* a transitional HUMUS type between MOR and MULL.

**moderator,** *n.* an essential component within the core of a NUCLEAR REACTOR, responsible for controlling the rate at which a NUCLEAR FISSION reaction takes place. The most commonly used materials used for the moderator are water, heavy water, and carbon (graphite). See CONTROL RODS; ADVANCED GAS-COOLED REACTOR.

**modified Mercalli scale,** *n.* a scale of EARTHQUAKE intensity originally devised by the Italian seismologist Mercalli in 1902 and adapted in 1931 to relate it more to the urbanized conditions of North America. The Mercalli scale utilizes twelve categories of intensity and subjectively describes the localized effect of earthquakes upon population, buildings and other engineering structures. The destructiveness of an earthquake is largely a qualitative measure as it is influenced by variable factors such as population density, the design and construction techniques used for engineering structures and the accuracy of the earthquake reports. The RICHTER SCALE is a more quantitative measure of earthquake magnitude and is usually preferred to the modified Mercalli scale.

**Mohorovičić discontinuity, Moho** or **M discontinuity,** *n.* the boundary between the EARTH'S CRUST and the MANTLE. The discontinuity occurs at an approximate depth of 8 km below the oceans and 40 km below the continents (see Fig. 28). See GUTENBERG DISCONTINUITY.

**molecule,** *n.* a group of atoms held together by strong chemical bonds that forms the smallest unit of a chemical compound retaining the characteristic properties of that compound. Molecules vary in size and composition from the very small hydrogen ($H_2$)

molecule, composed of two atoms of hydrogen, to proteins and starches, which have thousands of atoms in a single molecule.

**mollisol,** *n.* a CHERNOZEM type soil as listed in the American soil classification system. Of all the 10 orders in the American soil classification mollisols probably includes the greatest variety of SOIL PROFILES. This variation is due in no small part to the wide range of soil profile drainage which can range from waterlogged depressions to well-drained slopes.

**monoclimax theory,** see CLIMAX COMMUNITY.

**monocotyledon,** *n.* one of the two classes of flowering plants (ANGIOSPERMS), characterized by having embryos with one seed leaf (*cotyledon*), narrow leaves with a parallel vein system, scattered vascular bundles with no cambium, and flower parts arranged in multiples of three. Members of this class include the economically important grasses and grain crops. The only monocotyledon trees are palm trees. Compare DICOTYLEDON.

**monoculture,** *n.* **1.** the repeated cultivation of a single crop on a given acreage, often indefinitely. This may occur on PLANTATIONS, on farms whose only support is a fruit crop, and on large, highly mechanized farms growing wheat to the exclusion of other crops. This form of ARABLE FARMING developed first in the USA in the 19th century, spreading to Argentina, Australia and Canada by the 1930s, and to Europe in the post-war era as guaranteed prices under the COMMON AGRICULTURAL POLICY encouraged cereal production at the expense of MIXED FARMING.

**2.** the cultivation of only one crop during a particular season. In subsequent years an alternative crop may be grown, also in monoculture, to form a rotation of LEAFY CROPS and GRAIN CROPS.

**3.** permanent grain cropping in which the cereal variety is alternated in rotation.

Monocultures experience problems due to the exhaustion of nitrogen in the soil which causes yield reductions unless corrected by the application of FERTILIZERS, and to the build-up of pests and diseases that would normally be controlled by CROP ROTATION but require chemical insecticides, soil fumigation, or the development of disease-resistant plant varieties. Bush- and tree crops tolerate monoculture better than most FIELD CROPS as they require no annual crop rotation.

**monopoly,** *n.* a situation in which a single firm or individual supplies the entire output of a good or service. In practice, a firm is said to be monopolistic when it produces and sells a high enough proportion of a commodity to be able to control its price. In this respect, it is more likely, and more correctly, to be part of an *oligopoly*, a situation in which the output of a commodity is concentrated in the hands of a small number of firms.

**monsoon,** *n.* the seasonal reversal of winds and air pressure systems over continental landmasses and adjacent oceans. Monsoon climates largely result from the seasonal migration of the general circulation system although the differential heating and retention capacities of land and sea in addition to topographic factors also play an important role. Monsoon conditions most commonly occur on the eastern sides of continents in the tropics although less well developed monsoons are found in West Africa and northern Australia.

Probably the best developed and well known monsoon is the Indian Monsoon. Between October and March, the effects of the cold dry winds of the winter north-west monsoon blowing outwards from the high pressure system developed over central Asia are greatly ameliorated in India due to the protective barrier of the Himalayas and the Tibetan Plateau. Between April and September, the warm, rain-bearing winds of the summer south-east monsoon, which originates over the Indian Ocean, form a FRONT that gradually progresses over India. This front usually arrives at a location around the same date each year. The late arrival or failure of the rains can seriously affect agricultural production and in extreme circumstances results in famine.

**montane,** *adj.* (of an ecosystem) that characteristically found on high mountains in low latitudes. Increasing elevation causes a reduction in average temperature and an increase in precipitation which can result in a semi-permanent cloud or mist layer. Exposure values also increase with elevation and, as a result, the vegetation often becomes stunted in appearance. 'Elfin' forests occur in which trees grow horizontally instead of vertically (the *krummholz* condition). Because of the moist conditions, ferns, mosses and lichens may become abundant. See also ALPINE.

**moorland,** *n.* an open area of ground usually found above the level of enclosed or improved agricultural land. Moorland is commonly found on the extensive areas of peaty soils in the wetter and cooler areas of northwest Europe. The vegetation of moorlands is almost exclusively confined to low-growing evergreen shrubs typified by members of the Ericaceae family. Common species include *Erica cinerea*, *E.tetralix*, *Calluna vulgaris* and *Vaccinium myrtillus*. Grasses are usually infrequent although sedges, mosses and lichens may survive at ground level. Trees are rare.

Moorlands have been formed entirely as a result of human activity. FIRE is probably the main controlling factor in determining the existence of moorland. A burning interval of 5-10 years is common on moorlands which are managed as grouse moors or for sheep and/or deer grazing. Wherever burning has become infrequent then forest regeneration occurs. Birch, pine and juniper form the first tree colonizers and are followed by hazel, or alder in the damper areas. Vast areas of British moorland have been reafforested with coniferous trees since the early 1960s, for example, Keilder Forest in Northumberland.

**moorpan,** see HARDPAN.

**mor,** *n.* an acidic HUMUS layer (pH generally less than 4.5) common on poorly aerated soils. Mor is especially common in climatic regions with long, cold and wet winters and mild, damp summers and on older Palaeozoic rocks. The adverse soil climate precludes an inactive soil fauna, hence decomposition of plant litter is slow and incomplete. A thick organic layer (PEAT) can develop on the surface of the mineral soil which reduces the supply of oxygen and further diminishes the activity rate of the soil fauna. Typical soil types on which mor can be found include peaty BROWN EARTH, peaty GLEY SOILS, and peaty PODZOLS. Compare MULL, MODER.

**moraine,** *n.* the unstratified GLACIAL TILL transported on, within, or under, and deposited by, a GLACIER or ICESHEET. Several types of moraine are recognized.

(a) *terminal moraine* or *end moraine* forms a ridge up to 300 m high and is deposited at the maximum extent of a glacier or icesheet as it begins to waste. Terminal moraines often form natural dams.

(b) *recessional moraine* is a till deposit formed at the edge of a retreating glacier or ice sheet during a temporary halt in glacial retreat when its margins are relatively stable.

(c) *ground moraine* is the debris deposited by a steadily retreating glacier or icesheet which may result in an almost featureless plain.

(d) *lateral moraine* is an accumulation of rock debris found only along the margins of a valley glacier. As the glacier recedes the lateral moraine may be left as a ridge along the valley side.

(e) *medial moraine* results from the coalescence of lateral moraines as two valley glaciers unite. When the glacier retreats, a medial moraine is often left as an irregular ridge in the centre of the valley.

Steep slopes and poor quality stony soils often restrict the agricultural use of morainal areas to livestock farming. Areas of clay-rich moraines may often be utilized as dumps and LANDFILL sites as the impervious nature of clay prevents the contamination of groundwater supplies and land by toxic domestic and industrial waste. See GLACIAL DEPOSITION.

**morphological map,** see GEOMORPHOLOGICAL MAP.

**mortality rate,** see DEATH RATE.

**mortlake,** see OXBOW LAKE.

**moshav,** see AGRICULTURAL COOPERATIVE.

**mothballing,** see DECOMMISSIONING.

**mother ship,** see FACTORY SHIP

**motorway, autobahn, autostrada, highway** or **freeway,** *n.* any road built to high engineering standards and on which special restrictions of use usually apply. They usually consist of at least two carriageways each approximately 3.5 m in width, an emergency standing area or *hard shoulder*, and a central reservation which separates a similar number of lanes for traffic travelling in the opposite direction. Traffic enters and leaves motorways by means of separate junctions. The prime purpose of motorways is to permit large traffic flows to travel between key points quickly and safely. See also DUAL CARRIAGEWAY, TRAFFIC LANE, FLYOVER.

**mound cultivation,** *n.* the practice of growing crops on raised mounds or ridges, generally in tropical farming systems dependent on hand tools. Cut vegetation is buried in mounds and is either ignited and allowed to smoulder for several days (the heat having a beneficial effect on soil fertility as a result of increased phosphate availability) or else allowed to decompose as a green MANURE. Alternatively, vegetation is cut and fired where it lies, with the ashes subsequently buried in mounds or ridges into which are planted seeds or tubers. Mounds are sometimes covered with MULCH material to protect the soil from erosion.

A form of mound cultivation is found in the western highlands and islands of Scotland where *lazy-beds*, spade-dug ridges between 60 cm and 3 m wide divided by furrows up to 1 m wide, are used mostly for the cultivation of potatoes. Lazy-beds are not to be confused with PLAGEN SOILS.

**mountain climate,** *n.* the MESOCLIMATE developed over mountainous regions and high plateaus. Mountain climates are characterized by lower temperatures and increased precipitation patterns when compared to lowland areas of similar latitudes. ANABATIC WINDS and KATABATIC WINDS are common features of mountain climates.

**mountain glacier,** see GLACIER.

**mountain wind,** see KATABATIC WIND.

**mud flat,** *n.* a tract of low muddy land that is covered at high tide and exposed at low tide; mud flats are commonly found near ESTUARIES.

**mudflow,** *n.* the rapid downslope flow of saturated soil and rock debris under the influence of gravity. Mudflows occur most commonly in areas of sparse vegetation when torrential rain turns the surface soil into a flowing viscous mass. Mudflows can travel at up to 4 m/s and may be up to 2 m in depth.

The localized occurrence of mudflows can have catastrophic consequences, as seen in the destruction in 79 AD of the Roman city of Herculaneum. More recently, a mudflow in Canada transported several houses more than 500 m in just a few minutes. BOG-BURSTS are thought to be related to mudflows.

**mulch,** *n.* a layer of organic material such as straw, leaves, plant residue, sawdust, loose soil etc. applied to the soil surface to conserve moisture by reducing evaporation, and to suppress weed growth. A mulch may also be laid down to supply nutrients to plants as it decomposes, to prevent RUNOFF likely to lead to SOIL EROSION, and to prevent the soil surface from freezing.

**mull,** *n.* **1.** a nutrient-rich HUMUS usually found in well-aerated soils such as those in European lowland deciduous forests. The pH of mull is generally greater than 4.5 and the mineral soil which occurs beneath the humus layer can show a neutral or even slightly alkaline reaction (pH 7.0). Long hot summers and short cold winters encourages microbial activity in the soil and a large earthworm population ensures the mixing of organic material with the mineral layers. Mull-rich soils are usually very fertile and have long been used as agricultural sites. Typical soils associated with mull include deep BROWN EARTH, brown forest soil, and brown earth with surface GLEY SOILS. Compare MOR, MODER.

**multiple cropping,** *n.* the growing of more than one annual crop on the same land during one calendar year. Four main types of multiple cropping can be identified:

(a) *sequential cropping* or *successional cropping* in which a crop is planted and harvested, and is followed by further crops in the same year without recourse to fallowing. Double and triple cropping are common forms of sequential cropping, especially in the intensive SUBSISTENCE FARMING systems of China and South-East Asia, associated with WET-RICE CULTIVATION.

(b) *relay cropping* in which seedlings of a second crop are planted between plants or rows of an already established maturing annual or biennial crop. In the West African savanna where CONTINUOUS

CROPPING may last for several seasons relay planting becomes, in effect, a form of CROP ROTATION.

(c) *ratooning* in which the roots of a harvested crop produce a subsequent crop. It is a practice most associated with the cultivation of sugar-cane, and sometimes sorghum.

(d) MIXED CROPPING in which two or more crops are grown simultaneously.

Successful multiple cropping, especially sequential and relay planting is generally dependent on a favourable temperature regime and a reliable water supply, most often provided by IRRIGATION. The intensity of land use can be described by a *multiple cropping index (MCI)* in which the total crop area for one year is divided by the total cultivated land area for one year. When expressed as a percentage an MCI exceeding 150 indicates intensive land use and potentially high annual yields. Multiple cropping is widespread in the tropics. In the rice-growing areas of South and East Asia the MCI is rarely less than 150. Several crops are often only grown in association with others. It is estimated, for example, that in Africa 98% of cowpeas, the continent's major legume, and in Latin America 60% of maize, the major cereal, are grown in multiple cropping regimes.

**multiple cropping index,** see MULTIPLE CROPPING.

**multiple job holding,** see PART-TIME FARMING.

**multiple land use,** *n.* the use of land for more than one purpose. The term is not normally applied in circumstances where two or more agricultural activities are carried out on the same plot, as in MULTIPLE CROPPING, but where quite different activities make use of the same land, for example, AGRICULTURE or FORESTRY and recreation.

A critical task facing planning authorities is to harmonize different land or water use pressures; for example, open access to the land may conflict with agricultural uses leading to crop damage and the worrying of livestock, while the demands for unpolluted drinking water may be mutually exclusive with bathing, fishing, water sports and tranquil landscape appreciation. Multiple land use can only be successful if the uses are compatible, and then generally only with careful management, as with the work of the FORESTRY COMMISSION in the UK to open up forest plantations to the public by the provision of strategically located car parks, forest drives, trails and picnic sites.

**Multiple Use Sustained Yield Act (1960),** *n.* legislation applied to the Forest Service of the United States Department of

Agriculture under which FOREST MANAGEMENT must now operate under the twin guidelines of MULTIPLE LAND USE management and SUSTAINED YIELD

**multistorey cropping,** see MIXED CROPPING.

**muskeg,** see BOG.

# N

**national conservation strategy (NCS),** *n.* any CONSERVATION policy aimed specifically at national or regional problems. Thus in New Zealand a strategy has been developed to monitor fish stocks and regulate exploitation, while in Nepal, a strategy has been designed to maximize yields of fuelwood and minimize soil erosion. See also ECODEVELOPMENT

**National Environmental Policy Act (NEPA) (1969),** *n.* US legislation which charged federal government agencies with the responsibility to restore and maintain environmental quality throughout the USA. NEPA has been a landmark in environmental law as it required all federal government agencies to prepare ENVIROMENTAL IMPACT ASSESSMENTS on the predicted effects of any major project that may significantly affect the environment. The principles of the Act have also been extended to the state level in the USA and have been copied in similar legislation in at least 16 other nations. From March, 1988, member countries of the EC will be bound by similar controls.

The Act has been criticized for introducing costly and time-consuming bureaucratic controls and for promoting essentially cosmetic conservation in place of genuine environmental protection. However, there can be little doubt that without this legislation and the wider acceptance of EIA that it brought, many more fragile ecosystems and environmental amenities would have suffered from commercial developments than has been the case. See also WORLD CONSERVATION STRATEGY.

**National Environmental Satellite Service (NESS),** see NOAA.

**national grid,** *n.* **1.** a nationwide network of high-voltage power lines connecting power stations in the UK. The grid enables power to be transferred over long distances in quick response to variations in demand. National grids can also be built for water distribution and for natural gas.

**2.** a grid of metric coordinates printed on maps produced by the Ordnance Survey in the UK and Ireland, and by the New Zealand Lands and Survey Department.

**National Marine Fisheries Service,** see NOAA.

**national nature reserve (NNR),** *n.* any site selected by the NATURE CONSERVANCY COUNCIL as representative of the major types of flora, fauna, habitats and topography found in the UK. The range of ECOSYSTEMS covered by NNR include coastland, woodland, peatland, open waters, lowland and upland grassland, heaths, and scrubland. Potential sites are assessed against such criteria as size, diversity, rarity and fragility and if selected, are made the subject of scientific research and ecological management; the results of this work can often be applied to NATIONAL PARKS in order to maintain their quality.

**National Ocean Survey,** see NOAA.

**national park,** *n.* an extensive tract of land of outstanding natural beauty, reserved primarily for the conservation of its flora, fauna and scenery, and to a lesser extent for public recreation. The first area of land to be designated a national park was at Yellowstone, Montana in the USA in 1872. Since then about 100 countries have adopted the concept of the national park although their character and management varies widely from country to country. The INTERNATIONAL UNION FOR THE CONSERVATION OF NATURE AND NATURAL RESOURCES (IUCN) has defined a national park as an extensive natural area, largely unaltered by human exploitation and occupation, and which is state-owned and administered by national government; public recreational activities are permitted only so far as they are compatible with a primary objective of conserving the flora, fauna and landscape values of the park.

The national parks of the USA, Canada, Australia, the USSR, Poland and Czechoslovakia closely follow this definition whereas the national parks of more densely populated nations, in particular the UK, depart from the IUCN definition in a number of ways. Thus, in the UK, national parks comprise areas of great natural beauty which have been significantly altered by long periods of human occupancy; the land is frequently in private ownership and administered by local government. Greater emphasis is also placed on the maintenance of traditional cultural landscapes and on the provision of recreational facilities such as sailing, walking, climbing and fishing.

National parks can vary greatly in size and the largest occur in the Soviet Union, for example the Pechjora-Ilych National Park which covers 721 333 ha. Soviet national parks are under total state control and are designated as areas of intensive scientific research. Their success in terms of revitalising plant and animal species has been remarkable although this has been achieved in isolation from public involvement as access is strictly controlled.

The IUCN has urged all countries to allocate at least 10% of national land and water resources to conservation, preferably as national parks.

**National Park and Access to the Countryside Act (1949),** *n.* the legislation for England and Wales under which NATIONAL PARKS were established and are managed.

**National Weather Service,** see NOAA.

**National Wilderness Preservation System,** see WILDERNESS AREA.

**natural gas,** *n.* gaseous HYDROCARBONS (principally METHANE) trapped in underground rock reservoirs. It may occur alone but is more commonly associated with deposits of oil. Natural gas is the highest quality, cleanest burning FOSSIL FUEL and is heavily used as a domestic, commercial and industrial energy source; it is also an important raw material for the petrochemical industry. However, unless new, large-scale reservoirs of natural gas are discovered, all existing reserves may be exhausted by 2010.

**Nature Conservancy Council (NCC),** *n.* a British statutory body established in 1973 concerned with matters affecting nature conservation and which is charged with establishing and managing NATIONAL NATURE RESERVES in the UK. It undertakes research, provides information and advises government ministers on environmental issues.

**NCC,** see NATIONAL CONSERVANCY COUNCIL.

**NCS,** see NATIONAL CONSERVATION STRATEGY.

**neighbourhood centre,** *n.* a cluster of small shops and service outlets that serve the routine needs of the population living within short-journey distance, usually walking distance, from them. Social facilities, such as meeting rooms or youth-club premises may be located in such centres in order to exploit their accessibility. Neighbourhood centres develop at natural focal points in urban areas. The location of neighbourhood centres to service and thus to draw together socially the residents of wholly new areas of housing has been an integral part of the design of most British NEW TOWNS.

**nekton,** *n.* any aquatic organism capable of swimming or navigating independently of currents. Fish are the most important members of the nekton, which also include amphibians and large swimming insects. Compare PLANKTON.

**NEPA,** see NATIONAL ENVIRONMENTAL POLICY ACT.

**net primary productivity (NPP),** see PRIMARY PRODUCTIVITY.

**net residential density,** see RESIDENTIAL DENSITY.

**neve,** see SNOWFIELD.

**New International Economic Order (NIEO),** *n.* an economic and political concept enshrined in a set of proposals adopted by the United Nations in 1974, designed to restructure international economic relations and so reduce the inequalities between the DEVELOPED COUNTRIES of the 'North' and the DEVELOPING COUNTRIES of the 'South'.

The call for a NIEO was fuelled by the belief of many THIRD WORLD governments that the existing economic order perpetuated this inherent inequality, relegating primary COMMODITY exporters to an economically dependent status, and by a loss of faith in the ability of free market forces to close the gap between the North and the South.

It was argued in the report of the *Brandt Commission* (1980), an independent body examining international development issues, that an NIEO was in the interests of the North to ensure the harmonious growth of the world economic system, trading the South's need for economic equity against the North's desire for security in resource supply, and to avoid possible Third World debt defaulting.

Brandt recommended that to achieve a NIEO there should be a transfer of economic growth from the North to the South; greater cooperation in resource development with the developing countries gaining more control over the development of their own resources; an improvement in commodity trading arrangements through price stabilization measures, institutional reform and the setting up of commodity associations; a new LAW OF THE SEA; a code of conduct for multinational corporations; and a commitment by the governments of the developed countries towards increased foreign aid, international monetary reform, and the relocation of commodity processing and manufacturing activities in the raw material producing countries. The success of the Organization of Petroleum Exporting Countries in 1973-74 in wresting control of their indigenous petroleum reserves from Western oil companies encouraged many to advocate a NIEO in which Third World producers might secure a more equitable position in international commodity trade. However, in the 1980s, the prospects for a NIEO have receded with economic stagnation, and a preoccupation with domestic issues among leading industrial nations.

**new town,** *n.* a government-sponsored planned town, developed on former agricultural land surrounding or adjacent to an existing village or small town. The concept of the new town was pioneered in the UK, and between 1945 and 1950, 13 new towns were officially designated by central government; a further 13 were designated between 1950 and 1971. The new towns were seen as part of a broad national policy to disperse the population from the overcrowded conditions that existed in the older parts of the large cities and were carefully designed to provide high quality amenities for their new residents. The inhabitants of new towns were transferred from the inner city redevelopment schemes or voluntarily chose to migrate from cities.

The designs of new towns were based on the idea that the new urban areas should provide all the services people needed to make them independent communities. Thus the plans included shopping centres (at both neighbourhood and town-centre level), schools, recreation facilities and workplaces, as well as good transport facilities to link the town to the parent city and wider region.

The new town environment was also seen as an important basis for implementing policies for industrial renewal. It was felt that the pleasant surroundings and financial incentives would prove attractive both to new employers coming from overseas to seek sites in the UK and to those seeking relocation from unsuitable accommodation in the parent city.

The new towns were administered from the start by government-appointed development commissions which operated on an ad hoc basis outside the existing local government system, which was inappropriately structured to deal with the design and building of totally new towns.

The last British new town, Stonehouse, in Scotland, was both designated and cancelled within five years in the early 1970s. By this time the combination of major demolition and renewal in older urban areas, and the demographic downturn had removed the need for regional dispersal policies. Central- and local-government policy makers reallocated resources to the INNER CITY problem which was by then regarded as a quality-of-life problem rather than one of population congestion. See GREENFIELD SITE, OVERSPILL.

**NGO,** see NON-GOVERNMENT ORGANIZATION.

**NIEO,** see NEW INTERNATIONAL ECONOMIC ORDER.

**night soil,** *n.* human excrement collected nightly in towns and cities from cesspools and latrines, and transported to nearby farmland to be used as FERTILIZER. Until the improvement in sanitary arrangements in the 19th century the collection of night soil and horse droppings was commonplace throughout Europe. In

China, and in many parts of South-east Asia and Africa collection of night soil continues. Stored in earthenware pots and diluted with water it may be applied to individual plants or spread more widely. Although a high productivity fertilizer it may carry a variety of intestinal diseases. In some rural areas, pit latrines that have been sealed for several months may yield a useful dry MANURE with fewer public health dangers.

**nimbostratus,** *n.* a dense, dark-grey continuous layer of low CLOUD associated with the warm front of a DEPRESSION. Continuous PRECIPITATION usually falls from nimbostratus.

**nitrogen cycle,** *n.* the natural circulation of nitrogen by living organisms via the atmosphere, the oceans and the soil (see Fig. 43). Although the atmosphere comprises about 79% nitrogen, most of this is unavailable to living organisms. An intermediary NITROGEN FIXATION stage performed by soil organisms is necessary before atmospheric nitrogen becomes available. Nitrates in the soil, derived from the oxidation of ammonium compounds in dead organic material by soil bacteria, are absorbed and synthesized into complex organic compounds by green plants; these in turn return to the soil and are reduced to nitrates again when the plants (or animals feeding on them) die and decay. When they die, yet other organisms, the DECOMPOSERS begin a process of *denitrification* whereby nitrogen gas is released back into the atmosphere.

Shortage of available nitrogen was, prior to the 1860s, a major LIMITING FACTOR to agricultural productivity. Artificial production of nitrogen by the chemical industry has removed this restriction so much so that more artifical nitrogen is now manufactured than is provided by natural sources (about 32 million tonnes per annum). This is probably one of the most significant interventions by humans in the world's NUTRIENT CYCLES.

**nitrogen fixation,** *n.* the conversion of atmospheric nitrogen into organic nitrogen compounds. The fixation process is carried out by free-living soil bacteria and certain ALGAE, or by symbiotic organisms existing in the root nodules of leguminous plants.

Nitrogen also occurs in the oceans where non-symbiotic bacteria and blue-green algae are the prime agents of fixation. See NITROGEN CYCLE.

**NNI,** see NOISE AND NUMBER INDEX.

**NNR,** see NATIONAL NATURE RESERVE.

**NOAA,** *n. acronym for* the National Oceanic and Atmospheric Administration, an agency established by the US Department of

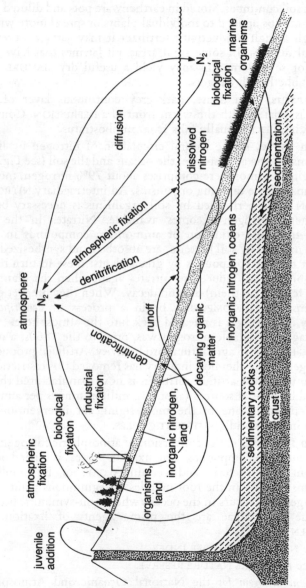

Fig. 43. **Nitrogen cycle.**

Commerce to encompass a number of previously existing services and to give them a common direction. The NOAA comprises:

(a) the *National Ocean Survey* (formerly the US Coast and Geodetic Survey), which creates nautical and air charts, conducts oceanographic surveys, makes tide and current predictions, and prepares navigational charts for American waters;

(b) the *National Weather Service* (formerly the US Weather Bureau), which forecasts the weather and conducts storm watches;

(c) the *National Marine Fisheries Service*, which researches economically important species of fish and marine organisms. This agency also oversees the Marine Mammals Protection Act and the Endangered Species Act;

(d) the *Environmental Data Service*, which studies the effects of environment on the nation's food energy supply. It operates the National Climatic Centre, the National Oceanographic Data Centre, and the National Geophysical and Solar Terrestrial Data Centre;

(e) the *National Environmental Satellite Service* (NESS), which maintains and collects data from the NOAA satellites;

(f) the *Environmental Research Laboratories*, which study ocean-atmosphere processes and the effect of pollution on coastal ecosystems;

(g) the *Office of Sea Grants*, which funds research institutions;

(h) the *Office of Coastal Management*, which exists to help maritime states in the development of management systems for their coasts.

**noise and number index (NNI),** *n.* an index for assessing air traffic noise based on the equation

$$NNI = PNdB + 15 \log N - 80$$

where PNdb = the logarithmic average of aircraft noise in perceived decibels and N = the number of aircraft heard on an average summer day between 0800 and 1800.

NNI contour values can be plotted around airports, and permitted land use patterns controlled accordingly. Thus, where NNI values exceed 60, land use is confined to warehouses, industry or transport. Between NNI values of 45 and 60, no new residential development is permitted and all existing houses, schools and hospitals must be fitted with sound insulation of the highest standard. Between NNI 35 and 45, new residential construction is permitted but only if the design includes full sound insulation. See NOISE POLLUTION.

# NOISE CONTROL

**noise control,** see NOISE POLLUTION.

**noise pollution,** *n.* noise from any source (such as from traffic, airports, industrial or manufacturing processes), that is deemed a nuisance to the residents of adjacent localities. To determine harmful levels of noise, sound pressure measurements in DECIBELS (dB) can be made using a decibel recorder. A mathematical equation is then used to convert sound pressure into a scale of loudness as perceived by the human ear. Usually, sound pressure is weighted towards higher pitched sounds and recorded in dBA units. Fig. 22 provides examples of some common noise levels. Constant exposure to noise of 75 dBA can be damaging to the human ear, physical pain occurs at 120 dBA and death can occur at 180 dBA.

Legislation to control noise is usually found within acts relating to road traffic or the management of civil aviation. For example, in the UK the problems associated with the landing and take-off of aircraft led to the *Air Navigation (Noise Certification) Order(1970)* which requires that aircraft comply with certain noise standards; consequently, recent trends in aircraft design have emphasized quieter engines. Road traffic noise, while of lower intensity, is more continuous and is acknowledged as detrimental to AMENITY for those living along busy streets. Recreational noise from, for example, transistor radios is an annoyance to many although, by its nature, it is sporadic and temporary and thus impossible to treat by legislation.

Households which are subject to unaccepatably high noise levels as determined by a NOISE AND NUMBER INDEX valuation may qualify for a government grant to install sound insulation. Alternatively, legally enforced maximum noise levels may be applied to the source of noise, thereby enforcing sound-deadening devices to be fitted to machinery and 'hush kits' to aero-engines.

**nomadism,** *n.* a way of life based on constant movement in search of sustenance and economic reward. The term is most often used in relation to pastoral peoples and their wanderings in search of grazing for their livestock. Several types of pastoral nomadism can be identified. *True or total nomadism* covers those people who have no permanent place of residence, do not practice regular cultivation, and move with their herds mostly within an established territory and along well-defined routes, for example, the camel nomads of the Sahara and Arabia.

*Semi-nomadism* is practised where families or communities have a fixed settlement around which there is some cultivation during the rainy season but in the dry season the men, if not the whole family, travel with their herds to distant areas in search of grazing, a system followed by the Fulani (Fulbe) in West Africa and the Maasai and Karimojong in East Africa.

While the rationale for a truly nomadic existence is largely due to the inability of the land to sustain sedentary activities it may also be a response to political pressures, or cultural traditions that hold the nomadic life in high esteem, as do the Tuareg of the Sahara and the Bedu of south-west Asia. See also TRANSHUMANCE.

**non-conforming user,** *n.* a building or site used for a purpose which is at odds with the uses of other sites which now surround it. The non-conforming user may have been operating on the site before the surrounding users arrived and before LAND USE PLANNING was adopted to regulate the relationship between the users of the area.

Classic examples of non-conforming users are found in housing areas where small industries established themselves in pre-planning days and continue to operate although adjacent houses may have been rehabilitated or rebuilt. These industries may be a nuisance because of their physical bulk which can obstruct the amount of light reaching the houses, the dirt and/or smell of their processes, the danger to passers-by from the volumes of heavy traffic generated or, in extreme cases, the toxic nature of their air or water discharges.

**non-frontal depression,** *n.* a type of DEPRESSION which does not originate from the more usual development of small wave-like irregularities along major FRONTS. Non-frontal depressions develop in a number of ways and include THERMAL DEPRESSIONS, LEE DEPRESSIONS and tropical depressions such as HURRICANES.

**non-government organization (NGO),** *n.* any non-party-political PRESSURE GROUP, advisory agency, aid charity or professional body which may list among its aims the protection of the BIOSPHERE and its inhabitants. It is estimated that in excess of 12 000 NGO now operate, ranging in size from the small natural history society intent on maintaining local plant and animal communities to the internationally active groups such as GREENPEACE and FRIENDS OF THE EARTH. Until about 1970, NGO were largely a developed world phenomenon but after the UN Conference on Environment at Stockholm (1972), NGO were established in many developed countries.

**non-renewable resource,** *n.* any naturally occuring finite resource which, in terms of the human time scale, cannot be renewed once it has been consumed. Most non-renewable resources can only be renewed over a geological time span and all the FOSSIL FUELS and the mineral resources fall into this category. In recent years, as resource depletion has become more common, the process of RECYCLING resources has reduced the reliance on virgin non-renewable resources.

**normal life zone,** see TOLERANCE.

**North Atlantic Drift,** *n.* a warm, poleward moving OCEAN CURRENT found in the North Atlantic Ocean. An extension of the GULF STREAM flowing northeastwards from the Gulf of Mexico, the North Atlantic Drift is responsible for the mild winters experienced throughout the oceanic regions of northwest Europe. Without its presence ports such as Rotterdam and Hamburg would be ice-bound during winter time.

**northern lights,** see AURORA BOREALIS.

**Nor'wester,** *n.* a warm, dry, ADIABATIC WIND which blows on the South Island of New Zealand.

**NPP,** see NET PRIMARY PRODUCTIVITY.

**nuclear fission,** *n.* a process in which an atom of an element is struck by a neutron causing the atomic nucleus to split apart and release other neutrons. A chain reaction results as these neutrons then strike other atomic nuclei. In so doing large amounts of energy are released. In NUCLEAR REACTORS the chain reaction caused by the fission of uranium-235 is regulated by the use of CONTROL RODS which absorb quantities of free neutrons. Compare NUCLEAR FUSION.

**nuclear fusion,** *n.* a process in which two nuclei of light elements such as hydrogen are combined to form a heavier nucleus such as helium with a substantial release of energy. This process was thought to occur only at exceedingly high temperatures (in excess of 100 million degrees Celsius) although in 1989 the preliminary results of experiments conducted in the USA produced controversial evidence that fusion was possible under the temperature and pressure available at room conditions.

Nuclear fusion power is seen as the ultimate source of abundant, cheap and pollution-free electricity with none of the present disadvantages of NUCLEAR FISSION or problems associated with the burning of FOSSIL FUELS. However, all efforts to harness fusion power for commerical energy production have so far been unseccessful.

**nuclear power** or **atomic power,** *n.* energy generated by a NUCLEAR REACTOR primarily by NUCLEAR FISSION, or experimentally by NUCLEAR FUSION. There is almost unanimous agreement that fission power is one of the most hazardous methods of producing power (usually in the form of electrical power) for use by industry, transport and domestic users.

The growth in nuclear power plants has been rapid. Whereas in the mid-1970s, developed nations such as the UK, Belgium and Japan were generating approximately 10-15% of their electricty supply from nuclear sources, by 1990, this figure is expected to have risen to 45-55%. France is the most significant producer of nuclear-derived electricity. However, on a world-wide basis only 8% of all electricty is generated from nuclear sources.

Nuclear power presents major technical, moral and ethical problems for humanity. The problems of NUCLEAR WASTE disposal, nuclear accidents, and possible terrorist attacks on reactors have caused some nations to rethink their nuclear power programme. Austria has 'moth-balled' its Zwentendorf nuclear power station, while in Sweden (where 40% of energy is nuclear-derived), the decision has been taken to DECOMMISSION all its nuclear power stations by the year 2000. Some countries, such as New Zealand, have declared themselves totally opposed to the use of nuclear power for energy generation.

**nuclear reactor,** *n.* a device within which a nuclear reaction can be induced and controlled for the production of vast quantities of heat energy which can be converted to other forms of energy, notably electricity. Reactors operate on the basis of NUCLEAR FISSION. Throughout the world, there are some 700 nuclear reactors in use for the generation of NUCLEAR POWER; 250 of these are in the USA and account for 230 000 million electron volts (meV) out of a world total of 500 000 meV generated by nuclear reactors. See FISSION REACTOR. See also MAGNOX REACTOR, ADVANCED GAS-COOLED REACTOR, WATER-COOLED REACTOR.

**nuclear waste,** *n.* radioactive waste materials produced as by-products of research and nuclear-power generation. Medical sources are responsible for a small percentage of waste produced, but the greatest amount results from the exhaustion and reprocessing of fissile fuel sources from NUCLEAR REACTORS, chiefly uranium-238 and plutonium-239. Most nuclear waste is produced by nuclear power stations and after a certain proportion of the FUEL RODS within the reactor core have been used, they are withdrawn for reprocessing. Typically, waste fuel will be stored at the power station in a pressurized buffer storage tube for 28 days following which the fuel rod will be broken down into its component parts in

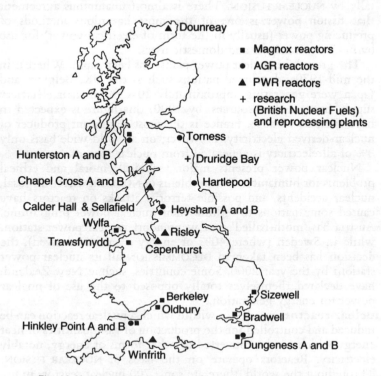

Fig. 44. **Nuclear reactor.** The location of nuclear reactors in the UK.

an automated dismantling facility and the irradiated fuel elements stored for 80 days in a water-filled pond to allow for further decay of fission product heating. The spent fuel is then placed in a special steel flask for transfer to a reprocessing plant, such as the Sellafield nuclear reprocessing plant in Cumbria, England. At the reprocessing plant up to 97.5% of the uranium and plutonium can be recovered and a small proportion recycled for new fuel. However, for the non-re-usable radioactive waste which remains no totally safe method of disposal exists. Nuclear waste is currently stored in slurry or liquid form in surface tanks protected by concrete and lead shields, buried in impermeable strata in deep bore holes as vitrified solids, or dumped in steel and concrete containers deep on the ocean bed.

Nuclear waste is classified into three categories of radioactivity:

(a) *high-level waste*, with a radioactivity greater than $3.7 \times 10^{10}$ becquerels (Bq) per gallon of slurry;

(b) *intermediate waste*, with a radioactivity of between $3.7 \times 10^4$ and $3.7 \times 10^{10}$ Bq per gallon of slurry;

(c) *low-level waste*, with a radioactivity of less than $3.7 \times 10^4$ Bq per gallon of slurry.

All radioactive materials have HALF-LIFE values which indicate the amount of time necessary for radioactivity levels to subside to non-dangerous values. Some half-life values are very short, for example, iodine-131 has an 8-day half-life and after 50 days its radioactivity is only 10% of its fresh value. However, plutonium-239 has a half-life of 240 000 years and retains a lethal radioactivity level for 500 000 years while uranium-238 has a half-life of $4.5 \times 10^6$ years.

**nuclear winter,** *n.* a theoretically deduced state which some scientists believe would exist after a nuclear war involving major atmospheric nuclear explosions. The devastation caused by the blast combined with the heat generated from the resulting fires, and the death of plants and animals as a result of radioactive FALLOUT would lead to a major increase in the PARTICULATE MATTER in the atmosphere. The lower atmosphere would become so opaque that the transmission of solar energy would be prevented, leading to a substantial lowering of the average air temperature. As the particulate matter was removed from the atmosphere by gravitational fallout then so the nuclear winter would retreat.

It should be stressed that not all physicists believe that a nuclear winter would follow a major atmospheric nuclear explosion.

**nuée ardente,** *n.* a rapidly moving incadescent cloud of hot volcanic gas, dust and steam formed by the sudden and explosive eruption of certain types of volcano. In 1902, Mont Pele on the Caribbean island of Martinique erupted and the resulting nuée ardente destroyed the city of Saint Pierre, killing 30 000 people.

**nutrient cycle or biogeochemical cycle,** *n.* the constant transfer of essential nutrients from living organisms to the physical environment and back to the organisms in a cyclical pathway. This sequence is achieved by physical processes such as WEATHERING and/or by biological processes such as decomposition. Due to the finite nature of the supply of nutrients they must constantly be re-used in order that organic life can continue. Of the approximately 100 elements in the EARTH'S CRUST about 30 are known to be necessary for living organisms and it is these which are cycled, at

# NUTRIENT CYCLE

varying speeds, between the BIOTIC and ABIOTIC components of an ecosystem. The remaining 70 elements undergo a simpler, and usually much slower *geological cycling* process within the abiotic cell.

# O

**oasis,** *n.* any location in an arid area where a regular and assured supply of water permits sustained plant growth, crop production and human settlement. Oases often depend on springs, wells, or surface or near-surface water. They vary in size and importance from small clusters of palm trees to vast fertile areas occupying broad valley floors, such as the Tafilalet in Morocco, and wide tracts of land supporting sedentary agriculture close to major waterways such as the Nile. Land holdings dependent on oases tend to be small and intensively cultivated, and are generally marked by a sharp transition from lush vegetation to the barren desert beyond.

**occluded front,** *n.* a type of FRONT formed by an advancing COLD FRONT overtaking a more slowly moving WARM FRONT and gradually raising the warm sector of a DEPRESSION off the earth's surface. Occluded fronts are commonly associated with mid-latitude depressions. When the air of the overtaking cold front is cooler than the cold air in front of the warm front then a *cold occlusion* results whilst if the cold front is warmer than the air ahead of the warm front then a *warm occlusion* occurs. The development of an occluded front indicates the decline of a depression.

**occupancy rate,** see RESIDENTIAL DENSITY.

**ocean,** *n.* a vast area of salt water surrounding the continents. There are five oceans: the *Atlantic*, the *Pacific*, the *Indian*, the *Arctic* and the *Southern Ocean*. They form an important part of the HYDROLOGICAL CYCLE and account for 71% of the earth's surface.

Oceans have numerous relief features and can be divided into the CONTINENTAL SHELF, CONTINENTAL SLOPE and the DEEP-SEA PLAIN. The salinity of ocean water is dependent upon water temperature, ocean movements, the amount of precipitation and the rate of evaporation. Surface temperature of ocean waters can also vary considerably, ranging from 35°C in shallow equatorial waters to the −1.9°C recorded in the Weddell Sea during the Antarctic winter.

Oceans are of great importance to humans. Coastal and deep-sea FISHING are important elements in the economies of many countries and disputes over fishing grounds have led to hostilities such as the

Cod Wars of the 1970s between the UK and Iceland. Dense populations in coastal areas such as Japan and parts of Africa are dependent upon fish as a source of protein but over-exploitation has led to a depletion of fish stocks in many fishing grounds.

Oceans are also of great economic importance. The extraction of bromine, magnesium and salts such as sodium chloride from sea water dates from early historical times and although recovery nowadays usually occurs in industrial chemical complexes, primitive panning methods are still used in the Mediterranean, in Brittany and in developing countries. Other exploitable economic resources found in the oceans include sands and gravels, as well as oil and gas reserves in the North Sea, the Tasmanian Gulf, and the Gulf of Mexico.

Pollution of the oceans is a serious and growing problem. MARINE POLLUTION in the form of oil spills, untreated human sewage and industrial waste can have catastrophic effects on oceanic environments which, in turn, can have repercussions for humans; for example, the contamination of shellfish by heavy-metals and other toxins has led to serious outbreaks of poisoning among human populations dependent on sea food. See also LAW OF THE SEA.

**ocean current,** *n.* a regular movement of ocean water following a well-defined route (see Fig. 45). Two types of ocean currents have been recognized:

(a) *wind-driven currents* are closely related to atmospheric circulation and their flow generally reflects the rotation of winds between the major oceanic pressure systems. In the northern hemisphere these currents circulate in a clockwise direction, whilst their motion is anticlockwise in the southern hemisphere. Currents moving from the equator to the pole, such as the East Australian Current and the North Atlantic Drift, are known as *warm currents*, while currents moving towards the equator such as the Humboldt or Peru Current and the Canary Current, are called *cold currents*;

(b) *density currents* involve the movement of ocean water due to the variations in water temperature and salinity. When surface water in high latitudes is cooled, it sinks and flows towards the equator beneath the warmer water of surface currents. Density currents may also result in areas of differing salinity; water with a low salinity is less dense than that with a high salt content and consequently flows in the upper zones of the seas and oceans. As it flows into areas of more saline water, there is a movement of the denser, more saline water below the surface flow in the opposite

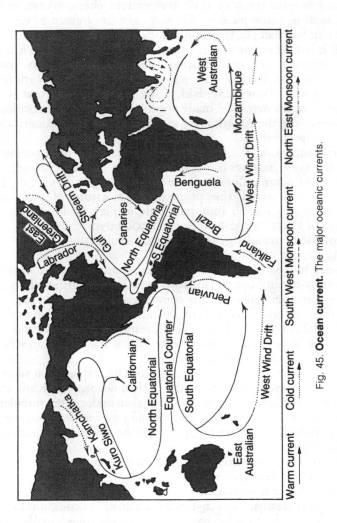

Fig. 45. **Ocean current.** The major oceanic currents.

direction. Such a current exists between the Mediterranean Sea and the Atlantic Ocean.

Currents can transmit their temperature characteristics to the surrounding atmosphere and may greatly influence regional climates; for example, the ameliorating effect of the North Atlantic Drift is largely responsible for the mild winters of western Europe. Where warm and cold currents meet, dense FOG may result. Following the *Kontiki* and *Ra* expeditions of the Norwegian archaeologist, Thor Heyerdahl, it has been suggested that wind-driven currents may have made possible the human colonization of Polynesia and the Americas by migrants from South America and Egypt respectively.

**oceanography,** *n.* the study of the physical, chemical, geological and biological features of oceans and ocean basins. It includes the analysis of oceanic flora and fauna, the topography of the ocean bed, as well as of the properties and nature of sea water. Oceanography has an important role to play in developing the theory of PLATE TECTONICS. See HYDROGRAPHY.

**ocean-floor spreading,** see SEA-FLOOR SPREADING.

**ODA,** see AID.

**offshore bar,** *n.* an offshore bank of sand and shingle that develops on gently sloping coasts. LAGOONS form behind offshore bars, and these may eventually silt up completely. LONGSHORE DRIFT can lengthen offshore bars to form islands such as Fire Island off the coast of Long Island, USA. These features are known as *barrier islands* or *barrier beaches*, which, especially if stabilized, can serve as protective barriers to the coast behind.

It is unclear how offshore bars form. One explanation suggests that storm waves erode material and then deposit the debris some distance offshore. Another contends that underwater turbulence from breaking waves causes the mounding of material on the sea bed. Changing sea levels may cause the onshore migration of offshore bars.

**Office of Coastal Management,** see NOAA.

**Office of Sea Grant,** see NOAA.

**O-horizon,** see H-HORIZON.

**oil** or **petroleum,** *n.* a naturally occuring viscous mixture of gaseous, liquid and solid hydrocarbon compounds usually found trapped deep underground beneath impermeable cap rock and above a lower dome of sedimentary rock such as shale. Most oil deposits are found in sedimentary rocks of marine, deltaic or estuarine origin.

OIL

It is thought that oil may take up to a million years to form; oil deposits have never been found in Precambrian or Recent rocks. The exact processes of oil formation are still debated but it is thought that oil results from the anaerobic decomposition of organic material during LITHIFICATION. The gradual compaction of the source rocks causes the migration of oil into other strata. Unless the oil becomes trapped beneath a cap rock the migration continues until the oil reaches the surface and the lighter hydrocarbons disperse or evaporate leaving the heavier, residual hydrocarbons to form *tar sands*.

The occurrence of oil reserves ranges from small isolated deposits to vast oil fields, such as the Gulf of Texas. Oil deposits are found throughout the world, with the major producers located in the Middle East (with 54% of all known reserves).

Oil deposits are detected by GEOPHYSICAL PROSPECTING methods and over the past 30 years, advances in drilling technology has allowed the exploration of offshore (marine) oil fields. If there is enough pressure from water and natural gas under the cap rock, some of the crude oil will rush to the surface when the well is drilled. Such wells, called *gushers*, are rare.

*Primary oil recovery* involves the pumping out of the oil, and is followed by the injection of water to force out further quantities of oil, a process called *secondary oil recovery*. Together, these processes recover only about 33% of the oil reserve, the thicker oils remaining. As the price of crude oil increases then so *enhanced oil recovery* methods (such as steam injection) may be used to remove a further 10% of the reserve. The enhanced oil recovery techniques currently require an energy input of the equivalent of one-third of a barrel of oil for each barrel actually extracted.

Onshore drilling operations can result in the subsidence of large areas causing damage to property, roads and railways. This problem can sometimes be overcome by injecting water to stabilize the underlying rock strata. Blow-outs and fires are a serious hazard whilst the accidental leakage of oil can contaminate agricultural land and groundwater supplies.

Marine oil pollution resulting from tanker accidents and leakage from pipelines is a serious problem which affects coastral tourism, fishing grounds and marine ecosystems. Most environmental intrusion from offshore operations is caused by ancillary onshore developments, such as petrochemical complexes to manufacture petrol, fuel oils, chemicals and plastics, and harbours and pipelines to service the offshore fields.

**oil field,** *n.* an area from which oil is produced in commercial quantities.

**oil-fired power station,** *n.* a type of power station in which oil is used to generate steam for use by electricity-generating turbines. During the era of cheap crude oil prior to the 1970s, the economics of electricity generation favoured oil as opposed to the traditional fuel of coal. In the UK, large oil-fired power stations were constructed at deep water coastal sites such as Milford Haven in Wales and Hunterston in Scotland to which supertankers brought Middle-Eastern oil. The first oil crisis of 1972–73 followed by the succesion of major oil price rises throughout the 1970s made the oil-fired stations uneconomic. Consideration was given to their conversion to natural gas but this too would have been only a short term expedient and economically unsustainable.

Most oil-fired power stations have now been 'mothballed' and are retained on a care-and-maintenance basis only. It is probable that even if oil prices had remained low this form of power generation would have become unacceptable due to rising environmental standards; this type of power station was designed to burn low grade oil which contained up to 4% sulphur thus making them major sources of sulphur dioxide pollution. See ACID DEPOSITION.

**oil shale,** *n.* a SHALE from which crude petroleum can be obtained through distillation. Oil shales are common throughout the world and yield locally important quantities of oil, for example,in the USA and USSR. During the recovery operation of oil from shale a 30% volume increase occurs thereby creating a major waste disposal problem.

**oil slick** or **oil spill,** *n.* the release of crude oil either by accident, design, or by natural extrusion into oceans, estuaries or rivers. Depending upon the prevailing wind speed and direction and tidal movements oil slicks can extend over a substantial area; for example the spillage from the supertanker *Exxon Valdez* in Alaska in 1989 covered an area of over 2500 km². Great damage can be done to marine life forms and if the slick is washed onshore then it can pollute the land between low and high water marks.

The average amount of oil released as slicks or spills into the oceans has been estimated at between 2 and 5 million tonnes p.a. with only 10% of this originating from natural outpourings. Despite the volume of publicity given to oil tanker disasters (such as the *Torrey Canyon* in 1967, the *Amoco Cadiz* in 1978 and the *Exxon Valdez*) and offshore oil platform blowouts (such as the Ekofisk blowout in 1977), less than 20% of oil slicks originate from such disasters. More than 66% of all marine oil pollution originates from the illegal dispersal of waste oil from industrial sources and from motor vehicle sumps. New technology now permits the resynthesis of waste oil into high grade oil thus removing the need to dispose of large quantities of waste oil in this manner.

**oil spill,** see OIL SLICK.

**oil terminal,** *n.* a collection place and junction point on an oil transport system. Oil is landed from pipes on the sea bed (as around the North Sea) or from land piplines in the case of landward oil fields (such as those in the Middle East) for temporary storage and onward transport by land and water to a point of further processing. Examples of terminals are Sullom Voe in Shetland, Scotland and Kharg Island in the Persian Gulf.

**old field,** see MEADOW.

**oligopoly,** see MONOPOLY.

**oligotrophic,** *adj.* (of a soil or water body) having a very low nutrient content. In soils this may be due to the paucity of nutrients in the parent rock; alternatively, excessive LEACHING can result in oligotrophic soils. Oligotroghic water bodies are unable to sustain active aquatic flora and fauna communities. Compare EUTROPHICA-TION.

**omnivore,** *n.* any animal species which gains its food supplies from a variety of sources, including plant material and animal matter. Compare CARNIVORE, HERBIVORE.

**onion skin weathering,** see PHYSICAL WEATHERING.

**ooze,** see DEEP-SEA DEPOSIT.

**opencast mining** or **open cut mining,** *n.* a method of mineral extraction for deposits lying at or near the surface and not requiring the use of shafts or tunnels. After the OVERBURDEN has been removed, the deposits are quarried. Opencast operations are usually extensive in nature and are used most often in the extraction of lignite, coal and iron ore. The economics of large-scale opencast mining allows the extraction of minerals of lower grade than would be possible in DEEP MINING. Normally, opencast mining is economic provided that the ratio between overburden and mineral does not exceed 15:1. As opencasting can cause extensive scarring of land, many countries only allow opencast operations to commence after the payment of RESTORATION BONDS. When mining ceases, the money is used to reinstate the land for productive uses such as agriculture and forestry. See STRIP MINING.

**open community,** *n.* any youthful plant COMMUNITY in which insufficient time has elapsed for all available living space to become colonized.

Openness within a plant community may be confined to above ground conditions only. Poor soils and/or aridity may result in closed conditions below ground.

Open communities can be strictly defined as areas of space not more than twice the diameters of the predominant vegetation clumps. If this value is exceeded then the vegetation is described as *sparse*. SUBSTRATUM, not vegetation dominates the landscape in this instance. Compare CLOSED COMMUNITY.

**open cut mining,** see OPENCAST MINING.

**open field,** see FIELD SYSTEM.

**opportunity cost,** *n.* the value of the opportunities foregone in carrying out a particular course of action. The cost of a particular action may be calculated either on the basis of its direct monetary outlay or according to the highest return that would have accrued from some alternative course it precluded, the two being the same only when the value of the outlay accurately reflects the value of the alternative. Thus, the opportunity cost of any given land use is the return that an alternative use might yield. In agriculture, a particular activity may yield a farmer a net profit, but if there is an alternative activity that would yield a greater return then the opportunity cost of the former is the difference between the return of the greater and smaller, that is, the cost of one alternative in terms of the alternatives not taken up.

**optimal range,** see ENVIRONMENTAL GRADIENT.

**optimum growth,** *n.* a condition in which a plant achieves maximum productivity through the particular balance of biological and environmental conditions, including day length, temperature, moisture and soil fertility. In CROP production, optimum growth may be attained by varying degrees of human and physical inputs, for example the application of FERTILIZER, the use of IRRIGATION, and more intensive forms of cultivation, sometimes compensating otherwise sub-optimal conditions. As there are specific conditions for each plant that are most favourable for its growth, it is possible to define areas of the world where yields will be highest and annual variability lowest. As environmental conditions deteriorate away from the optimum area, yields fall and variability increases, until the point is reached on the margins of production where the crop's minimum growth requirements are no longer met.

**optimum population,** *n*. the number of human inhabitants in an area that permits the most favourable return in accord with specified goals. Both the goals and the interpretation of what constitutes the most favourable return vary; an optimum population may be one which permits individuals to reach their full potential and to secure a reasonable standard of living, or one in which the population is adequate to exploit to the best advantage all the resources of an area while maximizing collective returns. In strictly economic terms, the optimum population is reached when total production or real income per capita is greatest. An optimum population may be reached somewhere between the conditions of UNDERPOPULATION and OVERPOPULATION, but the point of optimality is far from precise. See CARRYING CAPACITY.

**order,** *n*. any of the taxonomic groups into which a class is divided. For example, Carnivora and Rodentia are two orders of the class Mammalia. The order can comprise several or many related FAMILIES. See LINNEAN CLASSIFICATION.

**Ordovician period,** *n*. the geological PERIOD that followed the CAMBRIAN PERIOD about 500 million years ago and that ended about 440 million years ago when the SILURIAN PERIOD began. There was a major phase of mountain building and volcanic activity in Europe, North America and Australia during the Ordovician. A variety of invertebrate marine fossils are used to date and correlate Ordovician rocks, the most widespread being the graptolites. In North America, the first vertebrate fish appeared during this period. See EARTH HISTORY.

**ore,** *n*. an economically workable metalliferous mineral deposit. Ore deposits may occur in igneous, metamorphic or sedimentary rocks and may be magmatic, metasomatic, pneumatolytic, hydrothermal or residual in origin. Some gold and tin ores are found in PLACER DEPOSITS. Ore deposits are a non-renewable resource.

**organic sedimentary rock,** see SEDIMENTARY ROCK.

**organic weathering,** see BIOLOGICAL WEATHERING.

**orogenesis,** *n*. the process by which intensely folded and faulted mountain ranges are formed. Orogenesis results from the lateral movement of LITHOSPHERIC PLATES which causes the intense upward displacement of sections of the EARTH'S CRUST. Major periods of mountain formation (*orogenies*) last for tens of millions of years. The Alpine Orogeny, which saw the development of both the Alps and the Himalayas began around 35 million years ago and is still in progress. See FOLD MOUNTAINS. See also PLATE TECTONICS, EPEIROGENESIS, BATHOLITH.

**orogeny,** see OROGENESIS.

**orographic rain,** see RAIN.

**outcrop,** *n.* an area where bedrock occurs at the surface whether visibly exposed or underlying a veneer of soil and vegetation.

**out-of-danger species,** see RED DATA BOOK.

**out-of-town shopping centre,** *n.* a large-scale retail premises located on a GREENFIELD SITE between existing towns or at the outer edge of a town. Such centres are usually close to major roads and especially to intersections giving rapid access to the regional transport network. Sometimes the development consists of retail stores specializing in a particular product, for example in furniture, electrical goods or food. This is more common in the UK than in North America, where SHOPPING MALLS contain a range of other shops including discount stores and former city centre department stores, within a covered walking area.

Planning policy in the UK has opposed the development of North American-style malls on greenfield sites and instead has sought to locate them on highly accessible sites within existing cities in order to attract pedestrian users as well as car owners. It is considered that unrestricted development of such centres on greenfield sites would lead to an impoverishment of existing centres within cities and thus lead to a decline in the standards of service to less mobile members of the population. See also HYPERMARKET.

**outwash deposit,** see GLACIAL DEPOSITION.

**outwash plain** or **sandur,** *n.* an area of stratified glacial till deposited by meltwater streams sloping away from the margins of a glacier or icesheet. In an enclosed valley outwash plains are known as *valley trains*. In parts of northern Europe and the American Midwest outwash plains form productive agricultural land. Sand and gravel extraction is a common feature of many outwash plains. See GLACIAL DEPOSITION.

**outfield,** see INFIELD-OUTFIELD SYSTEM.

**overburden,** *n.* the soil and bedrock overlying mineral deposits that is removed prior to the commencement of OPENCAST MINING or STRIP MINING.

**overcrowding,** see CROWDING.

**overfishing,** see FACTORY FISHING.

**overland flow,** *n.* the movement of rain or meltwater over the ground surface in a thin unchannelled sheet. It usually develops during high-intensity rainstorms when the INFILTRATION capacity of the soil is exceeded. Overland flow is most frequently found in arid and semi-arid areas and is closely associated with SHEET EROSION. The gradual concentration of overland flow into RILLS may subsequently result in GULLY EROSION. See RUNOFF.

**overpopulation,** *n.* a situation in which the number of people in an area cannot be adequately supported by the available resources, leading to declining standards of living and a failure to realise human potential fully. Overpopulation may be said to occur whenever the OPTIMUM POPULATION level is exceeded.

The concept of overpopulation should not imply absolute limits, or necessarily that the CARRYING CAPACITY of an area has been outstripped by population growth, since the population that can be supported by the resource base may vary according to the level of technological development and the method of cultivation being practised. An area may be overpopulated, for example, if its inhabitants continue to practice an extensive form of SHIFTING CULTIVATION, but a progression to more intensive systems of ROTATIONAL BUSH FALLOW, perhaps in response to population growth, may yield higher returns and thus support the population adequately. Overpopulation may lead to excessive pressure on the land leading to environmental degradation and a further intensification of overpopulation as the productive capacity of the resource base deteriorates. See MALTHUSIAN MODEL.

**overseas development assistance,** see AID.

**overspill,** *n.* the planned displacement of population following slum clearance and redevelopment in an area of a city. In older parts of cities, the population density may be such that it is impossible to return the same number of people to the site after new buildings (governed in their designs and spacing by modern regulations regarding daylighting and accessibility) have been constructed. The difference in numbers between the original total numbers resident in an area and the expected new capacity of the rebuilt houses, is termed the *overspill population*. The old housing densities and planning policies of the 1960s, aimed at decongestion of cities, led to substantial overspill populations during the redevelopment of the inner parts of all the British CONURBATIONS. The need to accomodate overspill led, in turn, to the development of large areas of council housing on the edges of cities, the PERIPHERAL ESTATES, as well as to NEW TOWNS.

**owner occupation,** see LAND TENURE.

**oxbow lake, cutoff** or **mortlake,** *n.* a crescent-shaped lake formed when a river breaches the neck of a pronounced river MEANDER. As the river current forms a new course, silt is deposited

317

at the entrance of the meander which gradually becomes cut off from the main stream flow until it is completely separate from the river (see Fig. 40). If the water level in the oxbow is maintained by seepage from the river then a lake will remain; if not then the oxbow will eventually silt up and may no longer be a discernible topographic feature. Oxbow lakes are common on most large river FLOODPLAINS such as those of the Severn in England and the Mississippi in the USA. The straightening of rivers for navigation purposes may lead to the artificial creation of oxbow lakes.

**oxidation,** see CHEMICAL WEATHERING.

**oxisol,** *n.* the term used in the American soil classification system for a LATOSOL-type soil.

**ozone,** *n.* a highly reactive gas comprising triatomic oxygen ($O_3$) formed by the recombination of oxygen in the presence of ultraviolet radiation. It exists as a natural component of the atmosphere in concentrations of approximately 0.01 parts per million of air. An *ozone layer* exists in the upper atmosphere that protects life on earth by filtering out harmful ultraviolet radiation from the sun. In the lower atmosphere, OZONE POLLUTION from human sources is causing a build-up of ozone while in the upper atmosphere the average concentration of ozone is being depleted by a build-up of CHLOROFLUOROCARBONS (CFC). NASA has predicted a 10% depletion of the ozone layer by 2050 unless there is a dramatic decrease in the use of CFCs; scientists have already discovered 'holes' in the layer over the poles, particularly over Antarctica. Unless checked, it is feared that the increased solar radiation reaching the earth's surface result in a marked change in the distribution pattern of world climates and could bring about drastic changes in agricultural patterns.

**ozone layer,** see OZONE.

**ozone pollution,** *n.* atmospheric pollution resulting from a build-up of OZONE. Ozone is a natural component of the ATMOSPHERE and serves to filter out harmful incoming short-wave SOLAR RADIATION. However, it can also be formed by the reaction of hydrocarbons (the breakdown products of petroleum) and nitrogen oxides (released from high temperature and high pressure combustion processes, notably automoble engines) in the presence of sunlight. In this form, ozone forms an important component of PHOTOCHEMICAL SMOG. In very small concentrations (0.3 parts per million) it causes irritation to the human respiratory channels and at even lower concentrations may cause vegetation surfaces to become discoloured. Once affected, plant growth is substantially reduced and death may eventually occur. The citrus fruit industry in parts of California has been destroyed by ozone pollution.

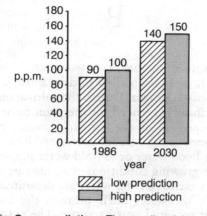

Fig. 45a **Ozone pollution**. The predicted increase in ozone 1986–2030.

# P

**paddock grazing,** see GRAZING MANAGEMENT.

**paddy (or padi),** *n.* **1.** also called **paddy field**, a small flooded field where wet-rice is grown. Due to the cultivation requirements of wet-rice, specifically the need for the plant to be submerged in water for three-quarters of the growing season, paddies have a relatively impermeable sub-soil and are ringed by embankments to impound rain or flood water or to hold water supplied by artificial IRRIGATION. The growing conditions of a paddy are especially rich: silt and nutrients carried in the water are deposited as the field is inundated, while the water itself protects the land from high temperatures and potential SOIL EROSION. The LEACHING of plant nutrients is limited by the high WATER TABLE, but the ability of WET-RICE CULTIVATION systems to maintain fertility, even under PERMANENT CROPPING, is due above all to the blue-green ALGAE and bacteria that proliferate in the slowly circulating warm waters of paddy-fields.

**2.** unmilled rice with the bran and husk intact.

**Palaeozoic era** or **Primary era,** *n.* the second oldest geological ERA that followed the PRECAMBRIAN ERA about 600 million years ago and ended about 225 million years ago when the MESOZOIC ERA began. The Palaeozoic can be divided into six periods in the UK and seven in the USA. Prior to the Paleozoic, the occurrence of fossils was limited. Lifeforms that developed during this era included many types of invertebrates, amphibians, reptiles and the first seed-bearing plants, the GYMNOSPERMS. See EARTH HISTORY.

**palynology,** see POLLEN ANALYSIS.

**pampas,** *n.* a long-established grassland community in Uruguay dominated by tall grasses of the Stipa and Melica species. Small thickets of trees occur in damper hollows and along river courses. Grasses occur in characteristic bunches, with substantial areas of bare soil between tussocks. Disagreement exists over the origin of the pampas: it is possible that they are ancient CLIMAX COMMUNITIES although it has been suggested that they are the result of regular burning caused by native peoples long before the arrival of European settlers.

Many areas of pampas have been converted into productive cereal production areas. Elsewhere, pampas have been improved by reseeding and application of fertilizers to form productive grazing areas for beef cattle.

**Pangaea,** *n.* the supercontinent believed to incorporate most of the continental crust of the earth and to have existed prior to the CARBONIFEROUS PERIOD. The concept of the Pangaea was formulated as part of the theory of CONTINENTAL DRIFT devised by the German climatologist, Alfred Wegener, in 1912. At some point the supercontinent fragmented into the protocontinents of GONDWANALAND and LAURASIA (see Fig. 19). See also PLATE TECTONICS.

**parabolic dune,** see DUNE.

**parasitism,** *n.* the relationship between two organisms, in which one (the *parasite*) lives upon, or within, the other (the *host*). The parasite is usually much smaller than its host and the effect of its presence can range from none at all to severe illness and possible death.

Parasitic relationships often display a range of forms of dependency of the parasite upon the host species. For example, *endoparasites* live within the host's body while *ectoparasites* live externally; *obligate parasites* can exist only as parasites while *facultative parasites* can live both as parasites and independently as SAPROPHYTES. Improved knowledge of species inter-relationships has suggested that parasitism is a much more common type of relationship than was previously thought. Compare SYMBIOSIS, COMMENSALISM.

**paramo** or **puna,** *n.* the vegetation of high tropical mountains in South America. Paramo is the term most frequently used in the northern Andes, while puna is most commonly used in the central Andes. Terms with similar meaning are *afro-alpine* in East Africa or *tropical-alpine* in Malaysia. No single name encompasses all the types of high tropical vegetation above the timberline although the use of the term paramo has been advocated for *all* vegetation of high tropical mountains. The greatest extent of paramo occurs in the Andes, in particular on the Altiplano of central and southern Peru, Bolivia, northern Chile and northern Argentina. Smaller patches occur in Venezuela, Columbia, Ecuador and northern Peru. In Africa it is confined to the Ethopian table land while in the Pacific area it is confined to the highest land of Borneo and New Guinea.

Paramo comprises high mountain grasslands, dominated by

bunch grasses but also containing shrub plants and rosette tree-like plants of the Compositae family. At the very highest elevation grasses are almost entirely replaced by the Compositae family. Because of the frequent severe frosts, the ground surface displays many features of arctic environments, most notably PATTERNED GROUND.

**parcellement,** see FRAGMENTATION.

**parent material,** *n.* the minerals from which a SOIL originates. Parent material may be derived from the WEATHERING of underlying bedrock or from sediments deposited by wind, water or ice. Only marginally altered parent material usually forms the C-HORIZON of a soil. The nature and composition of parent material is an important element in PEDOGENESIS and in particular determines the property of SOIL TEXTURE.

**park-and-ride system,** *n.* a transport arrangement which links car travel to rail or bus travel by provision of car-parking facilities at railway stations or bus terminals usually located at strategic points within the suburbs. This arrangement is one instrument of TRAFFIC MANAGEMENT, and its aim is to reduce the numbers of commuters or shoppers who attempt to take their cars into a city centre by offering free or subsidized parking at places where they might consider diverting to a public transport service.

**parkland,** *n.* an area of land characterized by open views and widely spaced trees. It is associated with the landscaped grounds of the large country estates and the mansions of post-Renaissance Europe, especially England.

**partial drought,** see DROUGHT.

**partible inheritance,** see FRAGMENTATION.

**particulate matter,** *n.* solid particles and liquid droplets present in the atmosphere. These pollutants have various sources, both natural and as a result of human activity. Examples include soot and ash emitted from factory and power station chimneys, dust raised by mining and agricultural practices, and dust and ash generated by volcanic eruptions.

Particulate matter provides the nuclei around which the various types of PRECIPITATION form and may sometimes be hygroscopic in nature, such as particles of salt. See HYGROSCOPIC NUCLEI. See also AIR POLLUTION, SMOG, FALLOUT.

**partnership authorities,** see URBAN AID.

**part-time farming,** *n.* an agricultural structure in which one or more members of a farm-based household are employed in work other than, or in addition to, farming the family's holding. Three main forms may be identified:

(a) *worker-peasant farming*, common throughout eastern and central Europe, in which a farmer adds industrial wage employment to small-holder agriculture, devoting only evenings and weekends to the farm. Formerly regarded as a transitional phase of employment leading out of agriculture altogether, it has become a growing and seemingly permanent sector of the rural economy in most European countries as small-scale farmers, wishing to retain their links with family farming traditions, seek a solution to low farm incomes. With recession in manufacturing industry, the farm may cushion the worst effects of enforced redundancy.

(b) *hobby farming* where an individual with a full-time and generally well-paid non-farm occupation purchases a farm as a diversion or hobby. Very often located in PERI-URBAN REGIONS, such farms tend to be smaller than average for the area and less intensively operated, reflecting the non-farm employment demands on the owner's time and resources. For some, hobby farming is a step towards full-time farming.

(c) farming that is combined with non-agricultural work (based on or in close proximity to the farm) that contributes significantly to family income. These may include: tourist-related activities (catering, the provision of accommodation or camping facilities, and recreational pursuits), the running of a small business from the farm, fishing and FISH FARMING. In such cases the farmer, unlike the worker-peasant, retains entreprenurial control and wholly self-employed status.

A particular form of part-time farming that may have elements of each of these defining characteristics is the Scottish system of CROFTING.

Most farm families have some non–agricultural income without it necessarily implying that the farm is a part-time enterprise, although in many parts of Europe and the USA such earnings form a significant proportion of family farm income, and in Japan on average exceed income from agriculture. The term part-time farming is being replaced by that of *multiple-job holding*, an occupational structure seen as beneficial in maintaining rural population densities and increasing overall rural standards of living. On the other hand, it makes few contributions to local employment, reducing land use intensities and imposing structural rigidity in FARM SIZES.

**pass,** see COL.

**pastoralism,** *n.* a type of farming concerned with the rearing of herbivorous livestock for food, clothing, shelter and artifacts useful to humans. There are many forms of pastoralism, ranging from the extensive commercial RANCHING of the sheep stations of northern Australia and the cattle ranches of the USA to the highly intensive dairy farming enterprises of western Europe. Capital investment may be limited or non-existent as is the case with pastoral NOMADISM, or considerable as with most modern forms of livestock farming in DEVELOPED COUNTRIES; on modern farms expensive equipment is required, for example, to facilitate GRAZING MANAGEMENT and to automate milking. See ANIMAL HUSBANDRY. See also FACTORY FARMING, FEEDLOTS, TRANSHUMANCE.

**pasture,** *n.* an area of grassland utilized by man in ANIMAL HUSBANDRY, and normally grazed by livestock rather than cut for fodder. The quality and condition of pasture influences, and is influenced by GRAZING MANAGEMENT.

Natural pasture, known as *rangeland* in North America, is a natural CLIMAX COMMUNITY consisting of native grasses, and herbaceous plants other than grasses and shrubs. Natural (or semi-natural) pastures include most of the Eurasian STEPPE, the North American PRAIRIE, the montane grasslands of the tropics, the MOORLANDS of the British Isles, and the mountain pastures of the Alps, Andes and Urals. Semi-natural pasture can be developed from a climax vegetation by natural or induced succession, by over-grazing, and by deliberate RANGE MANAGEMENT to encourage a higher proportion of useful, palatable and nutritious plants, and to restrict the growth of undesirable species. Most of the extensive tropical grasslands are essentially artificially created by the joint action of FIRE and grazing. In upland Britain fire is regularly used to improve pasture for sheep and to maintain grouse moors for sporting interests.

Permanent pasture is grassland maintained indefinitely for grazing. It is normally land that has been ploughed and sown with improved or introduced perennial grasses, and annual species able to propogate themselves by reseeding. Once established, permanent pasture forms a balanced grass sward and is very rarely ploughed for use in CROP ROTATIONS. With its condition improved by fertilizing, liming and controlled grazing, it can be maintained for long periods before the need for reseeding.

Permanent pasture is widespread in North America, Europe and New Zealand. With good management, permanent pasture may support stocking densities of 5-6 animals per hectare, as against

0.5-1.0 animals per hectare over much of the open-range, extensively managed natural pasture of the American prairies.

Temporary pasture, or *leys*, consist of grasses and legumes sown and kept down for a year or more, before being ploughed up again as part of an arable CROP ROTATION under LEY FARMING. Ley pasture generally contains more nutritious plant species than permanent pasture. Although leys may be grazed directly, they are commonly cropped for HAY and SILAGE, and thus strictly should be distinguished from pasture.

It is not possible to make precise distinctions between semi-natural and permanent pasture. Due to differences in definition and measurement international comparisons are not always valid, and should be treated with more caution than figures for CROPLAND or PERMANENT CROPS. In the EC in 1986 the percentage of the utilized agricultural area classed as permanent grassland ranged from 8% in Denmark to 81% in Ireland, with an average for the Community of 38% or some 49 million ha. International comparisons include: USA (56%), Canada (40%), Australia (90%), New Zealand (97%), and USSR (62%).

**patterned ground,** *n.* a series of distinct, geometric patterns formed on the soil surface of TUNDRA regions as a result of the natural sorting of stones and fine material. The regular freezing and thawing of the soil surface results in the finer materials being moved towards the centre of a roughly circular, or polygonal shape which may vary in size from 1 - 15 m in diameter. Patterned ground is best developed on undisturbed flat, or very gently sloping ground. On slopes with an angle greater than 5° the polygons become elongated and eventually form into *stone stripes*.

**PCB,** *abbrev. for* polychlorinated biphenyl, a group of at least 50 widely used compounds containing chlorine. PCBs can accumulate in FOOD CHAINS and are thought likely to produce a variety of harmful side-effects, particularly during the reproductive cycle of plants and animals. Many PCBs are non-BIODEGRADABLE.

**peasant,** *n.* any member of a farming system based on the family as the unit of production and consumption. The term is widely, if imprecisely, used to imply agriculturalists with a poor standard of living. Peasant farming is characterized by small farm holdings, land rights (whether owned or rented) being vested in the family rather than an individual, a dependence on family labour, production primarily but not exclusively to meet domestic requirements, and a minimal use of mechanization.

Peasants are distinguished from groups concerned solely with

SUBSISTENCE FARMING in that any surplus produced over and above family requirements can be sold at commercial outlets to provide a small income to pay rent to a landlord or, increasingly, to be used to purchase goods which the peasants cannot produce themselves.

The status of the peasantry worldwide is undergoing considerable change. While some countries, such as the USA and Australia have never had a peasant class, the 20th century has seen in much of Sub-Saharan Africa a shift from tribal farming to peasant farming structures. In Europe, peasant farming has been reduced by the demands of modern COMMERCIAL FARMING practices, leading to the emergence of worker-peasants (see PART-TIME FARMING), while changes in tenurial patterns (see LAND TENURE) have tended to de-emphasize family rights in favour of individual ownership.

**peat,** *n.* a black or dark brown mass of partially decomposed plant material formed under anaerobic conditions in a waterlogged environment. Peat can be formed from a wide variety of plant material including mosses and trees and the COMPACTION and CEMENTATION of peat sediments represent the first stage in the formation of COAL. Deposits of peat are found throughout the world, and can cover vast areas and develop to tens of metres in depth. Peat has a carbon content of 50% and is used as a domestic fuel in parts of Scotland, Scandinavia and the Netherlands. In the Republic of Ireland, commercial peat-fired power stations have been developed. Peat is also used as a soil conditioner.

**peat bog,** see BOG.

**ped,** *n.* a naturally formed AGGREGATE of soil particles. The ped is the smallest identifiable structural unit in a soil.

**pedalfer,** *n.* a group of soils which occur in humid regions where LEACHING, ELUVIATION and ILLUVIATION are the major pedogenic processes. Most pedalfers are characterized by the predominance of iron and aluminium minerals that remain after other soluble salts have been washed from the soil. Pedalfers, which incorporate soils such as PODZOLS and LATERITES, form one of the major soil groups, the other being the PEDOCALS.

**pedestrianized street,** *n.* a conventional street from which vehicular traffic is excluded between certain hours of the day. It differs from a *pedestrian precinct* in that the footpath and roadway remain differentiated from each other, and the closure to traffic is periodic and not total. Pedestrianization of a street can be used as an experiment to test the viability of a full precinct plan before committing money to structural works.

**pedocal,** *n.* a group of soils that occur in dry regions (especially where evaporation exceeds precipitation) and where CALCIFICA-TION, particularly in the B-HORIZON, is the dominant pedogenic process, leading to the characteristic predominance of calcium carbonate. LEACHING, ILLUVIATION and ELUVIATION occur only slightly in pedocals. Pedocals form one of the major soil groups, the other being the PEDALFERS.

**pedogenesis,** *n.* the combined effect of a number of interconnected processes which result in the formation and development of a SOIL. Climate, topography, PARENT MATERIAL, vegetation and the activities of animals (including humans) are all important factors in the creation of a soil.

**pedology,** *n.* the scientific study of the formation, characteristics, distribution and use of soils. See also AGRONOMY.

**pelagic fish,** *n.* fish which inhabit the surface waters of the oceans, such as herring and anchovy. Compare DEMERSAL FISH.

**peneplain,** *n.* an ancient land surface, the result of many millions of years during which erosion has taken place without disturbance from glaciation, uplift or volcanic activity. Peneplains are best developed over the ancient tablelands of southern Africa and in Brazil. SENILE SOILS are sometimes associated with peneplains.

**Pennsylvanian period,** see CARBONIFEROUS PERIOD.

**perceived environment,** see ENVIRONMENTAL PERCEPTION.

**perched block,** see ERRATIC.

**perched water table,** see WATER TABLE.

**percolating filter,** see TRICKLE FILTER.

**percolation,** *n.* the descent of water through a SOIL PROFILE due to the influence of gravity. See INFILTRATION, LEACHING.

**percoline,** see THROUGHFLOW.

**perennial,** *n.* a plant which extends its life cycle over many years. Two types of perennials can be recognised:

(a) *herbaceous perennials* in which the aerial shoots die back in autumn and the plant spends the climatically harsh part of the year hidden beneath the soil (often as a bulb, corm, rhizome or tuber).

(b) *woody perennials* which remain above ground at all times and survive unfavourable conditions through the construction of strong, resistant woody tissue in which lignin and cellulose have substantially strengthened the plant. Compare ANNUAL, BIENNIAL.

**periglacial zone,** *n.* the unglaciated area adjacent to an icesheet or GLACIER. The characteristics of periglacial areas vary considerably but low year-round temperatures and the dominance of freeze-thaw action are common to all (see PHYSICAL WEATHERING).

PERMAFROST, whether continuous or discontinuous, is frequently found in the periglacial zone. Wind and water action along with CHEMICAL WEATHERING are often subsidiary geomorphological processes.

During the PLEISTOCENE EPOCH, the periglacial zone was greatly extended and currently accounts for around 20% of the earth's surface, extending over large parts of Siberia, Iceland and northern North America. In the past, human activity in the periglacial environment has been limited but extensive areas in Siberia and northern Canada are now undergoing intensive prospecting for oil, gas and mineral resources.

**period,** *n.* the subdivision of EARTH HISTORY shorter than an ERA and and longer than EPOCH.

**periodic market,** *n.* an institution or event where people converge to buy and to sell, and to interact socially at regular intervals of less than daily occurence. The services performed by periodic markets may provide functions that are supplementary to fixed and permanent commercial enterprises, as in developed countries, or, as in many rural areas of the THIRD WORLD, the major or only such activities in the locality. In developing countries, they provide an essential wholesale and retail outlet for locally produced agricultural commodities and craft items, and the main point of consumer demand for externally manufactured goods. The most common interval between market meetings is weekly, but in West Africa, for example, the indigenous week is rarely one of seven days. Periodicity regimes based on a week of 3 to 10 days continue to operate largely unaffected by the conflicting demands of activities regulated by the modern seven-day week.

**peripheral estate,** *n.* any large area of PUBLIC SECTOR HOUSING situated at the edge of an existing urban area. Peripheral estates are most commonly found in northern Europe, and in the UK in particular. They emerged mostly during the 1950s and 1960s when many major cities in the UK attempted to deal with the great pressure of people who required rehousing from large areas of run-down housing stock. A massive house building programme was required to remedy the housing shortage and the prior need to clear the slums of the older inner areas aggravated the problem since demolition of these very high density areas created an OVERSPILL of people who had to be rehoused elsewhere. Under these combined pressures, and with the emphasis on quantity and speed, the quality of design and finish of the new estates suffered. Often the only land which could be found to rehouse the overspill was at the edge of the existing urban areas, a long way from the original home areas of the rehoused citizens. This proved to be one reason for the social discontent which quickly developed on the peripheral estates. The fabric of these estates has decayed rapidly and this, together with their monotonous design and the poorly maintained environment, has reinforced the social difficulties which have persisted in these areas because of the initial poverty of the population.

In the 1980s the so-called INNER CITY problems of social, economic and physical run-down are as likely to be found on the peripheral estates as in geographically central and older areas. The upgrading of the peripheral estates to late 20th century standards demands considerable political, economic and social effort.

**peri-urban region,** *n.* an area of land beyond the RURAL-URBAN FRINGE whose structure and activities are modified by the presence or extension of one or more urban agglomerations. Rural communities in the peri-urban region experience many of the problems faced by their counterparts in the rural-urban fringe brought about by the encroachment of urban land uses. However, such urban pressures are usually not so intensively felt as in those areas nearer the city. The loss of farmland to urban expansion is the most obvious feature of peri-urban regions. In England and Wales some 25 000 ha of farmland were lost annually in the 1930s, and although this slowed to less than 10 000 ha per annum in the late 1970s, not least as a result of GREEN BELT legislation, the trend has been most pressing in the more populous parts of southern England.

**permafrost,** *n.* permanently frozen, impermeable ground in PERIGLACIAL ZONES and ALPINE areas. Permafrost results when ground surface temperatures remain below freezing point for long periods. During the PLEISTOCENE EPOCH, the permafrost zone was greatly extended and at present around 20% of the earth's surface is affected by permafrost which may be continuous, discontinuous or sporadic in its occurrence.

Permafrost penetrates to depths of 400 m in Canada and Alaska and up to 700 m in Siberia. During the summer, the upper layers of the ground may thaw, forming an ACTIVE LAYER up to 4 m deep. This layer is highly susceptible to degradation and erosion resulting from human activity. For example, the removal of the protective insulating surface vegetation often enables THERMAL EROSION to occur; buildings, roads and railways may then become engulfed in mud-filled depressions as meltwater mixes with thawing soil. To overcome this problem, buildings must be constructed on pilings

PERMANENT CROP

allowing air to circulate underneath. Piped services, such as sewage, gas and electricity systems, are usually constructed above ground to prevent interference with the permafrost.

**permanent crop,** *n.* a perennial shrub or tree from which a useful fruit or sap is harvested regularly and continuously over a growth cycle of several years or decades.

Shrubs differ from trees in having branches near the soil surface while trees develop a single main trunk. *Shrub crops* such as coffee, tea and vines generally require greater inputs of labour for weeding, pruning and harvesting than *tree crops* such as olives, cocoa, oil-palms and rubber where the main labour demand is harvesting. The products of shrub crops tend to require early processing.

Permanent crops are grown by SMALLHOLDERS (generally as CASH CROPS) and on PLANTATIONS, and may in some tropical regions be harvested from trees in the wild. Very often they form MONOCULTURES, but in tropical and Mediterranean areas they may be part of complex, multistorey MIXED CROPPING systems.

The area devoted to permanent crops is very much less than that under permanent grassland or arable CROPLAND, comprising less than 1% of the utilized agricultural area (excluding woodland) in both USA and USSR, and 9% in the EC in 1986. In Mediterranean areas olive groves, vineyards and fruit orchards are more significant, and permanent crops occupy 18% of the utilized agricultural area in Spain, and 19% in Italy.

**permanent cropping,** *n.* **1.** the practice of crop cultivation without recourse to FALLOW years, although seasonal fallowing may be part of the AGRICULTURAL CYCLE. Permanent cropping is now widespread in temperate latitude developed countries where CROP ROTATION and high levels of chemical inputs replace lost nutrients and counter disease and pest depredations. In less developed tropical countries it is largely confined to especially fertile areas and the outskirts of towns where the application of NIGHT SOIL is an important contribution to the maintenance of soil fertility.

**2.** the practice of growing PERMANENT CROPS.

**Permian period,** *n.* the geological PERIOD that followed the CARBONIFEROUS PERIOD 280 million years ago and that ended 225 million years ago when the TRIASSIC PERIOD began. The Permian was the final period of the PALAEOZOIC ERA, although in many areas the Permian is followed by the TRIASSIC PERIOD with scarcely any evidence of lithological variation. There was a major phase of

330

mountain building and associated volcanic and igneous activity in Europe and North America during the Permian. The fossil record during this period is often localized and confused, with the result that the Permian and the later Triassic rocks are often combined into a *Permo-Triassic period*. Amphibians and reptiles continued to develop during the Permian. Substantial Permian coal deposits are found in parts of Australia, India and China. Other economic resources include sandstones and evaporites such as rock salt, potash and gypsum. See EARTH HISTORY.

**Permo-Triassic period,** see PERMIAN PERIOD.

**peroxyacetyl nitrates (PAN),** *n.* any of a series of complex chemical compounds that exist as SECONDARY POLLUTANTS and are major components of PHOTOCHEMICAL SMOG. PANs form when PRIMARY POLLUTANTS containing ozone, hydrocarbons and members of the nitrous oxide group ($NO_x$) are released into the air from vehicle exhausts and numerous industrial processes, and are recombined by the catalytic action of strong sunlight.

PANs are highly damaging to plant tissue and can cause eye irritation in humans. They are characteristic of a highly developed industrial and transport-dependent society. The first region to exhibit this form of pollution was the Los Angeles basin where vegetation damage due to PANs occurred in the early 1960s. The Californian citrus fruit industry was badly effected before control measures could be introduced. Considerable opposition from motor vehicle manufacturers had to be overcome before compulsory fitting of AFTERBURNERS or CATALYTIC CONVERTORS to the exhaust systems of all vehicles achieved a lowering of the PAN level. Evidence of PAN damage to vegetation may be found in many other large conglomerations, particularly those in sunny climes.

**pesticide,** *n.* any chemical designed to kill weeds, insects, fungi, rodents or other organisms that are considered by humans to be undesirable. See INSECTICIDE, FUNGICIDE, HERBICIDE, 2,4-T, 2,4,5-D, AGENT ORANGE.

**petrochemical complex,** *n.* an agglomeration of related industries based on the distillation of crude OIL into its component chemicals. Products typically produced by such industries include industrial chemicals, explosives, fertilizers, plastics, pesticides, synthetic fibres, paints, medicines and petroleum products.

Petrochemical plants generate an almost limitless variety of GASEOUS POLLUTANTS based mainly on hydrocarbons but also on terpines and oxides of nitrogen. Even when operated under the most stringent safety regulations, the accidental release of pollutants can occur, as well as industrial accidents which cause loss of human life and untold damage to the environment, as for example at Flixborough in Yorkshire in 1973 and at Bhophal in India in 1985 when a release of toxic gas into the atmosphere killed an estimated 2300 people and maimed 500 000.

**petroleum,** see OIL.

**phanerophyte,** see RAUNKIAER'S LIFEFORM CLASSIFICATION.

**phanerophyte climate,** see BIOCHORE.

**phased planting,** *n.* a technique, also known as successional cropping, often adopted by farmers practising SHIFTING CULTIVA-TION or ROTATIONAL BUSH FALLOW in which sections of a field are planted with the same crop or crop mixture but in a staggered or phased time sequence stretching over several weeks or months. In this way, demand for labour is spread more evenly, and fresh produce can be harvested over an extended period of time, thus reducing likely losses in storage. Losses from eschewing optimum planting and growing times may be considerable but where rainfall is unreliable this may be a soundly rational risk-spreading or risk-minimizing strategy.

**phenocryst,** see PORPHYRY.

**phenotype,** *n.* the physical constitution of an organism as determined by the interaction of its GENOTYPE and the environment in which the organism develops.

**phloem,** see CONDUCTING TISSUE.

**photic zone,** see EUPHOTIC ZONE

**photochemical smog,** *n.* an atmospheric HAZE sometimes found above large industrial and urban areas. Photochemical smog is the result of reactions between pollutants produced from most high temperature and pressurized combustion processes such as the high compression motor vehicle engine.

The primary ingredients of photochemical smog are often non-toxic and include certain hydrocarbons and oxides of nitrogen. The phytotoxic products of photochemical reactions are PEROXYACETYL NITRATES (PANs), OZONE and nitrogen. They are initiated by the energy of sunlight and are injurous to vegetation and some animals, even when present only in minute quantities (0.1 parts per million) and for a short duration (8 hours exposure).

The onset of a photochemical smog is marked by a deterioration in atmospheric visibility. This is followed by visible discolouration of the undersides of leaf surfaces and, if the smog persists, then eye irritation can occurs. Areas of photochemical smog often show a diurnal movement pattern dependent upon the movement of the

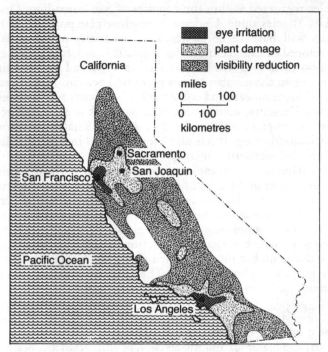

Fig. 46. **Photochemical smog.** The extent of pollution in California.

sun. At mid-day a pall of smog often lies over the city while at night it migrates outside the city limits. The problem of photochemical smog is greatest in California where 70% of the population suffer eye irritation, 80% of the vegetation shows characteristic discolouration or retarded or deformed growth while 97% of the state suffers reduction in visibility (see Fig. 46).

Photochemical smog has shown a spectacular increase in occurrence since its first appearance within a few square kilometres of Los Angeles County in 1942. By 1960 its occurrence had extended to 30 000 sq km throughout California. Stringent control measures, mainly the compulsory fitting of CATALYTIC CONVERTORS to the exhaust systems of motor vehicles from 1967 and the use of UNLEADED PETROL (gasoline) curbed the growth of the smog problem. However, by 1989, such had been the growth of motorized traffic in the Los Angeles area that smog levels were returning to their pre-1967 levels: there are 8 million motor vehicles registered within the counties of San Bernadino, Orange and Los Angeles alone. Under a three-phase plan proposed in 1989 all cars will have to be powered by electricity or other, as yet undeveloped 'clean technologies' by 2007. All free parking for petrol-powered cars will be abolished in 1989 and by 1994 limits will be set on the number of cars permitted per family, and petrol-powered lawnmowers and solid-fuelled barbecues will be banned. In other US states, control measures have not been as stringently applied although by 1988 some 25 of the 50 US states had recorded photochemical smog. It has also been recorded in Australia, India, Belgium and southern England.

**photosynthesis,** *n.* a complex process that occurs in the chlorophyll molecules in the cells of green plants whereby radiant energy from the sun provides the catalyst for the combination of carbon dioxide (from the atmosphere) and water and basic salts (from the soil) to produce simple sugars such as glucose. Oxygen is released as a by-product which is vital to respiration in plants and animals. Up to 100 different, but interconnected chemical changes are involved in the process of photosynthesis. They may be summarized in the equation:

$$6CO_2 + 6H_2O + \text{solar energy} \rightarrow C_6H_{12}O_6 + 6O_2$$

**phreatic zone,** see WATER TABLE.

**phylum,** *n.* a major taxonomic division of animals that contains one or more classes. The equivalent category in plant classification is the *division*.

**physical accessibility,** see ACCESSIBILITY.

**physical planning,** see LAND USE PLANNING.

**physical quality of life index (PQLI),** *n.* an index designed to measure the physical well-being of the population of a nation. It is calculated by the use of three indices, each of which is given equal weight: child mortality, literacy and life expectancy. Each country is rated on scale of 0-100 for each of the indicators and a mean value is then taken to give the PQLI rating. A PQLI rating of 77 has been set to represent basic human needs although approximately 66% of the world's population presently fall below this figure. The PQLI index provides a more useful indicator to the physical progress than the exclusively economic indicator of the gross national product figure. See QUALITY OF LIFE.

**physical weathering** or **mechanical weathering,** *n.* the loosening and disintegration of rocks and their constituent minerals by the physical action of frost, temperature change and salt crystallization.

Four main types of physical or mechanical weathering have been recognized:

(a) *freeze-thaw action* which entails the penetration of rock joints, bedding planes and pores by water which then freezes and expands. The surrounding rock is wedged apart and is then loosened as a thaw sets in. If the surface is on a slope, the loosened soil particles will be removed downslope by meltwater, leading to SOLIFLUC-TION. This *frost wedging* is often responsible for causing potholes in road surfaces. Freeze-thaw action involving the upward and downward movement of soil and rock (*frost heaving*) results from pressures caused by the freezing of groundwater in the upper REGOLITH; it is responsible for serious damage to roads, railways, underground pipes and the foundations of buildings. Freeze-thaw action is most prevalent in mountain areas, glacial and periglacial regions and during winter in temperate latitudes. See also PATTERNED GROUND;

(b) *thermal expansion and contraction* which may cause the disintegration of rock in areas of large diurnal temperature ranges such as deserts and mountain areas. The repetition of expansion and contraction weakens the outer layer of rock which eventually peel away, a process known as *exfoliation* or *onion skin weathering*. Doubts exist about the effectiveness of thermal expansion and contraction as a sole agent of physical erosion although it may serve to weaken the rock surfaces thus allowing other weathering processes to become more active;

(c) *salt weathering* which often causes the disintegration of rocks in the humid tropics, deserts and coastal areas and results from the evaporation of saline solutions within rocks and the subsequent crystallization of salts. In a process similar to frost wedging, salt weathering can lead to the honeycombing of rock surfaces. This process is also known as *cavernous weathering* and may cause the gradual destruction of the foundations and walls of buildings.

(d) *Unloading* or *dilation* involves the release of pressure on rocks found beneath the earth's surface and results in the formation of vertical and horizontal joints within the rock. This may lead to rock disintegration in itself or may allow other forms of weathering to operate.

Physical weathering often operates in conjunction with CHEMI-CAL WEATHERING and BIOLOGICAL WEATHERING.

**physiognomy,** *n.* (Ecology) the external appearance of vegetation including such features as colour, luxuriance, seasonality and

Planation surfaces have been widely used as indicators of landscape evolution, although their formation is frequently due to a wide range of localized as well as more universal factors. Many ancient planation surfaces, such as extensive areas of the Matto Grasso in Brazil, are of little agricultural value due to their SENILE SOILS .

**plankton,** *n.* organisms inhabiting the surface layer of a sea or lake, consisting of unicellular plant organisms (PHYTOPLANKTON) and animal organisms (ZOOPLANKTON). Zooplankton which feed mainly on phytoplankton, may be unicellular and solitary, unicellular and colonial, or multicelled. Plankton are less than 0.1 mm in size, and many are microscopic. They are entirely dependent on sunlight and can only flourish in the topmost layers of water at depths less than 100 m (see EUPHOTIC ZONE).

Plankton form the vital first step in aquatic FOOD CHAINS. Phytoplankton are estimated to generate 16 billion tonnes of carbon per year, and zooplankton a further 1.6 billion tonnes. However, they have undergone a drastic reduction in numbers in the last 25 years due to the POLLUTION of aquatic environments. Chemical residues from inorganic agricultural fertilizers, HERBICIDES and PESTICIDES, along with industrial wastes and oil spills have all contaminated the habitats of plankton. This has been particularly serious in shallow, fresh water bodies such as lakes.

**planning permission,** *n.* authorization which must be obtained from a local planning authority before any form of DEVELOPMENT can commence. The granting (or withholding) of permission is an effective method of DEVELOPMENT CONTROL. In the UK since 1947, the construction of new developments and the demolition or alteration of HISTORIC BUILDINGS have had to conform to specifications laid down by the TOWN AND COUNTRY PLANNING ACT, and by regional and LOCAL PLANS. Planning permission can be granted *unconditionally* whereupon development can take place according to the developer's plans, or *conditionally* whereby the planning authority imposes restrictions (as to the height, colour, right of access or construction material) upon the development. See GENERAL DEVELOPMENT ORDER, USE CLASSES ORDER.

**plantation,** *n.* **1.** a large-scale agricultural enterprise, generally located in tropical or sub-tropical regions, for the production and processing of crops, mostly for export markets in industrialized countries. Plantations are characterized by a tendency to MONOCULTURE or specialization in PERMANENT CROPS such as rubber, oil-palm, coconuts, cocoa, coffee and tea. Some may specialize in herbacious PERENNIALS such as bananas and sugar cane, or ANNUALS such as cotton, tobacco and jute. Plantations are generally in excess of 1000 ha, and are capital- and labour-intensive. Reflecting their colonial origin, plantations are sometimes managed by European staff, on behalf of multinational corporations, and employ indigenous wage labour.

The advantages of plantation agriculture lie in the easier application of modern farming practices; furthermore, production and processing, as well as quality, are more easily controlled than in contrasting SMALLHOLDING operations.

**2.** a stand of forest that has been deliberately planted for a particular purpose (see FORESTRY).

**plateau,** *n.* an upland area with an extensive, almost level summit which is frequently bounded by steep margins or ESCARPMENTS. Some plateau surfaces have been deeply dissected by downcutting rivers to form GORGES. Over time, lateral erosion may occur leaving only residual landforms known as MESAS and BUTTES to indicate the former plateau surface. Most plateaus are found above 600 m whilst the Colorado and Tibetan plateaus are over 1600 m above sea level.

Plateaus may be classified according to the nature of adjacent areas:

(a) *continental plateaus* rise sharply from plains or the sea and include much of Africa, the ice caps of Greenland and Antarctica and the tablelands of South Africa;

(b) *piedmont plateaus* are situated between lowland areas and mountain ranges. Examples include the Columbia River plateau in the north-west USA and the plateau of Patagonia;

(c) *intermontaine plateaus* are partially or completely enclosed by mountains. Examples include the highlands of northern and central Mexico.

Plateaus in the mid-latitudes are usually sparsely populated because of the low temperatures resulting from their high elevation. In contrast, the plateau regions of tropical areas are often heavily populated as the generally lower elevations provide conditions of temperature and humidity that are more conducive to human settlement. The resulting climate can also create ideal conditions for farming; coffee is a crop often associated with the plateaus of Kenya and Brazil. Besides their agricultural value, some plateaus are also rich in mineral deposits: in western Australia, Africa and the Brazilian Highlands reserves of gold, copper, iron and manganese have been exploited in plateau areas.

**plate tectonics,** *n.* a widely accepted theory that suggests that the LITHOSPHERE consists of a number of rigid plates the move on the partially molten ASTHENOSPHERE. Seven major lithospheric plates are recognized, together with numerous minor ones (see Fig. 47). CONTINENTAL DRIFT is thought to result from the movement of these plates.

Three types of plate boundary have been distinguished:

(a) *constructive* or *divergent boundaries*, which result when SEA-FLOOR SPREADING causes the formation of new lithospheric rocks and the gradual divergence of plates. Constructive boundaries are found at the MID-OCEANIC RIDGES in the Pacific, Atlantic and Indian Oceans. The rate of divergence may be up to 9 cm per year.

(b) *destructive* or *convergent boundaries* occur when plates move towards each other. One plate is overridden and forced downwards into the MANTLE to form a ZONE OF SUBDUCTION. Mountain formation due to intense folding, earthquakes and volcanic activity are usually associated with destructive boundaries. A destructive boundary is thought to exist along the west coast of South America.

(c) *conservative* or *shear boundaries*, which develop when plates move parallel to each other along transform FAULTS. Lithospheric rocks are neither formed nor destroyed along conservative boundaries. The San Andreas Fault in California, USA, is a conservative boundary.

Along divergent boundaries in the Red Sea area, upward moving convection currents in the mantle have led to the emplacement of major deposits of gold, silver, copper and iron ore. Other divergent boundaries are now being prospected for similar deposits. See also CONTINENTAL DRIFT.

**Pleistocene epoch,** *n.* the first geological EPOCH of the QUATER-NARY PERIOD that commenced about 2 million years ago and ended about 10 000 years ago when the RECENT EPOCH began. The Pleistocene was characterized by a global lowering of average temperature such that in northern latitudes snow and ice accumulation occurred with the formation of ice-caps, SNOWFIELDS and GLACIERS; FLUVIAL EROSION was replaced by GLACIAL EROSION. Conditions during the Pleistocene were not uniformly cold. Intervals of up to 10 000 years (INTERGLACIAL PERIODS) were free from ice which allowed deeply weathered soils and well-developed vegetation SUCCESSIONS to form. Towards the end of the Pleistocene, very rapid, short-lasting (200 years) improvements in the climate occurred (INTERSTADIAL PERIODS). In Europe, the Pleistocene has been traditionally divided into four main cold phases

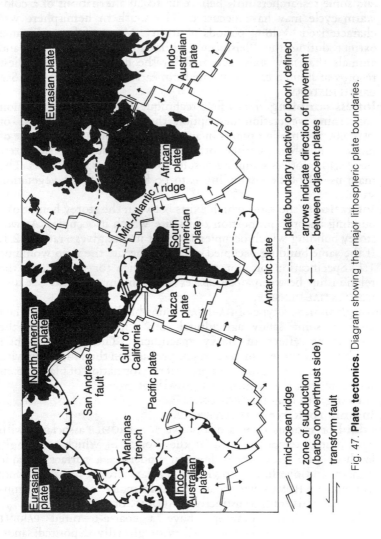

Fig. 47. **Plate tectonics.** Diagram showing the major lithospheric plate boundaries.

mid-ocean ridge

zone of subduction
(barbs on overthrust side)

transform fault

plate boundary inactive or poorly defined

arrows indicate direction of movement
between adjacent plates

# PLOTLESS SAMPLING

(Gunz, Mindel, Riss and Wurm) separated by warmer phases. Recent research has suggested such a division to be too simplistic and some researchers now believe up to 60 alternations of a cold-warm cycle may have occurred. The southern hemisphere was characterized by long periods of aridity. Many lifeforms became extinct during the Pleistocene whilst many of the plants and animals that now exist appeared for the first time. The earliest records of human existance date from early within this epoch. See EARTH HISTORY.

**plotless sampling,** *n.* any field technique used to gain information concerning vegetation description that does not make use of QUADRATS; sampling points in an area are determined by the use of random numbers. Unlike other techniques, such as TRANSECT SAMPLING, plotless sampling is suited for use over large areas. The main use of plotless sampling is to be found in forest vegetation studies.

**plot ratio,** *n.* the ratio which exists between the storey height of a building and the proportion of its site which it occupies. A two-storey building which occupies 100% of its site gives a ratio of 2:1. If the same building occupied only half the site the ratio would be 1:1. Specification of plot ratios is a device for controlling the relationship between open spaces and building height. See also RESIDENTIAL DENSITY.

**plough pan,** *n.* a type of HARDPAN found at the base of a ploughed layer in some sandy agricultural soils which results from the compacting effect of heavy machinery. The use of lighter machines, or those which exert less pressure on the soil, along with variable depth ploughing can prevent the formation of plough pans which restrict the normal root growth of crops.

**plucking,** see GLACIAL EROSION.

**plutonic rock,** see INTRUSIVE ROCK.

**podzol** or **spodosol,** *n.* a distinctive SOIL PROFILE associated with cool humid temperate area of the world, in which a strongly leached, pale grey upper horizon underlies a surface organic horizon and overlies a brown to very dark brown horizon where iron and/or HUMUS has accumulated. In many cases the accumulated materials form a distinctive HARDPAN or IRONPAN.

Podzols usually develop above a coarse-textured PARENT MATERIAL, often being associated with glacially deposited sands, gravels, tills and MORAINE. The natural vegetation associated with podzols is coniferous forest and ericaceous heath and moorland. The organic material produced by these vegetation types is acidic

MOR humus which assists in the processes of LEACHING. Podzols frequently record pH values of between 4.0 and 5.5. The high acidity and low nutrient levels associated with the A-HORIZON of podzols restrict their agricultural potential to rough grazing. Podzols can be converted to wider agricultural use only through the application of large quantities of inorganic FERTILIZERS. The process of conversion is slow, expensive and the results are not guaranteed. Podzols are best used for the growth of commercial coniferous forestry. The podzol soil type is widely recognized by all soil classifications although in the American system it is known as a spodosol.

**polar easterlies,** *n.* the cold winds which flow from the high pressure cells developed over the polar regions towards the lower pressure mid-latitudes. In the northern hemisphere, these winds, which blow from the north-east, occur only intermittently due to the limited seasonal extent of the Arctic high. When they do occur, they are usually located on the poleward side of migratory DEPRESSIONS over the northern Atlantic and Pacific. In the Southern hemisphere, the more permanent Antarctic ANTI-CYCLONE results in the persistent occurence of polar easterlies.

**polar front,** *n.* a large-scale FRONT in the North Atlantic and North Pacific oceans along which northward-moving tropical maritime AIR MASSES meet southward-moving polar maritime air masses (see Fig. 4). The polar front moves to the north during the winter and to the south in the summer. The formation of small wave-like irregularities in the polar front is responsible for development of temperate latitude DEPRESSIONS.

**polar jet stream,** see JET STREAM.

**polder,** *n.* any low-lying area in Holland reclaimed from the sea, a lake or an estuarine position by means of a manufactured DYKE. Excess water is removed by pumping, a process which is normally continued in perpetuity in order to prevent reimmersion of the area. Polder soils initially contain large quantities of marine-derived salts, in particular sodium and potassium, but these can be gradually washed out of the soil by repeated flooding of the soil with fresh water. Once free of the toxic salts polders form productive agricultural land.

**polje,** *n.* a steep-sided, flat-floored linear depression in KARST landscapes, up to 65 km in length and 10 km wide. The origin of polje is unclear but may be due to solution weathering (see CHEMICAL WEATHERING) along lines of weakness, the collapse of caverns formed by underground drainage, or, in some instances, may be the result of late glacial meltwater streams which carved out a channel. The floors of polje are often damp, clay-covered areas that can be used for agriculture in otherwise barren limestone regions. They can also provide valuable lines of communication in otherwise impenetrable terrain, for example through the northern limestone massif of Mallorca in the Ballearic Islands. See UVALA.

**pollen analysis** or **palynology,** *n.* the study of living and fossil pollen grains and plant spores. All flowering plants produce pollen while ferns and mosses produce spores as part of the process of reproduction. A resistant outer casing allows pollen and spores to be preserved in sediments which been deposited in lakes, peat bogs and soils. Pollen and spores from different species can be uniquely identified, so allowing the reconstruction of the vegetational history of a particular site or region. It is also possible to obtain information on such matters as variations in climate and the effects of human activity on vegetation cover. Pollen analysis can also be used as part of a RELATIVE DATING scheme by establishing the succession of deposits.

Pollen analysis often requires complex chemical and mechanical procedures to extract the pollen grains from within sediments in readiness for identification under a high-powered optical microscope. Such a process requires a specially equipped laboratory in which hazardous substances such as hydrofluoric acid can be safely handled.

**pollen zone,** *n.* an assemblage of particular pollen types, named after the dominant pollen type, classically considered to be suggestive of a particular climate. Changes between characteristic assemblages delimit the boundaries of different pollen zones. The pollen zone should have a regional distribution and so zonation should be based on diagrams where very local effects are excluded. Zones are best based upon so-called standard pollen diagrams. A complete pollen diagram for a north European site would have 8 pollen zones within it. See POLLEN ANALYSIS.

**polluter-must-pay principle,** *n.* the concept that the generator of industrial pollution should bear the costs of avoiding pollution or remedying its effects. In practice, the principle has been difficult to apply particularly with regard to atmospheric pollutants because the multiplicity of pollution sources often makes it hard to identify a specific polluter, as does the speed with which pollution can travel away from its source. However, the principle can be applied following a pollution disaster, such as a major oil spill from a tanker or from damage following a nuclear power station accident.

For more persistant pollution outputs, legislation such as that enforcing AIR QUALITY STANDARDS are more effective in controlling pollution levels.

Whichever approach is used, the cost of preventing or repairing environmental damage by industrial pollution must ultimately be met by the consumer. In the UK the cost of fitting gas-cleaning devices to 12 major coal-fired electricity generating stations was estimated by the government to cost £2000 million at 1986 prices. The price of electricity to the consumer would have to be increased by 4% to meet this cost.

**pollution,** *n.* the contamination of the BIOSPHERE with poisonous or harmful substances, usually domestic, industrial or chemical waste products. Pollution of the biosphere causes undesirable changes in the physical, chemical and biological infra structure and is reflected in the impaired performance, reduced growth, lowered reproductive capacity and ultimately the death of individual organisms. See ACID DEPOSITION, AIR POLLUTION, NOISE POLLUTION, MARINE POLLUTION, EFFLUENT, OIL SPILL, NUCLEAR WASTE.

**polyclimax theory,** see CLIMAX COMMUNITY.

**polyculture,** *n.* the raising of several different yet compatible species in the same area. A farming system based on polyculture is one in which PERMANENT CROPPING is the norm with two or more crops being grown simultaneously in the same field. In freshwater AQUACULTURE, a form of polyculture occurs whenever several species of fish, each with distinct feeding habits and occupying different ECOLOGICAL NICHES are raised in fish ponds. In South-East Asia, for example, harvests of tilapia, bream and several kinds of carp may be taken from the same pond. By maintaining a mixed-age, -size, and -species polyculture, a valuable source of protein is made available throughout the year.

**pools and riffles,** *n.* the alternations of deeper and shallower stretches of water in a river that are believed to be the forerunners of MEANDER formation. In the riffles, or shallows, the energy within the flowing water is used to overcome the high levels of friction imposed by the rough river bed, whereas in the pools, or deeper areas, the energy is diverted towards undercutting the river bank. Once started, meander formation appears to be a self-perpetuating process.

**population,** *n.* a group of individuals, usually of a single species, that inhabit a given area at a point in time. Population size will be determined by the CARRYING CAPACITY of the HABITAT.

**population crash,** *n.* a sudden and catastrophic reduction in the POPULATION size brought about by the inability of the HABITAT to support it. A population crash usually occurs after a population has grossly exceeded the CARRYING CAPACITY of the habitat in which it occurs and is due to the exhaustion of a primary requirement for life such as food, oxygen or living space. Population crashes are normally associated with simple algal populations in which control of the rate of increase is poorly developed. Compare POPULATION EXPLOSION.

**population explosion,** *n.* a sudden (and usually unpredictable) rapid increase in the population size of a species. Such an increase may be due to natural events, for example a seasonal changes in climate. More commonly, however, population explosions are linked to human factors, such as the introduction (accidental or deliberate) of species to geographical areas to which they had previously been absent. The best known examples are the population explosion of rabbits and prickly pear in Australia in the 19th century but many other examples exist, for example the accidental import of the Indian balsam plant (*Impatiens glandifera*) to southern England and its subsequent explosion along river banks throughout much of England and Wales. Compare POPULATION CRASH.

**population growth,** *n.* a permanent increase in POPULATION size brought about by an increase in births over deaths and/or immigration over emigration. In the human population, the maximum growth occurs during a period of DEMOGRAPHIC TRANSITION.

**pore space,** *n.* a minute space between soil particles and rock grains. Pore spaces are usually interconnected thus allowing the movement of SOIL WATER and SOIL ATMOSPHERE. See POROSITY

**porosity,** *n.* the amount of space between soil particles or rock grains. Porosity is usually expressed as the ratio of the volume of space between the soil particles to the total volume of a soil or rock sample, given as a percentage. See PORE SPACE.

**porphyry,** *n.* an IGNEOUS ROCK that contains large crystals (*phenocrysts*), within a finer GROUNDMASS of minerals.

**positive descrimination,** see DEPRIVATION, AREA OF.

**possibilism,** see ENVIRONMENTAL DETERMINISM.

**potash,** *n.* a SEDIMENTARY ROCK composed of potassium minerals precipitated by the evaporation of seawater. Potash is used by the chemical industry and in the manufacture of fertilizer.

**potassium-argon dating,** *n.* a technique for determining the age of fossil remains by measuring the quantity of the potassium RADIOISOTOPE $^{40}$K present in the sample. This isotope, which is present in the potassium occurring naturally in rocks and soil, has a HALF-LIFE of $1.28 \times 10^9$ years and decays to the stable isotope of argon, $^{40}$Ar. By calculating the ratio of the potassium isotope to the argon isotope present in the sample being tested, an age up to 10 million years may be estimated. See RADIO-CARBON DATING, DATING TECHNIQUES.

**PQLI,** see PHYSICAL QUALITY OF LIFE INDEX.

**prairie,** *n.* the vast natural grassland community extending across the interior of North America from the forest edge of Alberta and Saskatchewan to the Gulf Coast in south eastern Texas, and westwards to the foothills of the Rockies, an area now called the Great Plains. Prior to the changes achieved by the European settlers considerable regional diversity existed in the prairies. Three main subtypes have been distinguished:

(a) *true prairie* comprising sward grasses and bunch grasses such as Stipa species which grew to over 1 m in height. In summer, the true prairie had a tussocky appearance while in winter it assumed a smoother appearance;

(b) *mixed prairie* comprising grasses of two distinct types: *Stipa comata* and *Agropyron smithii* which grew to around 75 cm in height, and the dwarf grasses of less than 10 cm, the Buffalo grass, (*Buchloe dactyloides*) and Grama grass (*Bouteloua gracilis*). The increasing aridity to the west of the Great Plains was responsible for the shorter grasses. Overgrazing by cattle from 1870 onwards gradually eliminated most of the grass species except Grama grass;

(c) *Pacific* or *Palouse prairie* comprising a variety of bunch grasses.

Although prairies have been classified as natural grassland, the term should be applied with care as the prairies were probably the result of a long-term interaction between native Indian populations who burned the original forest in order to encourage grazing lands for the wild buffalo herds. Prairies should thus be considered as ancient PYROCLIMAX COMMUNITIES.

After 1870 the extension of railroads allowed the rapid settlement of Europeans to occur in the Great Plains. The introduction of beef cattle followed by sheep soon converted the native prairie to a species poor grassland which, in the driest and most heavily grazed areas, soon became eroded. The arrival of the mechanical corn reaper in the early 1900s encouraged rapid expansion of corn (maize) and wheat cultivation into areas with

PRAIRIE SOIL

less than 50 mm of rainfall per year. Massive SOIL EROSION produced the DUST BOWL problem and its associated economic hardships for the farmers of the mid- and far west regions of the Great Plains.

Gradually, the area has been restabilized through the use of more appropriate technology including the use of semi-permanent grassland, green fodder crops, contour strip planting, shelter belts of trees and the more judicious timing of ploughing the land so that the driest and windiest times can be avoided.

Small areas of 'old field' prairie have been conserved. Elsewhere the Great Plains comprise a vast agricultural store house capable of sustaining not only the population of North America but also permitting export of grain to the USSR and providing FOOD AID to THIRD WORLD countries.

**prairie soil,** see CHERNOZEM.

**Precambrian era,** *n.* the earliest geological ERA that commenced with the consolidation of the EARTH'S CRUST about 4600 million years ago and ended about 600 million ago when the PALAEOZOIC ERA began. The Precambrian spans more than 85% of the earth's history and is sometimes divided into the older *Archaeozoic era* that ended about 2500 million years ago, and the younger *Proterozoic era.* Most Archaeozoic rocks are highly metamorphosed sedimentary and volcanic rocks with granite intrusions. Proterozoic rocks contain less metamorphic and igneous strata and more sedimentary rocks. There appear to have been several periods of mountain building, glaciation and volcanic activity during the Precambrian. Rocks of this era outcrop in parts of Scotland, England, France, Canada, Africa and the USSR. Proterozoic rocks often contain substantial deposits of the ores of gold, silver, nickel, copper iron and cobalt. See EARTH HISTORY.

**precipitation,** *n.* **1.** the deposition of atmospheric moisture as rain, hail, sleet, snow, frost and dew.

**2.** the deposition of minerals in solution due to EVAPORATION. See also ACID DEPOSITION.

**predator,** *n.* any organism (typically an animal) which gains its food by killing and eating other organisms. The relationship between the predator and its prey is often an intricate one and forms part of a complex FOOD CHAIN. See also TROPHIC LEVEL.

**present value,** see LAND VALUE.

**preservation,** *n.* the prevention of destruction; in the contemporary context, preservation is applied to extremely rare, and hence valuable objects. The traditional preservationist approach can be seen in museums where ancient artefacts are displayed behind glass and in a controlled environment. The preserved object is inevitably an inert (dead or lifeless) object. More recently, preservation has been extended to inanimate objects such as historic buildings, costume, vintage motor cars and to language, for example the original Swiss language of Romansh. Compare CONSERVATION.

**pressure gradient** or **barometric gradient,** *n.* the rate of horizontal change in air pressure from a particular point in any direction, as indicated by the location of ISOBARS on a weather chart. The maximum gradient occurs at right angles to the isobars. Strong winds are usually associated with steep gradients, which are portrayed by closely spaced isobars.

**pressure group,** *n.* any group of people united by common intent and who actively seek publicity, occassionally through spectacular and dangerous action, in order to draw attention to their cause thereby bringing pressure to bear upon decision makers (politicians and planners) and influencing public opinion. Environmental issues frequently attract the attention of pressure groups, for example groups such as GREENPEACE and FRIENDS OF THE EARTH. Action has been taken on issues such as the culling of seals in Canada, the removal of native tree species and their replacement by EXOTIC PLANTS in upland Britain and the exploding of atmospheric nuclear bombs in the Pacific Ocean by the French government.

**pressurized water reactor (PWR)** see WATER-COOLED REACTOR.

**primary consumer,** see FOOD CHAIN.

**Primary era,** see PALAEOZOIC ERA.

**primary forest,** *n.* any original, virgin forest unmodified by deletirious human activity. It is probable that no primary forest now exists as even the remotest areas of forest, such as that in the Amazon basin, have been subjected to air-borne pollutants and chemical residues from agricultural regions. The decline of primary forest began about 5000 years BP when humans became numerically and technologically significant as an agent of change in the biosphere (see ANTHROPOGENIC FACTOR).

**primary oil recovery,** see OIL.

**primary pollution,** *n.* pollution resulting directly from any substance which has been released into the atmosphere in harmful concentrations. The primary pollutant may already be present in the atmosphere, such as carbon dioxide, but only becomes toxic when additions to it causes it to rise above its natural concentration (see GREENHOUSE EFFECT); it may also involve the direct release into the atmosphere of a chemical substance not normally present in the air, for example, hydrogen fluoride released from large coal-fired furnaces. Compare SECONDARY POLLUTION. See AIR POLLUTION.

**primary producer,** see FOOD CHAIN.

**primary productivity,** *n.* the rate at which chemical energy is manufactured by green plants per unit area and in a given time period (usually one year or one GROWING SEASON) through the action of PHOTOSYNTHESIS. This energy is usually in the form of organic substances capable of being used as food material. *Gross primary productivity* (GPP) is the total amount of energy assimilated by the primary producers per unit of time while the *net primary productivity* (NPP) is equal to GPP minus losses due to respiration. NPP is usually about half the GPP value (for example, GPP biosphere average = 8372 kJ/m$^2$/year, while NPP = 4186 kJ/m$^2$/year). Primary productivity levels of the major BIOMES of the world show great variation due to latitudinal position, elevation and climate. Fig. 48 shows NPP and BIOMASS values for a selection of the main biomes.

| Vegetation unit | Net primary production (g/m$^2$ p.a.) | | Biomass (t/ha) | | Leaf area index (m$^2$/m$^2$) |
|---|---|---|---|---|---|
| | Range | Mean | Range | Mean | |
| Tropical rain forest | 1000–5000 | 2000 | 450–800 | 450 | 6–16.6 |
| Seasonal rain forest | 450–2500 | 1000 | 420–600 | | |
| Savanna | 200–2000 | 700 | 20–600 | 366 | |
| Arid desert zone | 10–250 | 70 | 1–40 | 7 | |
| Mediterranean zone | 250–1600 | | 170–350 | | |
| Mid-latitude deciduous forest | 200–3000 | 800 | 60–700 | 300 | 4–8 |
| Boreal forest | 200–2000 | 650 | 80–520 | 200 | 7–16 |
| Tundra | 3–400 | 140 | 1–30 | 6 | |
| Agriculture | 100–4000 | 650 | 4–120 | 10 | |

Fig. 48 **Primary productivity.** Summary chart of vegetation productivity for the major vegetation units and agriculture.

**primary succession,** *n.* the natural sequence of vegetation development on a site which has not previously borne vegetation. Such sites are rare but include ground left bare by deglaciation, avalanches, slumps and slides, land eroded by floodwater or covered by silt after a flood. Compare SECONDARY SUCCESSION. See PRISERE.

**primary tillage,** see TILLAGE.

**prime agricultural land units,** see LAND USE CLASSIFICATION.

**primitive area,** see FOREST RESERVE.

**priority treatment area,** *n.* a sector of a city designated for special resource allocation and improvement schemes due to its social and economic deprivation and/or its physical degradation. The use of this designation is a latter-day expression of the principle of positive discrimination by area. Areas so defined receive special allocations from within their local authority's budget and often attract additional help in the form of special managerial arrangements such as area teams of professional officers working on site, and/or special grants from central goverment or the EC. See also INNER CITY, DEPRIVATON, AREA OF.

**prisere,** *n.* the complete natural sequence of vegetation development from PIONEER COMMUNITY to CLIMAX COMMUNITY. It is unlikely that an undisturbed priseral development still occurs due to the constant modifications to the soils, vegetation and climate as a result of human activity. See SERE. See also PRIMARY SUCCESSION.

**probablism,** see ENVIRONMENTAL DETERMINISM.

**production rate,** *n.* the rate at which organic matter is produced by PRIMARY PRODUCTIVITY over a given time period and usually expressed as grams per square metre. Production rate is the most common measure of plant growth for agricultural crops and forest products. However, care must be taken when comparing production rates calculated by different centres as many variations are possible in the method used to arrive at a final production rate value. For example, variations include whether production has been measured at a single point in time, or is the accrued production between two points in time, or whether the production rate includes animal material or not.

**productivity,** *n.* a measure of output per unit of input of land, labour or capital needed to operate a productive process. Agricultural productivity measured in output per unit of land area is more commonly referred to as the YIELD. Labour productivity, which can be increased either by saving labour or by producing more with the existing labour supply, is the average value product per MAN-DAY.

The gross labour productivity of a farm enterprise may be given by the gross output per man-year, while the net labour productivity is the farm income per man-year. In evolutionary terms, the development of farming systems in a progression from SHIFTING CULTIVATION towards PERMANENT CROPPING enabled the production of greater food supplies but at the expense of labour productivity. Only with the increased MECHANIZATION and the use of chemical inputs has the trend been redressed.

A feature of modern COMMERCIAL FARMING has been the preoccupation with improving labour productivity, but it has been gained at the expense of other inputs, especially support energy, that is, the sum of all the non-renewable energy inputs needed to produce a crop both directly as fuel costs and indirectly in the manufacture of farm machinery and agrochemical inputs. A modern energy-intensive AGRIBUSINESS may use 5 calories of energy input to produce 1 calorie of food whereas the self-sufficient shifting cultivator in the tropics may use just 1 calorie of work energy to produce 4 calories or more of food. Relying on an abundance of solar energy for rapid photosynthesis, systems with low labour productivity may function with greater energy productivity or efficiency. In extractive and industrial activities, as in agriculture, productivity gains have been achieved at the expense of fossil fuel energy resources. Should these resources become a major LIMITING FACTOR there may be a need to re-appraise productivity measures.

**programme aid,** see FOOD AID.

**project aid,** see FOOD AID.

**Proterozoic era,** see PRECAMBRIAN ERA.

**profile of equilibrium,** see RIVER PROFILE.

**psammosere,** *n.* a vegetation succession which begins on sand dunes or dry, weathered rock particles. Psammosere species show extreme specialization to withstand drought and extremes of nutrient availability. See PRISERE.

**psychological accessibility,** see ACCESSIBILITY.

**pteridophyta,** *n.* a division of the plant kingdom which comprises contemporary ferns, club mosses and horsetails. The plants are flowerless although distinct roots, stems and leaves can be distinguished; internally, a well developed vascular system exists. A complex two-stage reproductive cycle has usually been developed in which an asexual spore stage preceeds a water-dependent sexual (gametophyte) stage. Pteridophytes first appeared in the DEVONIAN PERIOD (405-345 million years BP) and developed during the

CARBONIFEROUS PERIOD (345-280 million years BP) to dominate the vegetation and form the economically important COAL MEASURES. Most members of this group became extinct and surviving species now form only a very minor proportion of the contemporary flora.

**public sector housing,** *n.* houses built and managed by a public body for rent. Most public sector housing is owned by local authorities such as district councils (in the UK) or similar authorities which administer cities and counties elsewhere. Some public housing is provided by agencies of central governments, for example, Scottish Homes or the Northern Ireland Housing Executive.

Public housing is provided for the citizens of each authority on a basis of proven need and eligibility which is determined according to criteria fixed by each authority; such criteria might include length of residence in the area, special needs because of number of children in a family or priority rating by reason of disability or having been displaced from previous home by REDEVELOPMENT.

The council acts as the landlord, allocating tenants to houses which accord with their expressed needs and locational preferences as far as posssible. Rental levels are fixed by the authority and may contain an element of subsidy from the public purse. The council manages the public housing stock and is responsible for structural maintenance and repair.

**pulp,** *n.* the soft fibrous mass used in the production of paper and rayon formed by the mechanical and chemical processing of rags, straw and especially wood. Coniferous SOFTWOODS are more suitable than tropical HARDWOODS for the production of woodpulp for paper manufacture. The distribution of such forests in the northern hemisphere and their relative proximity to the main centres of demand for paper and paper board has confined pulp and paper production largely to DEVELOPED COUNTRIES. Of 135 million tonnes of woodpulp produced in 1984 the USA accounted for 37%, Canada 15%, and the USSR, Japan, and Sweden each about 7%. Paper consumption has been rising by 2% per annum in developed countries but by 5.5% per annum in DEVELOPING COUNTRIES whose share of world consumption rose from 11% to 16% during the 1970s. However, average per capita consumption amounts to only 5 kg per annum as against 272 kg in the USA and 122 kg in the UK.

**pumped storage scheme,** *n.* a system of generating peak-load electricity using turbines powered by water dropping from a high-level reservoir to a lower one; the water is raised initially by pumping from the lower reservoir during times of off-peak loading on the electricity grid. Surplus electrical energy available at night time from nuclear power stations is frequently used to pump water to the high-level reservoir. A pumped storage scheme can deliver its full electrical output within about 4 minutes of start-up and can sustain a peak load for several hours. It is expensive to generate electricity by this method which only becomes economic during high, but short-lasting, peak load periods.

The largest examples in the UK are at Dinorwig in North Wales and at Ben Cruachan in Scotland. In the latter case the water is raised from Loch Awe to a newly constructed dam on the mountain side and returns to the loch via a generating station in a chamber built within Ben Cruachan.

**puna,** see PARAMO

**PWR,** see WATER-COOLED REACTOR.

**PYO,** see PICK-YOUR-OWN FARMING.

**pyroclastic material,** *n.* the fragmental rock, ash and dust material blown into the atmosphere by the explosive activity of a VOLCANO. Animals and people caught in volcanic ash falls are often asphixiated due to oxygen depletion and poisonous fumes. Heavy falls of volcanic ash and other pyroclastic material can result in the partial or complete burial and loss of settlements, lines of communication and agricultural land. Light falls of pyroclastic material can contaminate water supplies, clog up surface drainage systems and damage crops and grazing land. Pyroclastic material often weathers to form fertile and highly productive soils.

**pyroclimax community,** *n.* a distinct unit of vegetation which has developed in response to frequent burning of the ground. The effects of fire on vegetation may be totally destructive (as in major forest fires) although periodic light fires can encourage the selective development of fire-resistant species. The technique of burning the landscape to promote growth has been practised by agriculturalists for many centuries and some vegetation communities are entirely dependent on fire for their existance, for example the African SAVANNA grasslands and the HEATHLANDS of the north European plain. See SLASH-AND-BURN, FIRE.

# Q

**QC,** see QUALITY CLASS.

**quadrat,** *n.* the basic sampling unit used in the field study of vegetation. Quadrats have been widely used since early in the 20th century to provide data on the frequency, abundance, cover value and density of species in plant COMMUNITIES.

A frequently used quadrat size is 1 m² although the choice of quadrat size is dependent upon the form and size of the vegetation unit under study; the size used must ensure that the species composition of the plant community under examination is adequately represented. This *minimal area* can only be calculated with certainty when the community is homogeneous and has not been broken into distinct parts by grazing. Fig. 49. gives typical quadrat sizes used with temperate zone vegetation.

| Vegetation unit | Quadrat size (m²) |
|---|---|
| Forests (including tree stratum) | 200–500 |
| Forests (undergrowth only) | 50–200 |
| Dry grassland | 50–100 |
| Dwarf shrub heath | 10–25 |
| Hay meadow | 10–25 |
| Agricultural weed communities | 25–100 |
| Moss communities | 1–4 |
| Lichen communities | 0.1–1.0 |

Fig. 49. **Quadrat**. Typical quadrat sizes for field studies in temperate zone plant communities.

**quality class (QC),** *n.* (Silviculture) a method of indicating the general growth performance of a STAND of trees. The QC value is calculated from the relationship between the average height of 40 trees of the same type and their age. Generally, the faster the rate of growth the higher the quality class. Many QC calculations have been made for specific tree types and 'average' growth rates calculated. In the British forestry system such calculations have been made for all the main commercial species and five QC divisions established, ranging from QCI (best) to QCV (poorest). QC values are not comparable between different species and are of only limited use as indicators of future timber yield from a forest. In the UK the QC was replaced in 1966 by the YIELD CLASS measurement system.

**quality of life,** *n.* a complex set of indices which, taken as a whole, provide a definition of the general state or condition of a human population in a given area. Many interpretations can be placed upon the term 'quality of life'. The traditional measure of well-being has been the economist's index of gross national product (GNP), the total value of all goods and services produced anually by a nation. However, this index does not include factors such as the cleanliness of the environment, peaceful co-existance with other human societies and with nature nor the medical health facilities available to a population.

Environmentalists have suggested that a new social indicator of *gross national quality* (GNQ) is required, where all the undesirable features of a society would be costed and subtracted from the GNP to provide a GNQ value which would be a more accurate reflection of quality of life. The task of placing a value on the 'environmental bads' has not yet been achieved, hence a GNQ index remains an unmet target. See PHYSICAL QUALITY OF LIFE INDEX.

**quarrying,** *n.* **1.** the open surface excavation of mineral resources such as slate, granite, sand and gravel.

**2.** see GLACIAL EROSION.

**quartzite,** *n.* a medium grained METAMORPHIC ROCK formed by the regional or contact METAMORPHISM of sandstone. Quartzite is sometimes used for road metal.

**Quaternary period,** *n.* the present geological PERIOD that followed the TERTIARY PERIOD and began around 2 million years ago. The Quaternary can be divided into the PLEISTOCENE EPOCH and the RECENT EPOCH. Humans appeared for the first time during the Quaternary. See EARTH HISTORY.

# R

**rad,** *n.* a unit of absorbed RADIATION DOSE now superseded by the gray but still in widespread common use. 1 rad = 0.01 Gy.

**radial drainage,** see DRAINAGE PATTERN.

**radiation,** *n.* **1.** the transfer of heat and other energy by means of electromagnetic waves, as in the transfer of SOLAR ENERGY from the sun to the earth. This radiation enters the earth's atmosphere in short wavelength forms ( as gamma, ultraviolet and X-rays) whilst the loss of heat from the atmosphere occurs via the longer wavelength forms (as infra-red and microwaves. See INSOLATION, ALBEDO.

**2.** (Nuclear physics) the emission of alpha-, beta- and gamma particles from radioactive particles. Prolonged exposure to ALPHA PARTICLES and BETA PARTICLES can cause serious damage to living tissue, while gamma particles (commonly associated with nuclear reactors and bombs) can lead to *radiation sickness* in humans involving skin cancers, loss of hair, destruction of bone marrow and ultimately, death. See IONIZING RADIATION.

**3.** (Evolution) *adaptive radiation* occurs when primitive ancestors evolve into many divergent forms, each adapted for survival under specific conditons. Adaptive radiation is often found to have occurred on isolated islands such as the Galapagos Islands where for example, an ancestral finch has evolved to fill 14 separate modes of life.

**radiation dose,** *n.* the quantity of IONIZING RADIATION received by an object. It can be measured in a number of ways:

(i) the *absorbed dose* is a measure of the energy given to a unit mass of matter such as protoplasm. It is measured in units of gray (symbol: Gy) where 1 Gy = 1 joule per kilogram (J/kg). The absorbed dose was formerly expressed in rads where 1 rad = 0.01 Gy.

(ii) the *dose equivalent* is the absorbed dose of ionizing radiation multiplied by a numerical factor, the *quality factor*, which allows for the different effectiveness of the various types of RADIATION in producing damage to tissues. Dose equivalence is expressed in

357

sieverts (symbol: Sv) which have replaced the previously used REM unit.

(iii) *effective dose equivalent*, the quantity of radiation dose obtained by multiplying the dose equivalent to various tissues by a risk-weighting factor appropriate to each organ and summing the product. The unit of measurement is the sievert.

(iv) *collective effective dose equivalent*, or the *collective dose* is the quantity of radiation dose obtained by multiplying the average effective dose equivalent by the number of individuals exposed to a given source of radiation. The unit of measurement is the person-sievert.

**radiation fog,** *n.* a type of FOG formed by the rapid nocturnal cooling of the ground which chills the overlying layers of air below their DEWPOINT and leads to the condensation of atmospheric water vapour. Radiation fogs frequently form in valley bottoms and are associated with a moist atmosphere, calm conditions and cloudless skies. This type of fog occurs in most latitudes and can develop at any time of year. In summer, early morning radiation fogs are quickly evaporated by the heat of the sun but in winter, such fogs may persist for several days. Compare ADVECTION FOG, STEAM FOG.

**radiation sickness,** see RADIATION.

**radioactive waste,** see NUCLEAR WASTE.

**radio-carbon dating** or **¹⁴C-dating,** *n.* a technique by which organic material can be accurately dated. Developed in the 1940s at Chicago University, the method is based on the principle that elemental carbon is incorporated into green plants from the atmosphere via PHOTOSYNTHESIS and passes up the food chain until it becomes incorporated in the bones and teeth of animals. The death of an organism terminates the incorporation of carbon, an ISOTOPE of which, ¹⁴C, is unstable and decays at a predictable rate. When a comparison of the ¹⁴C isotope concentration remaining in a sample is made against a calibration curve (see Fig. 50), an accurate age can be calculated. ¹⁴C has a HALF-LIFE of 5730 years and with careful laboratory analysis, ¹⁴C-dating can be used to date objects of up to 70 000 years of age. When the technique was first formulated it was assumed that there was a constant amount of ¹⁴C in the atmosphere; however, it was later shown that this amount varied and thus a correction factor had to be applied to any calculation made using radio-carbon dating. See DENDROCHRONO-LOGY, POLLEN ANALYSIS, DATING TECHNIQUES.

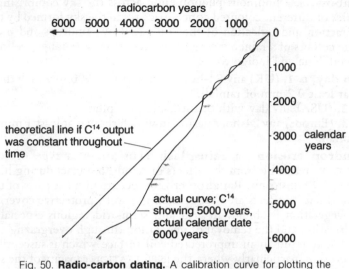

radiocarbon years

6000  5000  4000  3000  2000  1000

theoretical line if C¹⁴ output was constant throughout time

calendar years

1000
2000
3000
4000
5000
6000

actual curve; C¹⁴ showing 5000 years, actual calendar date 6000 years

Fig. 50. **Radio-carbon dating.** A calibration curve for plotting the age of a sample showing the correction factor to be applied to ¹⁴C dates.

**rain,** *n.* PRECIPITATION from clouds in the form of drops of water, formed as a result of the condensation of atmospheric water vapour in rising air masses. Raindrops usually have a diameter of between 0.5 and 2.5 mm and form in a number of ways:

(a) *orographic rain* results when moisture-laden air is deflected upwards by a physical barrier such as a mountain range. The water vapour in the air cools and condenses and heavy rain falls on the windward side, whilst a RAINSHADOW develops on the leeward side. The mountainous west coast of Scotland receives heavy orographic rainfall whilst the east coast lies in a rainshadow.

(b) *cyclonic* or *frontal rain* is associated with the passage of a DEPRESSION or a TROPICAL CYCLONE. Much of the rainfall on the east coast of North America and in the UK is frontal in origin.

(c) *convectional rain* is caused by the rapid heating of the moisture-laden lower layers of the atmosphere due to an intensely heated land surface. This air rises under conditions of atmospheric INSTABILITY and the water vapour condenses to generate heavy rainfall. THUNDERSTORMS in temperate and tropical regions are formed by convectional heating.

**rainbow,** *n.* a luminous phenomenon across the sky comprising a series of different coloured concentric arcs in the sky formed by the refraction and reflection of the sun's rays by raindrops and mist. The colours of a rainbow range from red, through orange, yellow, green, blue and indigo to violet.

**rain day,** *n.* **1.** (UK) any 24-hour period from 0900 in which there is at least 0.2 mm of rainfall.

**2.** (USA) any day with quantifiable precipitation.

**3.** (Europe) any 24-hour period in which there is at least 1 mm of rainfall.

**raindrop erosion,** or **rainsplash erosion,** *n.* a type of SOIL EROSION resulting from the effects of raindrop impact during high intensity rainstorms. Raindrop erosion occurs in most parts of the world but its effects are exaggerated if a soil's protective covering of vegetation is absent. In arid and semi-arid regions especially, cultivation and the removal of vegetation through overgrazng and FIRE may result in an unprotected soil surface which is susceptible to erosion, particularly where the INFILTRATION capacity of the soil has been reduced by the washing-in of fine soil particles which block the surface PORE SPACES. See also SHEET EROSION.

**rainfall,** *n.* the amount of PRECIPITATION recorded in an area during a defined period, and measured in a standard RAIN GAUGE. Rainfall is usually measured in millimetres per annum. See also RAIN DAY.

**rainfall reliability,** *n.* an expression of the variability of rainfall from one year to the next. Mean monthly and annual statistics may disguise the interannual variation in the amount of PRECIPITATION and its seasonal distribution. The reliability of rainfall is expressed either as a mean or as an extreme percentage departure from the normal; for example, a location may have a mean annual rainfall of 500 mm with a ±35% interannual variability. In areas of very low rainfall where a year's precipitation or more may fall in 24 hours the percentage variability is naturally greatest (see Fig. 51).

**rain gauge,** *n.* an instrument that measures the amount of PRECIPITATION that falls in a given time period. The dimensions, shape and location of a rain gauge vary slightly from country to country. In the UK, for example, a rain gauge is a copper or plastic container at the top of which is a funnel, 125 mm in diameter and protruding 300 mm above a grass surface. The rain gauge should be positioned so that it is clear of turbulence caused by vegetation or trees. The horizontal distance between the rain gauge and the obstacle should be at least twice the height of the impediment.

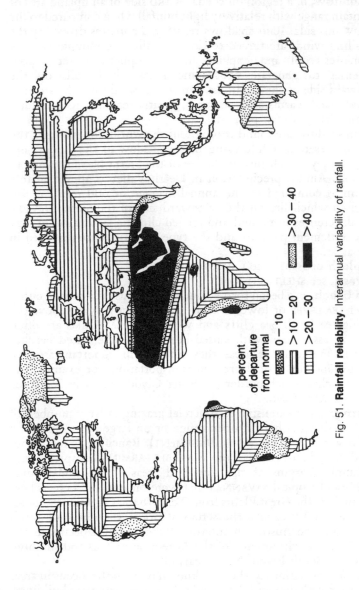

Fig. 51. **Rainfall reliability**. Interannual variability of rainfall.

percent
of departure
from normal

0 — 10    >10 — 20    >20 — 30    >30 — 40    >40

**rain shadow,** *n.* a region on the LEEWARD side of an upland area or mountain range with relatively light rainfall when compared to the WINDWARD side. Rain shadows result as air masses driven by the prevailing winds are forced to rise over hills and mountains. The air masses cool and expand and atmospheric water vapour condenses to generate orographic RAIN which falls on the windward side of the landmass. On the leeward side, the air mass contracts and warms as it descends, greatly reducing its humidity and associated rainfall potential.

Rain shadows are found throughout the world, for example, the Atacama desert in Chile lying in the shadow of the Andes, and Death Valley which lies in the lee of the Sierra Nevada range in the USA. Annual precipitation in Death Valley is approximately 40 mm as compared to the annual average of 600 mm for San Francisco which lies on the windward side of the Sierra Nevada. Such differences in rainfall and related climatic factors can greatly affect settlement patterns and the type of agriculture practised in these areas.

**rainsplash erosion,** see RAINDROP EROSION.

**rainwash,** see SHEET EROSION.

**raised beach,** *n.* a beach formed during previous occasions of either higher SEA LEVEL or lowered land level and now lying above the existing sea level. Sea cliffs and WAVE-CUT PLATFORMS are often found in association with raised beaches. Where raised beaches occur in mountainous areas they often form important areas of level ground for agriculture and transportation, for example the raised beaches which border the River Clyde estuary in Scotland.

**ranch,** see RANCHING.

**ranching,** *n.* the extensive commercial grazing of large numbers of cattle or sheep either on open range or on large areas fenced by barbed wire (see GRAZING MANAGEMENT). Ranching is less widespread now than a century ago as ARABLE FARMING has spread onto the humid margins of the semi-arid areas of the mid-latitude STEPPES and tropical SAVANNAS that were once its preserve. These areas include the Great Plains from Texas to the Canadian PRAIRIES, the llanos of Venezuela, the sertao of Brazil, the Karoo of South Africa, and the Australian interior. Ranching also occurs in the more humid environments of the Argentine PAMPAS and the high country of South Island, New Zealand.

*Ranches,* or stations as they are known in Australia, occur in areas of sparse population with low land values, have few farm buildings, but may feature numerous windmills to pump groundwater to the surface. Stock CARRYING CAPACITIES are low, at best perhaps 3 hectares per head of cattle, but in the more arid areas sometimes no more than 2 cattle per 100 hectares. Ranches in western Texas may exceed 8000 hectares, while some in Arizona are twice that size. Australian sheep stations average 8000 hectares, while some northern cattle stations cover 20 000 hectares or more. In recent times, however, ranching has tended towards smaller and more diversified units with some cultivation.

**rangeland,** see PASTURE.

**range management,** *n.* the organization and control of the use of open PASTURE or *rangeland* to achieve maximum livestock production commensurate with maintaining a stable ECOSYSTEM. During the late 19th century, overstocking on many of the RANCHING enterprises of the American Great Plains led to degraded natural pasture and a much reduced CARRYING CAPACITY, especially after the severe winter of 1886-87. The introduction of barbed wire enabled GRAZING MANAGEMENT schemes to be applied to formerly free-range areas. More recently, range management practices have been extended to integrate the demands of stock-rearing and wildlife conservation, often in the light of conflicting pressures from recreational uses and other forms of natural resource exploitation, such as timber and mineral extraction.

**ranker, regosol** or **lithic haplumbrept,** *n.* a SOIL PROFILE displaying poor horizonation and in which the A-HORIZON alone can be marginally determined from the PARENT MATERIAL; no other distinct horizons can been seen. Ranker soils form on a wide variety of parent material, under differing climates, topography and under different vegetation. They do not form soils suitable for agricultural use. The term is used only in certain soil classification systems.

**rapid,** *n.* a section of a river displaying broken, rapidly flowing water resulting from a sudden increase in the gradient of the river. Rapids are usually caused by the varied rates of erosion on different types of underlying bedrock. River navigation is often impeded by rapids unless a series of LOCKS are constructed to bypass the feature.

**rare organism,** see RED DATA BOOK.

**ratooning,** see MULTIPLE CROPPING.

**Raunkiaer's lifeform classification,** *n.* a functional classification of plants, based on their ability to adapt to survive unfavourable climatic conditions. The Danish ecologist, Carl Raunkiaer chose the position on the plant of the vegetative buds as the best expression of adapatation to survival. He recognized 15 main types of life forms which he placed into 5 main categories:

(a) *phanerophytes*, comprising plants which bear their buds in an exposed aerial position and are thus ill-suited for survival in arid or cold climates. Four sub-groups are recognized depending upon the height above ground of the buds;

(b) *chamaephytes*, comprising woody or herbaceous, low-growing plants with buds produced on aerial branches close to the soil. Four sub-groups are again recognized, all of which are well represented in areas characterized by drier or cold climates;

(c) *hemicryptophytes*, comprising plants whose shoots degenerate to the level of the ground at the beginning of the unfavourable period, so that only the lower aerial parts of the plant remain alive. Most of the flowering herbs and grasses belong to this group;

(d) *cryptophytes*, comprising plants which bear buds at varying depths below ground (*geophytes*) or in soil saturated with water (*helophytes*) or in water itself (*hydrophytes*);

(e) *therophytes*, comprising plants which survive winter as seed, that is, are ANNUALS, and have a life cycle extending from spring to autumn.

From extensive statistical investigation of world flora, Raunkiaer established that 46% of the world flora comprised phanerophytes; 9%, chamaephytes; 26%, hemicryptophytes; 6%, cryptophytes and 13%, therophytes. Later work has shown that plant climatic boundaries can be determined through the proportion of species in the main categories. For example, in cold temperate zones one finds 10% of the flora composed of chamaephytes, in boreal zones, 20% and in arctic zones, 30%.

**reafforestation,** *n.* the planting of trees on land that has carried forest within the previous 50 years, or within living memory, but which has been removed by natural or human agency.

Some sources distinguish between reafforestation and *reforestation*, the latter being the establishment of forest plantations by means of either or both AFFORESTATION and reafforestation, but excluding regeneration of forest crops naturally or through FOREST MANAGEMENT. The FAO estimated that by 1980 some 14.5 million hectares of forest land were being reforested or renewed worldwide each year, mostly in China, the USSR, the USA, Canada, Brazil and Japan. Compare DEFORESTATION.

**Recent epoch** or **Holocene epoch,** *n.* the present geological EPOCH that followed the PLEISTOCENE EPOCH around 10 000 years ago. See EARTH HISTORY.

**recessional moraine,** see MORAINE.

**recycling,** *n.* the recovery and re-use of materials from wastes. Recycling can comprise:

(a) *natural recycling* where all the finite substances necessary for life such as water, nitrogen and carbon constantly undergo recycling (see CARBON CYCLE, HYDROLOGICAL CYCLE, NITROGEN CYCLE);

(b) the *recycling of obsolete manufactured goods* from which valuable mineral resources (such as steel, copper, plastic or fibres) can be recovered and re-used. The extent to which the recycling of such goods takes place depends upon (i) the cost and scarcity of producing or extracting such resources; (ii) the cost and ease with which obsolete items can be recycled, and (iii) the political and social climate which can determine the acceptibility of recycling.

In recent years increasing material scarcity has led to an upsurge in recycling of materials. Recycling falls into three classes:

(a) *re-use* in which an item can be easily reused with a minimum of reprocessing, as for example, a milk bottle.

(b) *direct recycling* in which an item fails to pass a production quality control check and is internally recycled by the producer without first being used. Surplus or off-cut materials can also be directly recycled for example, paper, wood, iron and steel.

(c) *indirect recycling* which involves the collection of material after use in commerce or domestic situations, followed by its sorting, cleaning and reprocessing. Because of the number of stages involved, this form of recycling is expensive in terms of energy and manpower. Even after substantial reprocessing, the recycled commodity is often of low quality and cannot be remanufactured for its previous use, for example, glass may be ground into pellets and incorporated into skid-resistant road surfaces, while plastics can be made into sheeting, pallets or fence posts where appearance and structure are not primary considerations.

**Red Data books,** *n.* a series of source books containing data on the threat to and rarity of plant and animal species. Red Data books are produced on behalf of the INTERNATIONAL UNION FOR THE CONSERVATION OF NATURE AND NATURAL RESOURCES at both national (country) and continental levels level and cover mammals, birds, amphibians and retiles, fish and angiosperms (higher plants); regional cover is incomplete particularly for the developing world. For each species data is provided on status, geographical distribution, population size, habitat, breeding rate and any conservation measures taken to protect the species.

A species qualifies for inclusion in the Red Data book only if it can be justified as 'rare' although different interpretations of rarity exist. In the UK, a rare plant is one which occurs in 15 or fewer of 10 km² national grid squares which cover the country. Some 18% (321 species) of the native British flora can be classified as rare on this basis.

Each species is assigned a rarity status within a range of five categories:

(a) *endangered species*, which applies to those species or subspecies most likely to become extinct if current trends persist;

(b) *vulnerable organisms*, which are severely exploited at the present time or which inhabit areas of major environmental disturbance and which are unlikely to adapt to those changes. Species in this category will probably move into the endangered status if current trends persist;

(c) *rare organisms*, which are at risk because of their small total world population. Rare taxa usually have a very restricted geographical distribution, for example exiting solely on islands, or mountain summits;

(d) *out-of-danger species*, which were formerly in categories (a)-(c), but which have responded to conservation measures and the threat to their survival has been removed.

(e) *indeterminate species* which comprise plant and animals which probably belong to groups (a)-(c) but insufficient knowledge prevents an adequate assessment of their status.

**redevelopment,** *n.* the process of reconstructing the physical fabric of a run-down area, usually within a city or major urban settlement. The term implies the total replacement of the old environment by a newly created one, and is frequently used to describe the major reconstruction carried out in the UK in the 1960s following demolition of slum areas and the reclamation of derelict industrial sites. The process, which was very expensive in both money and social disruption, was gradually replaced by a less sweeping approach involving the rehabilitation of INFILL SITES by new construction. See PRIORITY TREATMENT AREAS, OVERSPILL, DEPRIVATION, AREA OF, REPLACEMENT RATE, HOUSE CONDITION SURVEY.

**reef,** *n.* **1.** a ridge of rock usually lying just below the surface of the sea but which may be visible at low tide. The rock type and method of reef formation varies considerably. Reefs are found in coastal areas and open oceans throughout the world, and if uncharted, may represent a hazard to shipping. See CORAL REEF.

**2.** a gold-bearing conglomerate in Africa.

**3.** a gold-bearing quartz vein in Australia.

**reforestation,** see REAFFORESTATION.

**refuge site** or **refugium,** *n.* any area of pedologic, topographic or climatic uniqueness, into which organisms have migrated, either as a result of competion from more adabtable species, or through their inability to adapt to changing habitat conditions of surrounding areas. Habitat conditions are thus suitable for the support of ancestral or *relic populations.* Refuge sites of this type can become 'graveyards' as more species become forced to occupy successively smaller living areas, eventually secumbing to species extinction.

**regional metamorphism,** see METAMORPHISM.

**regolith** or **mantle rock,** *n.* the layer of weathered rock debris which overlies unweathered BEDROCK. The regolith incorporates SKELETAL SOILS and superficial deposits of alluvial, aeolian, glacial and organic materials.

**regosol,** see RANKER.

**regulator organism,** *n.* any organism capable of maintaining a relatively stable internal metabolism despite changes in the external, physical environment. The main species which can be classed as regulators are the mammals which, by means of sophisticated heat-regulation circulatory systems, can maintain a constant internal body temperature across a wide range of external environments. See CONFORMER ORGANISM.

**rehabilitation,** *n.* the restoration, or reclamation for use, of derelict land, houses or locations. See also DERELICT LAND, HOUSE CONDITION SURVEY, IMPROVEMENT GRANT.

**rejuvenation,** *n.* the process by which the erosive capacity of a river is renewed. When a river's BASE LEVEL is lowered then the river will begin to regrade its profile to the new base level. During the regrading process, there is often a marked change of slope, known as a KNICKPOINT, where the newly graded profile intersects

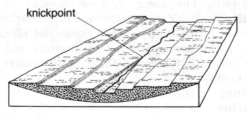

Fig. 52. **Rejuvenation.** River terraces formed as a result of rejuvenation.

367

# RELATIVE DATING

with the former. Knickpoints are often marked by WATERFALLS or RAPIDS (see Fig. 52).

This recurrence of river downcutting can take two forms:

(a) *dynamic rejuvenation*, resulting from a eustatic fall in sea level or from the uplift of land masses (see EUSTASY);

(b) *static rejuvenation*, resulting from the increase in the river's load discharge (see FLUVIAL TRANSPORTATION). This can occur following RIVER CAPTURE or changes in climate which cause an increase in both discharge and load. As a result, the river's drainage system has greater eroding potential and a new phase of downcutting commences.

Several landforms are associated with rejuvenation including INCISED MEANDERS and RIVER TERRACES.

**relative dating,** *n.* the determination of the age of an object or event in history relative to other objects or events. A relative dating scheme allows events to be labelled as occurring earlier, later or synchronous to other events and is usually based on the principle that objects located deeper in the stratigraphical layering pre-date items which occur higher in the column. The *post-glacial pollen zone sequence* is a widely used example of a relative dating scheme. It is usually impossible for a relative dating scheme to provide a precise datum for the occurrence of events although wherever a great many events from one site can be chronicled in rank order then the relative dating technique can become very accurate. The method gains considerably if ABSOLUTE DATING points can be established at key points in the chronosequence. See POLLEN ANALYSIS. See also DENDROCHRONOLOGY, RADIO-CARBON DATING.

**relative humidity,** *n.* the ratio (expressed as a percentage) between the WATER VAPOUR content of a given amount of air and the moisture it could contain if the air had been saturated at the same temperature and pressure. For example, if the air holds two-thirds of the water vapour it could when saturated, then the relative humidity is 66%. The water retention capacity of air usually increases with a rise in air temperature and high relative humidities are common in many tropical and equatorial climates. These climates are frequently oppressive to humans and introduced animals, such as cattle. Compare ABSOLUTE HUMIDITY. See HUMIDITY, SATURATION.

**relay cropping,** see MULTIPLE CROPPING.

**relic population,** see REFUGE SITE.

**rem,** *n.* *acronym for* roentgen equivalent man, a unit of dose equivalence now replaced by the sievert (Sv), but still in widespread common use where 1 rem = 0.01 Sv. See RADIATION DOSE.

**remote sensing,** *n.* the collection and analysis of scientific data about phenomena at, above, or below the earth's surface and the oceans without coming into physical contact with them. Information can be gathered in a variety of ways, using conventional aerial photography, radar, or airborne electronic-scanning devices. Satellites are being increasingly used for remote sensing and a series of LANDSAT satellites were launched from 1972 onwards specifically to provide an inventory of land resources and to monitor environmental changes. Remote sensing techniques can be used to provide data on such matters as the development and movement of meteorological disturbances, the existence of certain mineral deposits and the build-up of pressure along faults in the earth's crust prior to an earthquake. See also FALSE-COLOUR IMAGERY.

**rendoll,** *n.* a RENDZINA-like soil as listed in the American soil classification system.

**rendzina, calcareous rego black soil, rendoll** or **derncarbonate soil,** *n.* a grey, brown or black soil of medium texture developed above PARENT MATERIAL with a high calcium carbonate content such as limestone. Rendzinas are usually shallow (no more than 30 cm), are excessively drained and contain fragments of poorly weathered calcium carbonate. The typically dark colour of the rendzina is due to the abundance of HUMUS. The combination of shallowness and excessive drainage makes this soil type unsuited for agricultural use.

**renewable resource,** *n.* any commodity which theoretically cannot be totally consumed due to its ability to reproduce (biologically) or regenerate (physically) in number. Renewable resources originate either as inexhaustable sources (such as SOLAR ENERGY), as a major physical cycle (such as the HYDROLOGICAL CYCLE), or as a biological system (such as all plants and animals which replicate themselves). In recent years, human activity has seriously depleted some resources previously classed as renewable, for example, North Sea fish stocks and numerous forest resources. This has occurred when the resource has been harvested at a rate faster than it can renew itself. See NON-RENEWABLE RESOURCE.

**replacement rate,** *n.* the proportion of the original residential population that can be accommodated in a REDEVELOPMENT site. In slum clearance areas in the UK during the 1960s, the replacement rate was typically between 35 and 45%. The assumed replacement rate is an important factor in assessing the OVERSPILL population that may need to be accommodated elsewhere in the region.

**Report on Community Forest Policy,** see GATTO REPORT.

**reptile,** *n.* any member of the Reptilia class of vertebrates. Ancestral reptiles (dinosaurs) can be found in the fossil record, first appearing at the end of the CARBONIFEROUS PERIOD, 280 million years ago. They dominated the land during the MESOZOIC ERA (230-60 million years ago) developing a wide variety of walking, running and swimmimg forms; it is possible that they also formed the ancestors of modern birds. Most reptiles are cold blooded (*poikilothermic*) and posses dry, scaly skins. Offspring are produced in large, hard-shelled amniotic eggs. Modern reptiles include tortoises, turtles, lizards and snakes.

**reserve,** *n.* a presently unexploited MINERAL RESOURCE.

**residential density,** *n.* a measure of the intensity of use of a residential area by the people who live in it. The residential density value is a function of:

(a) *gross residential density*: the ratio of the total population to the total land area of the unit being used for spatially grouping the data, for example, the city ward or the census enumeration district.

(b) *net residential density*: the ratio of the total population in the areal unit to the net area in residential use, which includes each house, its immediate curtilage and half the width of the adjoining road.

(c) *occupancy rate*: the ratio of residents to the habitable rooms within the dwellings studied; a related expression of density of dwelling units is the calculation of number of habitable rooms per acre or hectacre.

The measures of density, when mapped, give an indication of those areas of a city which are deemed to be overcrowded. Standards of acceptable density levels change as public perceptions of minimum living standards become more exacting. In the UK, occupany rates of more than 1.5 persons per room are now regarded as undesirable. See CROWDING.

**residential segregation,** *n.* the clustering of ethnic, religious or social class groupings in particular residential areas of a city. In modern cities segregation is most marked in areas of lower socio-economic status where there is the additional factor of racial difference. The sharply defined segregation of working-class Catholics from working-class Protestants in Belfast is an extreme example of this and contrasts with the liberal religious mix in the more socially diverse areas of the same city. Segregation of this sort is voluntary and driven by the perception of an external threat to the group involved. See also GHETTO.

**residual mineral deposit,** *n.* the residual mineral and rock debris resulting from CHEMICAL WEATHERING processes, such as TERRA ROSA in limestone regions. Some residual deposits have considerable economic value, such as BAUXITE and LATERITE. Bauxite ores are mined in Surinam, Jamaica, Guyana, Australia and several parts of Africa whilst laterites are exploited in Brazil and Venezuela.

**resilience,** *n.* (Ecology) the robustness or durability displayed by an ECOSYSTEM in the event of changes in climate, pollution, competition from other organisms or, increasingly, as a result of human activity (see DISTURBANCE). Resilience is a measure of the probability of the extinction of an ecosystem and is dependent upon:

(a) the extent of natural changes within the ecosystem. Some ecosystems show a natural tendency to fluctuate between widely spaced extremes and in so doing are well placed to accommodate changes imposed by humans. Conversely, those ecosytems which show very small fluctuations have poor resiliance to anthropogenic change;

(b) the rate at which the system can recover from disturbance.

**resource,** *n.* **1.** any commodity such as water, oxygen, chemical nutrients and territory, required by living organisms in order to complete their life cycle. During the life history of a species, many different resources will be required and in differing proportions. For most living organisms resources are used solely to sustain life, although for humans, resources have assumed an additional role as they have been used to sustain and advance the material well-being of our species.

**2.** any source of economic wealth, for example, land, minerals, fossil fuel, labour etc. The utility of a resource is dependent upon contemporary levels of technological progress and what is regarded as a resource at a particular place or time may be thought of as useless elsewhere or at a different time. For example, crude oil deposits became valuable resources only when suitable technology existed to produce the automobile engine. See RENEWABLE RESOURCES, NON-RENEWABLE RESOURCES, RESOURCE MANAGEMENT.

**resource management,** *n.* the planned use of RESOURCES in order that long-term stability of both economic and ecological systems can be maintained. Over-use or mis-use of a resource can result in increased production costs, scarcity, physical erosion of the environment, and/or human hardship. Many national governments as well as non-political, international agencies such as the INTERNATIONAL UNION FOR THE CONSERVATION OF NATURE AND

NATURAL RESOURCES are now dedicated to sophisticated resource management techniques in order to sustain the quality of the biosphere. Examples of these techniques include: REAFFORESTATION; SOIL CONSERVATION; RECYCLING of wastes; LAND USE PLANNING, and population control. See also ENVIRONMENTAL MANAGEMENT.

**rest area,** see SERVICE AREA.

**restoration bond,** *n.* a cash sum deposited by a developer to pay for the rehabilitation of land disfigured during an exploitive phase of a development. Restoration bonds are common with open-cast mining developments particularly when located on the RURAL-URBAN FRINGE or in agricultural areas.

**ria,** *n.* a long, narrow and often irregular coastal inlet that decreases in width and depth as it extends inland. Usually lying at right angles to the coast, rias are formed by the post glacial rises in sea level drowning the lower part of a river valley. Rias are found along the coasts of Brittany, north-west Spain and south-west Ireland and their value as natural harbours has often led to the development of ports where they occur.

**Richter scale,** *n.* a logarithmic scale of earthquake magnitude devised by the American seismologist C.F. Richter in 1935. The scale indicates the amount of energy generated by an earthquake and ranges along a progression from 0 to 9. The most severe earthquakes register over 8.0. The use of the Richter Scale to measure earthquakes is preferred to the more subjective MODIFIED MERCALLI SCALE.

**ridge of high pressure,** *n.* the narrow elongated extension of an ANTICYCLONE which is contained between two areas of lower pressure. The passage of a ridge of high pressure usually results in a short bright and dry spell within the unsettled weather associated with the DEPRESSIONS (see Fig. 53).

**rift valley** or **graben,** *n.* an elongated trough-like depression in the earth's surface bounded by two parallel FAULTS and formed by the downthrow of the central block or the upthrow of the adjacent blocks. Examples of rift valleys include Death Valley in California, USA and the Central Valley of Scotland.

**right of way, 1.** the legal right of someone to pass over another's land, acquired by grant or by long usage.

**2.** any FOOTPATH or road established by grant or long usage. Sometimes organized walks are needed to establish a right of way which has been obstructed by, for example, the erection of a fence.

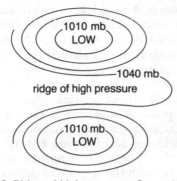

Fig. 53. **Ridge of high pressure.** See main entry.

**rill,** *n.* a shallow channel often occurring in the middle of a slope which is formed by the concentration of thin sheets of surface water into flows capable of eroding the soil (see SHEET EROSION). Rills can reach up to 30 cm in width and depth and may gradually coalesce and enlarge downslope as the increased concentration of surface RUNOFF results in GULLY EROSION.

**rime,** *n.* an accumulation of ice formed on the WINDWARD side of objects such as cars, fences and trees. Rime results when super-cooled water droplets freeze when blown by a slight wind into contact with any exposed surface. See FROST, SUPERCOOLING.

**Ring of Fire,** see EARTHQUAKE.

**ring road,** *n.* any road specifically planned to convey motor vehicles around the perimeter of a town or city thus preventing the flow of vehicles (especially heavy goods vehicles) into the urban centre. Ring roads may make use of the existing highway system or more commonly, they may comprise purpose-built dual carria-geway roads or MOTORWAYS. Such has been the growth both of urban areas and of the volume of traffic that many old ring-roads have become superseded by more modern structures, for example the original 'North Circular' routeway in London by the M25 orbital motorway.

**river,** *n.* a natural stream of fresh water flowing along a definite course, usually into the sea, being fed by tributaries. Rivers may originate from the coalescence of small water channels, from lakes or from natural springs. In humid areas, rivers may be permanent features of the landscape while in arid areas they may flow only intermittently. Most continental landmasses have major river systems, for example the Nile in Africa, the Mississippi in North America, the Ganges in Asia and the Rhine in Europe, all of which have CATCHMENT AREAS extending over thousands of square kilometres. See RIVER PROFILE, RIVER CAPTURE, REJUVENATION.

**river capture, abstraction** or **beheading,** *n.* the breaching of a WATERSHED by a rapidly eroding river to acquire the headwaters of another stream and incorporate them within its own drainage system (see Fig. 54). River capture results from HEADWARD EROSION and the point of diversion is known as the ELBOW OF CAPTURE. The captured river is deprived of a considerable part of the headwater catchment and becomes a MISFIT RIVER in its own valley. The misfit river's new source may be some distance below the elbow of capture, with the abandoned dry section of the valley forming a WIND GAP. If several streams are captured then the dominant river can have a very extensive CATCHMENT AREA. The catchment area of the River Humber in England provides a good example of the development of a river system through river capture.

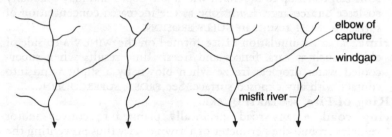

Fig. 54. **River capture.** Diagram showing a river capture by a rapidly eroding stream

**river profile** or **thalweg,** *n.* a longitudinal section of a river's course. The profile generally follows an ascending concave curve, with the steepest gradients located close to the stream's source and the gentlest near the mouth as the river approaches its BASE LEVEL.

Along its course, a river is constantly eroding, transporting and depositing material and if these processes are balanced then the river is said to have achieved a *profile of equilibrium* and become a *graded river*. This equilibrium is of a dynamic nature as changes in a river's profile may result from factors such as an increased load and from REJUVENATION. Compare CROSS-SECTION. See FLUVIAL EROSION, FLUVIAL TRANSPORTATION, FLUVIAL DEPOSITION.

**river regime,** *n.* the average daily or monthly discharge of a river, usually expressed in cubic metres of water passing a measuring point. The river regime value can be used in the modelling of water yield from a CATCHMENT AREA and can provide information vital in the forecasting of flood and/or low-flow situations.

**river terrace,** *n.* a flat or slightly inclined alluvial platform located on the sides of a river valley. Terraces may be developed through REJUVENATION, as the river cuts through its own deposits into the underlying bedrock (see Fig. 52). Given time, the course of the new valley widens and a second flood plain forms within the first. The margins of the original valley floor are then left as terraces, above the new level of the river. Subsequent rejuvenation may cause further terraces to form, giving a stepped appearance. As the terraces on either side of the valley occur at the same level, the resulting CROSS-SECTION is symmetrical.

River terraces may also form as a meandering river erodes vertically into its FLOODPLAIN. However, the terraces formed by this method are unpaired and provide an asymmetrical cross section.

River terraces are found throughout the world; examples may be found on the Fraser River in British Columbia, the River Findhorn in Scotland and on the Clutha River in Otago, New Zealand.

Terraces have several features that make them of use to humans. Their surface is generally above flood levels, ALLUVIAL SOILS are often of great agricultural value and the ALLUVIUM itself is well-drained, flat and stable enough for settlements and lines of communication to be constructed. The alluvium can also provide a source of sand and gravel for construction purposes. Due to these factors, many settlement patterns in eastern and central USA can be linked to the occurrence of river terraces.

**Roaring Forties,** *n.* the expanse of ocean between 40° and 50°S where the prevailing WESTERLIES flow with great strength and regularity. The weather is usually stormy, wet and mild, and the region is characterized by the constant eastward tracking of DEPRESSIONS.

**rocking stone,** see ERRATIC.

**rock salt** or **halite,** *n.* a SEDIMENTARY ROCK composed of a sodium chloride mineral precipitated by evaporating seawater. Rock salt is used by the chemical and metallurgical industries and in the manufacture of domestic salt, glass and soap. Rock salt is often used mixed with sand for the treatment of icy roads.

**rock flour,** see GLACIAL TRANSPORTATION.

**room occupancy,** *n.* the degree of crowding within a house, expressed as the ratio of the number of people in the household to the number of habitable rooms (excluding kitchen and bathroom). In the UK, an occupancy rate of 1.5 persons or more per room is regarded as an unacceptable degree of overcrowding. See CROWDING, RESIDENTIAL DENSITY.

**root nodules (of leguminous plants),** *n.* tumour like swellings of the root tissue of clover, beans, peas, vetch and other legumes. They contain endosymbiotic nitrogen-fixing bacteria which allow the plant to survive in nitrogen-deficient soils. The roots of legumes appear to secrete a substance which encourages the multiplication of nodule-forming bacteria, of which Rhizobium species are most common. Nitrogen-fixing nodules also occur in some non-leguminous plants, notably the alder tree.

**rotational bush fallow,** *n.* any tropical farming practice based on a FALLOW SYSTEM in which the cultivation period is longer or the FALLOW is shorter, or both, than under true SHIFTING CULTIVATION, but where the years of cultivation remain less than the time in fallow. This more intensive use of land may have evolved from longer-fallow shifting cultivation in response to population pressure or a more general pressure of needs. Rotational bush fallow or recurrent cultivation covers a wide range of practices through variation in the cropping and fallow regime, and ecological and cultural diversity.

Several features of rotational bush fallow are similar to those of shifting cultivation, particularly in the use of FIRE to clear the land, in the tools employed, and in the practice of MULTIPLE CROPPING. However, to compensate for the inability of shorter fallow periods adequately to restore soil fertility, the application of MANURE is more likely, as is the need for more cultivation, both in seed bed preparation (for example, by MOUND CULTIVATION), and in weeding, due to the plot's longer cropping period. Households are unlikely to be oriented solely to SUBSISTENCE FARMING, and are in general permanently settled. Although rights to land may still be usufructory (see LAND TENURE), holdings are usually clearly marked as the property or responsibility of particular families. In regions with harsh dry seasons, bare soil may be exposed between growing seasons, increasing the risk of SOIL EROSION. As the aim of rotational bush fallowing is to create a regular system of fallows which are prevented from reverting back to their natural state, as in true shifting cultivation, the spread of these practices in the face of increasing population pressure and the intensification of tropical agriculture has contributed to DEFORESTATION.

**rotational grazing,** see GRAZING MANAGEMENT.

**rotational period,** see CROP ROTATION.

**r-species,** *n.* any species of plant or animal characterized by a small body size, rapid development, a high innate capacity for increase, a reproductive phase which begins early in the life history and, once completed does not occur again, and a complete life history which lasts less than one year. Examples of r-species are ANNUAL plants or insects. Compare K-SPECIES.

**rudaceous rock,** see SEDIMENTARY ROCK.

**runoff,** *n.* water that moves across the surface of the land into streams rather than being absorbed by the soil. Runoff is typically a product of rainfall or melting snow and its volume is affected by such factors as rock and soil permeability, vegetation cover, ground temperature and groundslope. See OVERLAND FLOW, INFILTRATION.

**rural development,** *n.* any process for promoting the economic growth, modernization, increased agricultural productivity, and provision of basic needs and services such as education, health care and water supply in rural areas. The achievement of these aims generally depends upon the nature of the administrative systems through which a variety of programmes are implemented, and on national policy towards such issues as LAND TENURE, AGRARIAN REFORM, the disbursement of AID, and FOOD POLICY. See also INTEGRATED RURAL DEVELOPMENT PLANNING.

**rural-urban continuum,** *n.* a concept associated with the sociological typology of communities relating their social characteristics to the size, density and environment of a settlement. The term implies that in each kind of settlement in a continuum from the rural village to the urban city there is a characteristic way of life with different degrees of social stratification, stability, homogeneity and integration, and that the continuum indicates a process of social change.

The view that settlement patterns are associated with social characteristics and lifestyles has been attacked for failing to recognize that societies with essentially rural characteristics may be found in cities while those with urban values or attributes may occur in villages. Similarly, the process of social change may be evident within rural communities without any significant increase in settlement size. With only partial empirical support for the concept, it has taken on a wider if looser geographical meaning to represent the physical difficulty in recognizing a clear distinction between town and country. Thus, the idea of a rural-urban continuum supports the notion of a RURAL-URBAN FRINGE in which essentially urban ways of life mingle with, and merge into, rural ones.

**rural-urban fringe,** *n.* a zone of mixed rural and urban LAND USES on the edge of a city or built-up area, with distinctive social and demographic characteristics. The rural-urban fringe comprises a zone on the immediate periphery of the built-up area where rural land uses are on the point of conversion to urban functions, and an outer fringe which is predominately rural in character but contains urban elements such as airports, hospitals, INDUSTRIAL ESTATES and HYPERMARKETS.

A rural-urban fringe exists around most cities, its width varying according to the intensity of urban pressures on the rural environment. Major forces promoting urban expansion into rural areas include the increased demand for land for urban development, the loss of agricultural labour to urban employment, and the GENTRIFICATION of rural villages. Further problems for farmers in this region include uncertainty about city growth and urban policy, VANDALISM, unintended trespass, POLLUTION, waste disposal, and the worrying of livestock by domestic pets. Such pressures on rural areas are offset by government GREENBELT policies but inevitably, the rural-urban fringe is often a major zone of conflict over land use.

# S

**saddle-point,** see GAME THEORY.

**safety rods,** see CONTROL RODS.

**Sahel,** *n.* a semi-arid area of Africa lying between the Sahara desert and the moister SAVANNA and forest zones to the south. Stretching across the countries of Senegal, Mauritania, Mali, Burkina Faso, Niger and Chad, northern parts of Nigeria and Cameroon, and eastwards towards Sudan, the term is derived from the Arabic word meaning border or fringe, as between the desert and the lands of cultivation.

The Sahel is characterized by annual rainfall of less than 500 mm, and high temperatures with a high diurnal range. Much of the region experiences an interannual variability of rainfall that departs from the mean by ±40% (see RAINFALL RELIABILITY). Since the late 1960s, DROUGHT has been a recurring problem, with many meteorological stations recording rainfall totals of 80% or less of the mean annual amounts. A natural vegetation ranging from thorn-wooded grassland to tussocky grasses with large patches of bare earth between has in many areas been overgrazed by cattle, and overpopulated by humans both processes contributing to the DESERTIFICATION of the area.

The area was formerly one of considerable wealth, with a succession of powerful empires between the 8th and 17th centuries regulating a trans-Saharan trade in gold, ivory, salt, slaves and kola nuts. The fertile area of the Niger inland delta supported large and prosperous urban centres such as Timbuktu, for a time one of the world's largest cities. Today, the countries of the Sahel are among the poorest in the world, ravaged by drought, food shortages and periodic FAMINE, a deepening FUELWOOD CRISIS, poor communications, and, in the case of Chad, by military conflict.

**salinity,** *n.* the level of dissolved salts in rivers, lakes, seas and oceans, expressed in parts per thousand. All waters contain some dissolved salts although the OCEANS contain notably more salt than rvers. Inland lakes in arid areas show highest salinity levels due to high EVAPORATION rates.

**salinization,** *n.* the accumulation of highly soluble sodium, magnesium and potassium salts in a soil. Salinization usually occurs in arid and semi-arid areas where evaporation rates exceed those of precipitation, especially in coastal regions and areas with underlying EVAPORITE deposits. The limited effects of LEACHING in these environments causes an increase in salinity due to the retention of highly mobile sodium and potassium salts within the soil. Salinization is greatly accelerated by the upward movement of saline groundwater through CAPILLARITY. High air and ground temperatures cause evaporation and the deposition of salts from the groundwater.

Salinization can also result from the excessive and wrongly timed application of IRRIGATION water. Irrigation followed by a lengthy hot, dry phase can result in a substantial upward movement of salts which accumulate as a *salt pan*. In the worst examples (as occurred soon after the expansion of irrigation in Egypt following the construction of the Aswan High Dam) the salt pan forms a toxic layer which kills agricultural crops. The process of salinization causes the formation of SOLONCHAK soils. See also ALKALIZATION, CALCIFICATION.

**SALR,** see SATURATED ADIABATIC LAPSE RATE.

**saltation,** see FLUVIAL TRANSPORTATION.

**salt marsh,** *n.* **1.** a coastal marsh found in a backwater of an estuary or in the lee of a SPIT. As marine sediments become trapped by vegetation the marsh increases in height and inundation occurs only on the occasions of highest tides. Eventually, salt marsh can be reclaimed for pasture, as occurred with Dungeness Marsh in Kent, England.

**2.** an inland marsh of arid areas where the water has a high saline content brought about by EVAPORATION.

**salt pan,** see SALINIZATION.

**salt weathering,** see PHYSICAL WEATHERING.

**sampling,** *n.* the process of selecting an unbiased representative sample for examination from within a total population. In environmental studies it is not possible to study total populations and instead, a small proportion (often as small as 5-10%) must be selected to represent the whole. It is essential that the sample is drawn at random from the total population. See PLOTLESS SAMPLE, QUADRAT, TRANSECT.

**San Andreas Fault,** *n.* a series of FAULTS along the coast of California where the American and the Pacific LITHOSPHERIC PLATES move past each other. It is analogous to other faults in the

Pacific basin. The San Andreas line leads from the Gulf of California, where the East Pacific Rise dips under North America, northwest past San Fransisco. See EARTHQUAKE.

**sand,** *n.* a coarse granular mineral comprised mainly of quartz grains. Sand is derived from the chemical and physical weathering of rocks rich in quartz, notably sandstone and granite. The diameter of sand grains vary according to the soil texture classification being used. See SOIL TEXTURE and Fig. 57. Soils with a high sand component are well aerated, warm quickly in spring time and are easily cultivated; however, their high POROSITY often results in drought-prone soils with an impoverished nutrient content due to the rapid LEACHING process which operates within the SOIL PROFILE.

Sand is an important economic resource. It is a component of concrete, is used to filter contaminated water and forms an important ingredient in the manufacture of glass, abrasives and electrical equipment.

**sandstone,** *n.* a SEDIMENTARY ROCK formed by the COMPACTION and CEMENTATION of sand grains. Sandstone can be classified according to the mineral composition of the sand and cement. Sandstone is often used as a building material, in glass-making and in the manufacture of cement.

**sandstorm,** *n.* a windblown cloud of sand rising to only a few metres off the ground, reducing visibility, scouring paintwork, dessicating vegetation and causing severe irritation of respiratory surfaces in animals. Sandstorms are usually confined to arid areas. See also DUSTSTORM.

**sandur,** see OUTWASH PLAIN.

**sanitary landfill,** see CONTROLLED TIPPING.

**saprophage,** *n.* any animal species which obtains its food from dead or decaying plant or animal material. See SAPROPHYTE.

**saprophyte,** *n.* any plant or microorganism which obtains its nutrition in solution from dead or decaying organic matter. FUNGI and BACTERIA are saprophytes and as such carry out the essential process of releasing chemical substances which would otherwise remain locked up in the cells of plants and animals. Once released, the chemical nutrients are available for re-use by other organisms, chiefly green plants. Saprophytes are responsible for mobilizing up to 90% of the nutrients in some ecosystems. See SAPROPHAGE. See also CARBON CYCLE, NITROGEN CYCLE, DETRITUS FOOD CHAIN.

**saturated adiabatic lapse rate (SALR),** *n.* the rate of heat loss which occurs from a rising, saturated air mass. The SALR can vary depending upon the temperature of the air mass and the associated water vapour content. A minimum value of 0.4°C per 100 m to a maximum of 0.9°C per 100 m can be recorded. Compare DRY ADIABATIC LAPSE RATE, ENVIRONMENTAL LAPSE RATE. See SATURATION.

**saturation,** *n.* the condition of a body of air when it can hold no more WATER VAPOUR at its particular temperature and pressure, the RELATIVE HUMIDITY then being 100%. The amount of water vapour required to saturate a parcel of air rises as the air temperature increases. Unless particulate matter is present in the body of air for water droplets to form around then SUPERSATURATION may result at temperatures below the usual DEWPOINT.

**savanna,** *n.* any of a number of tropical vegetation COMMUNITIES in which grasses (Graminea) and sedges (Cyperaceae) predominate. In practice, savannas exhibit a great range of species composition, ranging from completely treeless grasslands to woodlands in which trees and shrubs form an almost continuous cover beneath which there is a grass-dominated undergrowth. Areas covered by savanna usually exhibit a marked seasonal climate with distinct wet and dry seasons. Annual rainfall varies between 500 and 2000 mm while the daily minimum temperature rarely falls below 20°C.

Savanna can be classified in a number of ways. One of the most convenient comprises three simple categories:

(a) a luxuriant high grass/low tree savanna found only in Africa and in which elephant grasses (Pennisetum species) grow up to 3 m during the wet season. Deciduous trees are interspersed in the grassland. This category of savanna grows in close proximity to TROPICAL RAINFOREST;

(b) acacia tall-grass savanna in which tussock grasses form a continuous ground cover of between 0.75 and 1.75 m in height. In Africa the most commonly found tree is of the Acacia species; in Australia the eucalypts are commonplace while in South America a wide range of tall, evergreen trees occur;

(c) a discontinuous cover of highly xerophyllous grasses with scattered thorn bushes. Large patches of bare ground exist. The term *desert grass savanna* has been applied to these areas. They occur most frequently along the fringes of true hot deserts, for example, along the southern fringes of the Sahara, around the edges of the Great Australian Desert and in India, around the western and southeastern edge of the Deccan plateau.

The origins of savanna are complex and are due to specific combinations of climate, soils, geomorphology, burning and biotic interaction (including ANTHROPOGENIC FACTORS). FIRE undoubtedly has played a major role in the formation of savanna. Repeated burning can totally eradicate woody species and also alters the grass species composition towards fire-tolerant varieties. At sites which have not been burnt for more than 30 years then tree regeneration can become common. Once formed, savannas have been maintained by the pressures of grazing imposed by the vast herds of wild animals (deer, buck, zebra, giraffe, kangaroos, wallabies, guanaco); and more recently by domesticated species (cattle, goats).

**scarp,** see ESCARPMENT.

**schist,** *n.* a medium-grained METAMORPHIC ROCK formed by the regional METAMORPHISM of slate. Schists have well developed foliation.

**science park,** *n.* an area developed for the use of firms engaged in research and development in advanced technology. Commercial development is closely linked to fundamental research taking place in universities and the provision of science parks in university towns has become common since the late 1970s.

The provision of these parks has been an attempt to facilitate a trend observed in the USA since the early 1970s (especially in 'Silicon Valley' in California) for high-tech industries to cluster near advanced research establishments. In the UK, science parks developed spontaneously in areas of high growth such as Cambridge well before financial incentives in the form of government grants were made available to encourage the setting up of science parks in less favoured locations. See also INDUSTRIAL ESTATE, ADVANCE FACTORY.

**sclerophyllous,** *adj.* (of a plant) having small, grey green leaves often with a down-like covering of hairs; such adaptations allow the plant to survive the hot, dry summers of MEDITERRANEAN-type climates in which this type of vegetation flourishes. Most sclerophyllous species are evergreen with a thick cuticle covering the leaves. In extreme examples the leaves become reduced to small blades or spines. Stems are often corky or calloused; root systems comprise both a single deep tap root and a dense mat of surface roots. Many sclerophyllous plants show further adaptation by remaining dormant throughout the long, hot and dry summers and grow in the cooler, moister winter and spring. See FYNBOS, GARRIGUE, MAQUIS.

**scree** or **talus,** *n.* an accumulation of angular rock debris of variable size at the base of a cliff or steep slope. Scree deposits may be tens of metres thick and form when the PHYSICAL WEATHERING of rock faces causes rock debris to disintegrate and fall. Scree may form steep-sided accumulations known as *talus cone* and if these cones coalesce then a *scree slope* or *talus slope* forms.

The angle of repose for scree deposits is usually between 25° and 35°. Accumulations of scree are subject to a gradual downhill movement known as *talus creep* which results from various processes including the movement of rock debris through daily temperature changes, freeze-thaw action, and the impact of falling rock fragments.

Many fresh scree slopes are highly unstable and even a minor rockfall can cause scree to slide and destroy roads, railways or settlements within the vicinity. To prevent such loss of life and damage to property, scree slopes may be fixed through the planting of vegetation.

**scree slope,** see SCREE.

**scrub,** *n.* any low-growing woody, or thorny vegetation cover, that results from the clearance of vegetation for commercial use, (such as agriculture or forestry) that is followed by a period of regeneration.

**scrubber,** *n.* an apparatus used to remove pollutants from exhaust gases produced by industrial processes. Most scrubbers use a fine liquid spray to bombard effluent gases and so physically remove solid particles. Another commonly used technique passes the gas through a wetted packing, the liquid in which strips out solid or liquid pollutants. See CYCLONE DUST SCRUBBER, ELECTROSTATIC PRECIPITATOR.

**sea breeze,** *n.* any light wind which occurs during the day in coastal areas and flows onshore due to the slight decrease in air pressure caused by the more rapid heating of the land surface.

**sea-floor spreading** or **ocean-floor spreading,** *n.* the lateral expansion of the ocean floor from the point of the MID-OCEANIC RIDGES. These ridges indicate constructive LITHOSPHERIC PLATE boundaries where upward convection currents in the MANTLE cause the extrusion of lava and the formation of new oceanic crust. The theory underlying sea-floor spreading is supported by PLATE TECTONICS.

**sea lane,** *n.* a recognized pathway in busy areas of sea that enables dense concentrations of shipping to navigate safely.

**sea level,** *n.* **1.** the average level of the surface of the sea uninfluenced by tidal movement or by waves. An average sea level can only be calculated after many years of measurement at a point and must take account of the rise in the level of the land relative to the sea (see ISOSTASY). In the UK, the mean sea level is calculated from measurements made at Newlyn in Cornwall and this level forms the datum point for the construction of all contour lines and spot heights used by the Ordnance Survey in its production of charts and maps.

Sea level is undergoing constant change depending upon the the moon's gravitational influence, on the amount of water held within GLACIERS, and in the speed with which water moves through the HYDROLOGICAL CYCLE. Some scientists believe that the world sea level will rise during the next century due to the gradual warming of the atmosphere and the consequent melting of the ice caps (see GREENHOUSE EFFECT).

**2.** the mean elevation of the sea between the levels of high and low tides, used as a standard base for measuring altitudes and depths.

**sea level change,** *n.* the rise and fall of SEA LEVEL due to EUSTASY or tectonic movement (see PLATE TECTONICS). Sea level changes have played an important role in the shaping of many coastal areas as well as influencing river drainage systems through REJUVENATION.

**sea loch,** *n.* the Scottish term for FIORD.

**seam,** *n.* a STRATUM of ore or mineral material such as coal.

**seamount,** *n.* a steep-sided underwater mountain with a conical peak of volcanic origin. Sea mounts may rise to 3700 m above the DEEP-SEA PLAIN. Several ranges of seamounts and GUYOTS are located in the Pacific Ocean, including the Mid-Pacific Range and the Emperor Range.

**sea wall,** *n.* an artificial wall-like structure usually made of large boulders or concrete and designed to prevent the damage and destruction of harbours and settlements by MARINE EROSION or the flooding of low-lying coastal areas.

**secondary consumer,** see FOOD CHAIN.

**Secondary era,** see MESOZOIC ERA.

**secondary forest,** *n.* the vegetation type which usually replaces PRIMARY FOREST following its removal. When forest disturbance has been small the secondary forest may bear a strong visual resemblence to the original forest although species diversity, layering, energy flow patterns and faunal diversity will inevitably have been simplified. In other instances when vegetation is allowed to recolonize an area of former agriculture (as in tropical SLASH-AND-BURN clearings) the resulting vegetation reveals considerable lack of diversity and far fewer animal species (particularly bird and arboreal species of monkey). At best, those areas of forest now called 'natural' are merely very old secondary woodlands.

**secondary oil recovery,** see OIL.

**secondary pollution,** *n.* atmospheric pollution resulting from a chemical reaction between two or more sources of PRIMARY POLLUTION or between primary pollutants and naturally occurring atmospheric elements. The most notorious secondary pollutant is PHOTOCHEMICAL SMOG. See ACID DEPOSITION, SYNERGISM.

**secondary succession,** *n.* any plant COMMUNITY which develops on sites from which a previous community has been removed. ANTHROPOGENIC FACTORS are the most common agents of this change, in particular DEFORESTATION, ploughing and FIRE, although it can also result from natural causes such as land slips, flooding, storm damage, and disease. The speed with which a secondary succession can progress depends upon the severity of DISTURBANCE, the presence of suitable seed sources from which succession can re-commence, and the lack of further site disturbance. Compare PRIMARY SUCCESSION.

**secondary tillage,** see TILLAGE.

**secured landfill,** see CONTROLLED TIPPING.

**sediment,** *n.* the soil particles and rock fragments transported and deposited by the action of rivers, glaciers, sea and wind. See SEDIMENTATION, SEDIMENTARY ROCK. See also ALLUVIUM, ALLUVIAL FAN.

**sedimentary rock,** *n.* a rock formed by the LITHIFICATION of sediment. Three main types of sedimentary rock are recognized:

(a) *clastic* or *mechanically formed sedimentary rocks*, derived from fragments of existing igneous, metamorphic and sedimentary rocks that have been eroded, transported and deposited by wind, water or ice. Clastic rocks can be classified according to sediment size:

  (i) *rudaceous rocks* are composed of coarse sediments, such as breccias and conglomerates;

 (ii) *arenaceous rocks* consist of cemented medium-sized sediments, such as sandstones;

(iii) *argillaceous rocks* result from fine sediments, such as shales;

(b) *organic sedimentary rocks* are formed from the remains of plant and animal life. Carbonaceous rocks include the major economic resources of coal and the associated by-products of oil and natural gas. Many calcareous and siliceous rocks have organic origins,

(c) *chemical sedimentary rocks* are derived from the precipitation of minerals, usually by evaporation, from both fresh water and sea water. Evaporates as well as some calcareous and siliceous rocks are formed in this way.

Most sedimentary rocks were formed in a marine environment although some are continental or transitional in origin. STRATIFICA- TION occurs in sedimentary rocks due to variations in the nature of the deposition environment and in the material being deposited. Each rock stratum is separated along a BEDDING PLANE. Many important economic resources are associated with sedimentary rocks including building stone, gold, tin, chromite, platinum and diamonds, coal, oil, and natural gas.

**sedimentation,,** *n.* the deposition of SEDIMENT. See GLACIAL DEPOSITION, MARINE DEPOSITION, WIND DEPOSITION.

**seine netting,** *n.* a method of FISHING in which a net is suspended vertically between two points in the water by means of floats at its top and weights at the bottom; the net is gradually drawn together thus entrapping fish. The seine net is used for catching PELAGIC fish. The use of seine nets has often been declared illegal due to its effectiveness in catching all the fish which attempt to pass through it, including immature and breeding female fish thus leading to the over-exploitation and exhaustion of the fish stock. This is particularly so when net with a small mesh size is used. Compare TRAWLING, LINE FISHING.

**Seismic Sea Wave Warning System,** see TSUNAMI.

**seismic wave** or **shock wave,** *n.* the tremor that results from the dissipation of energy at the FOCUS of an EARTHQUAKE.

**seismograph,** *n.* a sensitive instrument that measures and records the earth tremors caused by SEISMIC WAVES.

**selective breeding,** *n.* the process of developing new genetic lines of plants and animals with desirable heritable characteristics by means of cross-breeding and fertilization. Selective breeding involves the application of genetic principles to produce new and improved farm animals and crops that better serve the needs of farmers, the suppliers of agricultural inputs, food processors, and consumer demand. See HYBRID SEED, GENETIC ENGINEERING.

**selective herbicide,** *n.* a chemical substance which, when applied to soils, vegetation or crops, is effective in killing specific plant types, notably WEED species. See HERBICIDE.

**self-sufficiency,** *n.* a way of life in which all the basic needs – food, clothing and shelter – are met and sustained by the efforts of the immediate family group or local community without recourse to goods and services produced by others.

Traditional agricultural systems based on SUBSISTENCE FARMING and pastoral NOMADISM, and supplemented by barter exchange, might reasonably qualify as self-sufficient in that the immediate environment, solar radiation, precipitation and domestic labour provide the necessary inputs to sustain life, and the productive output is consumed by the particular unit of social organization. In the modern world, outside the realm of HUNTER-GATHERERS and remote tribal groups practicing SHIFTING CULTIVATION there are few communities that are fully self-sufficient.

In developed temperate zones until recent times a goal of MIXED FARMING was a diversified output which might provide the farmer with a self-sufficient food base for his family, if not a fully self-sufficient way of life. Modern farming practices have promoted greater farm specialization and a growing dependence on external inputs to maintain the productive processes, thus losing any claim to self-efficiency.

**selva,** see TROPICAL RAINFOREST.

**semi-arid climate,** *n.* a climate which forms a zone of transition between true DESERT climate, and SAVANNA and MEDITERRANEAN climates. Examples of semi-arid climates are found in southwestern USA and on the margins of the Australian deserts. Semi-arid areas usually have sparse vegetation due to the lack of moisture. The potential for agriculture is severly limited unless IRRIGATION schemes can be developed. See ARID CLIMATE.

**semi-arid region,** see DESERT.

**semi-desert scrub,** *n.* a type of vegetation characterized by low-growing (2 m), widely spaced shrubs, bushes and succulents. Unlike plants of the MICROPHYLLOUS FOREST, the roots form a dense sub-surface mat, an adaptation probably made to exploit EPHEMERAL rains. Grasses are uncommon. Semi-desert scrub occurs in Mexico, on the margins of the Atacama desert, around the Sahara and Arabian deserts, the Namib desert and around the Great Australian desert. Productivity of this vegetation type is very low and varies between average net PRIMARY PRODUCTIVITY values of 50-200 $g/m^2/year$.

**senile soil,** *n.* any SOIL which has remained undisturbed by natural or human activities for extremely long periods of time, in some cases for millions of years. Such soils are thought to exist only on the ancient PENEPLAINS of southern Africa and South America. Extensive, deep WEATHERING has resulted in a thick layer of debris but of which only the top metre or so is involved with the cycling of nutrients between plants and the soil. As a result, the reservoir of available nutrients is often limited and can become reduced due to erosion, removal of vegetation and loss of nutrients via LEACHING and FIRE. Senile soils are usually incapable of supporting luxuriant vegetation (such as forest) nor are they suitable for agriculture.

**sequential cropping,** see MULTIPLE CROPPING.

**sere,** *n.* any stage in the development of a vegetation type between denudation and the stabilization of a habitat. The complete succession sequence fron bare ground to the CLIMAX COMMUNITY involving many seral stages is called a PRISERE.

**service area** or **rest area,** *n.* a group of buildings and car parks adjacent to a MOTORWAY or major highway which provide a variety of services required by road travellers. Facilities may range from a simple cafe and petrol (gas) station to highly sophisticated restaurant, shopping, car repair and fuelling facilities for long-distance travellers, as well as overnight accommodation.

**set aside policy,** see FALLOW.

**sewage** or **sludge,** *n.* any viscous, semi-solid mixture of bacteria- and virus-laden organic matter, toxic metals, synthetic organic chemicals and settled solids removed from domestic and industrial waste water at sewage treatment plants. The tanks containing the sludge must be periodically emptied and the accepted practice is to dump it in coastal waters. However, as a result of more stringent legislation relating to the release of raw sewage into rivers and oceans more sewage treatment plants have been built, which in turn has led to a massive increase in the amount of sludge requiring disposal. Between 1975 and 1985 the quantity of sludge dumped in coastal waters of the USA increased by 60% and after 60 years of continuous sludge dumping, 105 km² of the sea bed in the New York Bight is covered with black sludge. During storms this sludge is washed back onto beaches; it has also contaminated shell-fish beds which has led to outbreaks of food poisoning.

The safe disposal of sludge presents a major problem for public health engineers. It is accepted that transferring the dumping grounds beyond the CONTINENTAL SHELF is not a long-term solution. Modern sewage contains many TOXIC WASTES that contain slowly degrading or non-BIODEGRADABLE substances such as PCBS, pesticides, radioactive ISOTOPES and toxic mercury compounds that can accumulate in ocean FOOD CHAINS.

**shale,** *n.* a dark fine-grained SEDIMENTARY ROCK formed by the COMPACTION of mud, clay or silt. Carbonaceous shales may contain substantial deposits of coal and oil whilst calcareous shales may be used in the manufacture of cement

**sharecropping,** *n.* a form of LAND TENURE in which the tenant pays the landlord a share of the crop produced rather than a monetary payment for the right to use the land. The landlord generally provides much of the movable capital such as farm equipment, seed and stock, as well as the fixed capital of the land and buildings, while the tenant farmer provides labour.

Efforts to reform or regulate sharecropping have focused on the issues of security of tenure and limiting the share paid to the landowner. The system is characterized by short-term leases, or as in many parts of the THIRD WORLD, by no more than oral agreements; tenants thus have little incentive to invest in the land or engage in long-term LAND USE PLANNING. Maximum shares have become fixed in law in most developed countries and in areas where effective LAND REFORM has been implemented. In Taiwan for example, legislation in 1951 reduced shares from 50-70% of all crops, payable in rice, to 37.5% of the standard annual yield of the main crop. See also FARM RENT.

**shear boundary,** see PLATE TECTONICS.

**shear strength,** *n.* the degree to which soil and rock can resist MASS MOVEMENTS. See SHEAR STRESS.

**shear stress,** *n.* any disturbing force that may result in the MASS MOVEMENT of rock and soil, such as undercutting by a river, accumulation of water after a thaw or heavy rainfall, EARTH-QUAKES, and vibrations caused by earth tremors, blasting, or the passage of heavy rail traffic. See SHEAR STRENGTH.

**sheet erosion, sheetwash** or **rainwash,** *n.* a type of SOIL EROSION characterized by the downslope removal of soil particles within a thin sheet of water. Sheet erosion usually occurs near the top of a slope and occurs during heavy rainstorms when the INFILTRATION capacity of the soil is exceeded and soil particles displaced by raindrop impact are borne away by sheets of surface water. Sheet erosion may occur in most parts of the world but is most frequently found in arid and semi-arid regions. In such areas especially, cultivation and the removal of vegetation through overgrazing and FIRE may leave an unprotected soil surface which is susceptible to this type of soil erosion following high intensity rainstorms. See also RAINDROP EROSION, RUNOFF.

**sheetwash,** see SHEET EROSION.

**Shelford's law of tolerance,** *n.* a concept which states that an organism will respond to every individual ecological factor on a scale which varies from a minimum to a maximum intensity and the distance between these minimum and maximum values is called the zone of TOLERANCE. See LIEBIG'S LAW OF THE MINIMUM.

**shelter belt,** *n.* a STAND of trees deliberately planted to modify local climatic conditions in windswept areas. Acting as a windbreak, a shelter belt is designed to reduce SOIL EROSION and to protect crops from the direct effects of high-velocity winds. The effectiveness of protection depends on the height, density and shape of the plantation. Other microclimatic effects include the reduction in EVAPORATION rates, the control of snow-drifting, small increases in daytime temperature, and decreases in night temperatures with a commensurate risk of frost. Shelter belts may be no longer than the field to be protected, or extend, as in parts of the USSR, over several kilometres or in staggered rows over considerably greater distances.

**sheltered housing,** *n.* accommodation designed especially for elderly or handicapped occupants. Individual houses are normally designed on a single level and are organized in groups, in order that a number of facilities may be shared. Sheltered houses are usually supervised through a system of electronic devices which enable the residents to call for the help of caretaking staff in times of emergency. In planning such houses, care is taken to ensure that they are located close to local services yet protected from main traffic flows.

**shield volcano,** see VOLCANO.

**shifting cultivation,** *n.* a type of tropical farming system in which short periods of cropping are alternated with longer periods of fallowing. There are, however, great variations in the type of crops grown, in the methods of cultivation, and most critically, in the lengths of the cropping and FALLOW periods.

True shifting cultivation, otherwise known as the classical long-fallow or forest FALLOW SYSTEM, is ecologically well adapted to forested areas of low population density such as the TROPICAL RAINFOREST of the Amazon Basin, west-central Africa, and parts of South-East Asia. A site or *swidden* is cleared of natural vegetation by felling and the use of FIRE (hence the reference to shifting cultivation as SLASH-AND-BURN agriculture). Soil disturbance is normally slight, with crops sown or planted in the warm ash. After a season or two of cropping, the site is then abandoned to regain its fertility under SECONDARY FOREST regrowth over a period up to 30 years. Cultivation and village settlement meanwhile shift to other areas. Cultivation may be said to shift within an area that is otherwise covered by natural vegetation. Less extensive systems, or those where farms move in a regular sequence about a permanent settlement, are better defined as ROTATIONAL BUSH FALLOW.

Shifting cultivation is further characterized by the general absence of draught animals; virtually no application of MANURE; the use of simple hand tools such as the hoe and digging- or dibble-stick; the practice of MULTIPLE CROPPING; the low expenditure of labour per unit of production; low output per hectare; and production oriented largely towards SUBSISTENCE FARMING goals. In many SAVANNA or grassland areas, shifting cultivation may be combined with other field rotational systems, and even PERMANENT CROPPING. For example, permanently and intensively cropped infields may be contrasted with outfields under shifting cultivation (see INFIELD-OUTFIELD SYSTEM). Elsewhere, uplands may be farmed on a long-fallow regime while valley bottoms are kept in near permanent cultivation.

**shock wave,** see SEISMIC WAVE.

**shopping mall,** *n.* a covered shopping centre consisting of indoor 'streets' or arcades often on more than one level with shops opening off them. Shopping malls provide comfortable shopping environments secure from unfavourable weather and usually contain the full range of shops associated with the traditional city centre shopping area; however, the shopping mall can claim greater convenience than the traditional shopping area due to its 'one-stop-shopping' facility and ease of parking close to the shops.

They are usually built at the edges of towns or near MOTORWAY or highway intersections as they are designed for use by car-borne shoppers. The retail area is often dwarfed by the area devoted to car parking. The car parks may be so large that a free bus service is provided to drop off points at the entrance to the shopping mall.

In the UK, the development of large OUT-OF-TOWN SHOPPING CENTRES is not as great as in North America due to the existence of better public transport and to planning styles which have favoured the development of local district centres and city-centre redevelopment.

**shore,** *n.* the area of land between the low tide watermark and the highest point reached by high tides and storm waves or, on a steep coast, the base of the cliffs. In layman's terms, it is also taken to mean the edge of the land surface visible from the sea or a lake. See also SHORELINE.

**shoreline,** *n.* the area marking the points of contact between land and water bodies, such as the sea or a lake. The term *coastline* is generally taken to refer only to the boundary between the land and seas and oceans.

**shrub crop,** see PERMANENT CROP.

**sial,** *n.* the upper layer of the EARTH'S CRUST found under the continental masses (see Fig. 28). The sial is largely composed of granitic-type rocks which are less dense and lighter in colour than rocks in the SIMA. Sial rocks are rich in silica and alumina.

**siderite,** see METEORITE.

**siderolite,** see METEORITE.

**sidewalk,** see FOOTPATH.

**sidewalk farming,** *n.* a farming system found in the USA in which the farmer, although owning the holding, lives in an urban area greater than 50 km from it and only returns to the farm at critical points in the crop's AGRICULTURAL CYCLE. Sidewalk farming is generally only feasible with cereal cultivation, as GRAIN CROPS such as wheat and barley require only minimal attention between sowing and harvesting. Sidewalk farming emerged in the post-war era as a result of improved transportation facilities. See SUITCASE FARMING. See also EXTENSIVE AGRICULTURE.

**Sierra Club,** *n.* a society founded in 1892 by influential American scientists, businessmen and politicians and devoted to the furtherance of all conservation issues. See AUDUBON SOCIETY.

**silage,** *n.* any green fodder preserved by excluding air from the storage site. After cutting, plant material continues respiration and loses much of its nutritional value unless kept from the air. Under proper storage conditions in silos or bunkers, silage made from green FODDER CROPS, for example, maize, legumes, grasses, kale and rape, ferments slightly and keeps for several months. Silage is best with a moisture content of 50-65% and sufficient packing to exclude air. Some crops, notably grass, oats, peas, beans and vetch may be treated with treacle and water. In the UK, fresh-cut grasses may be stored in open silos after treatment with formic acid to prevent microbial putrefactors.

**silica,** *n.* the chemical compound silicon dioxide. Quartz is one of the most important forms of silica and is used in the manufacture of glass, electronic equipment, lenses and prisms. Coloured varieties of quartz may be valued as semi-precious gems.

**silicates,** *n.* a group of chemical compounds consisting of silicon, oxygen and at least one metal. Silicates are the most important group of rock-forming compounds in the earth's crust, accounting for around 95% of the crust by weight. There are a variety of silicate minerals including the FELDSPARS, the MICAS and SILICA.

SILICEOUS

Certain types of feldspar are used in the manufacture of ceramics, glass, abrasives and paint whilst some micas are used to manufacture lubricants, paint and electrical equipment.

**siliceous,** *adj.* composed of or containing silica or a silicate compound.

**sill,** *n.* a concordant sheet-like mass of INTRUSIVE ROCK formed by the injection and solidification of MAGMA along the BEDDING PLANES of country rocks. The thickness of sills may range from a few centimetres to hundreds of metres but the horizontal extent is always much greater than the thickness. The rock that forms sills is often harder than the surrounding rock and differential erosion may result in outcropping sills forming plateaus. Rock is often quarried from sills for use as road metal.

**silt,** *n.* fine-grained mineral material formed from the erosion of rock fragments and deposited by rivers and lakes. The accumulation of silt in dams and reservoirs can effectively reduce their storage capacity and in the worst instances render them useless for the storage of water. The defined diameter of silt particles varies according to the soil texture classification used (see SOIL TEXTURE and Fig. 57).

**Silurian period,** *n.* the geological PERIOD that followed the ORDOVICIAN PERIOD about 440 million years ago and that ended 395 million years ago when the DEVONIAN PERIOD began. There was a major phase of mountain building and volcanic activity in North America while land plants and animals appeared for the first time in the Silurian along with vertebrate fish in Europe. A variety of invertebrate marine fossils are used to date and correlate Silurian rocks. See EARTH HISTORY.

**silviculture,** *n.* the cultivation of a tree crop primarily for economic profit. In silviculture, the complete sequence of forest development will be subject to a management plan which will determine pre-planting drainage and fencing, the choice of species to be planted based upon the maximum YIELD CLASS attainable at the site, fertilizer applications, the dates of thinning and final felling, and the marketing of the crop.

Silvicultural practices are typically found in coniferous plantations in the developed countries of the northern hemisphere, notably those of Western Europe. Except in West Germany, silvicultural land use is only occasionally combined with agriculture. Such an approach to tree cultivation has been much criticized by ecologists and conservationists due to the lack of concern shown to traditional forestry values such as ecological diversity and the

AMENITY value of woodland. However, the rate of deforestation is such that without the intensive management practices associated with silivicluture, it is estimated that timber supplies will become extremely scarce by 2020 AD. See FOREST MANAGEMENT. See also AFFORESTATION, DEFORESTATION, FORESTRY.

**sima,** *n.* the lower layer of the EARTH'S CRUST found under both continental masses and the oceans (see Fig. 28). The sima is largely composed of basaltic rocks which are darker and denser than the granitic-type rocks of the SIAL. Sima rocks are rich in silica, magnesium and iron.

**simulation study,** *n.* an analytical technique involving the construction of simplified, mathematical MODELS of complex, real world systems. Simulation studies are of particular use whenever the behaviour of a system is the result of *stochastic processes* (that is controlled by the rules of chance), or where due to conditions of size, rarity or slowness of operation of the natural system, it is impractical to work on real-world examples. Thus, simulation studies are of value in the study of forest ECOSYSTEMS, in particular to examine the impact of DEFORESTATION on soils and forest animals.

The advent of the computer has allowed considerable advances to be made in the realism of simulation studies. Their use allows the storage and use of large data sets relevant to the system under study while their speed of computation allows complex ALGORITHMS to be used as controlling mechanisms for the study.

**sink area,** see BENTHIC ZONE.

**sink hole,** see SWALLOW HOLE.

**Site of Special Scientific Interest (SSSI),** *n.* any area in the UK which has been identified by the NATURE CONSERVANCY COUNCIL (NCC) as containing flora, fauna, geology and/or physiographic features worthy of conservation. Approximately 3500 SSSIs have been established in the UK since 1949 and are graded according to a four-point scale based on the scientific value of the site, Grade I sites being equivalent to NATIONAL NATURE RESERVES.

Any proposed change in land use which may affect the long-term stability of the SSSI must be identified by the local planning authority and notification of the proposed changes given to the NCC. Objections to land use change can be lodged by the NCC and unless a compromise can be reached between the land developer, the planning authority and the NCC then a public enquiry must be held to determine the future of the site.

**skeletal soil,** *n.* a newly formed SOIL in which only limited amounts of HUMUS material is found and in which SOIL HORIZONS have had insufficient time to form. The particle size of the minerals is predominantly that of coarse sand and clay-sized materials are noticeably absent. Plant nutrient levels are usually abundant due to the freshly weathered nature of the PARENT MATERIAL. See also IMMATURE SOIL, SOIL TEXTURE.

**slash-and-burn,** *n.* a method of land clearance in which vegetation is cut, allowed to dry, and then fired prior to cultivation of the soil and the planting of crops. It is common in tropical regions where SHIFTING CULTIVATION and ROTATIONAL BUSH FALLOW are practised. See also FIRE.

**sleet,** *n.* **1.** (UK), a form of PRECIPITATION consisting of partly melted snow or a mixture of rain and snow.

**2.** (USA) small ice particles produced by freezing of raindrops formed in an upper, warmer layer, but falling through an underlying cold air layer, causing them to partially melt.

**slope retreat,** *n.* the retreat of a slope due to the processes of WEATHERING, MASS MOVEMENT and EROSION.

**sludge,** see SEWAGE.

**slump,** *n.* the downward sliding of soil and rock debris under the influence of gravity. Slumps are usually considered a form of LANDSLIDE and are characterized by their rotational movement on a curved slip plane.

**smallholding,** *n.* a small farm. In DEVELOPED COUNTRIES the term often carries the implication that the holding is capable of supporting only PART-TIME FARMING, but in most DEVELOPING COUNTRIES where a majority of farms may be less than 5 ha, smallholdings have to support the SUBSISTENCE FARMING needs of the bulk of the rural population. The term has further distinction in countries where gross inequalities in land distribution prevail, the contrast being drawn between smallholdings on the one hand and PLANTATIONS and LATIFUNDIA on the other in respect of size, operating conditions, etc.

In the UK a smallholding is officially defined as a holding of less than 20.25 ha and with a rental value below a certain level. In England and Wales, local authorities are empowered to establish statutory smallholdings for rent to individuals with some degree of farming experience. Originally established to provide employment for landless labourers and ex-servicemen after the First World War these holdings must be in excess of 0.4 ha but not so large as to require more than 900 standard MAN-DAYS of labour per annum.

**smog,** *n.* a dense, discoloured RADIATION FOG containing large quantities of soot, ash and GASEOUS POLLUTANTS such as sulphur dioxide and carbon dioxide. Until recently, smog was a serious environmental problem in many industrialized urban areas in the developed world. For example, in London, during December 1952, over 4000 people died from respiratory ailments attributed to an infamous 'pea soup' smog.

The clear links between many health problems and smog has led to many governments in developed countries introducing legislation to promote the use of smokeless fuel and reduce the emission of noxious and toxic gases into the atmosphere, such as the British Clean Air Act of 1956 and the American Federal Clean Air Act of 1970. The rapid industrialization of the developing world, with minimum environmental protection legislation, has led to serious smog problems in many major cities such as Bangkok, Calcutta, Cairo, Lagos and Sao Paulo. See also AUTOMOBILE EMMISSIONS, SMOKE CONTROL ZONE, PHOTOCHEMICAL SMOG, AIR POLLUTION.

**smoke control zone** or **smokeless zone,** *n.* any geographically defined region, usually an urban area, within which it is illegal to produce smoke from chimneys. The UK was the first country to introduce a CLEAN AIR ACT in 1956 specifically to reduce the quantity of PARTICULATE MATTER in the atmosphere and thus reduce the frequency of SMOGS which were known to be a major health hazard. The creation of smoke control zones within which smoke control orders apply have done much to improve the atmosphere in urban areas. See also AIR POLLUTION.

**smokeless zone,** see SMOKE CONTROL ZONE.

**snood,** see LINE FISHING.

**snow,** *n.* a form of PRECIPITATION from clouds in the form of ice crystals formed in the upper atmosphere following the condensation of atmospheric water vapour whenever the air temperature is below freezing. Snowfalls are common in high latitudes and some mid-latitude areas, with the heaviest snowfalls usually being recorded in mountainous and high latitude areas. The Rocky Mountains in the USA have an average annual snowfall depth of 10 m.

**snowfield,** *n.* an area of permanent snow in high latitudes or mountainous regions from which GLACIERS originate. As snow accumulates, it freezes and compacts into ice known as *firn* or *neve*. Subsequent snowfalls compresses the firn further until glacier ice forms. When deposits of glacier ice are large enough, they begin to move under the influence of gravity. The lower edge of a snowfield is known as the SNOWLINE.

**snowline,** *n.* the lower edge of a SNOWFIELD below which the winter accumulations of snow melt during the summer. The level of the snowline is dictated by latitude, altitude and aspect and generally drops from over 5000 m above sea level at the equator to sea level at the Poles.

**social forestry,** see AGROFORESTRY.

**social infrastructure,** see INFRASTRUCTURE.

**soft chain,** see DETERGENT.

**softwood,** *n.* the wood of coniferous trees such as pine or cedar. Softwood timber is particularly suitable for conversion to cellulose and paper pulp; otherwise its use is usually confined to rough joinery work, although the timber from Scots Pine can be used for furniture making. The North American Western Red Cedar is particularly valued as external cladding on houses. Compare HARDWOOD.

**soil,** *n.* the loose weathered rock debris and decayed organic material which covers large areas of the continental landmasses and is able to support plant growth. A soil differs from a REGOLITH in that the latter is devoid of HUMUS and cannot support plant and animal life. The nature and characteristics of different soils are largely determined by the climatic regime, the underlying PARENT MATERIAL, local ecology, topography and the age of the soil. The organic and mineral material may account for only half the volume of some soils; the remaining space is taken up by PORE SPACES which are occupied by the SOIL ATMOSPHERE and SOIL WATER.

A soil takes many thousands of years to develop, the exact amount of time being dependent upon the speed at which WEATHERING of the parent material occurs. On a newly exposed surface PEDOGENESIS will typically pursue the following sequence:

(a) unweathered material,
(b) regolith,
(c) SKELETAL SOIL,
(d) MATURE SOIL,
(e) old, or SENILE SOIL.

At least 10 000 years without disturbance is usually necessary for the formation of mature soils and many millions of years will elapse before a soil becomes senile. By contrast, an eroded soil may form within the space of 50 years if the natural vegetation cover is removed and leaves the soil surface exposed to the erosive processes of wind, frost, rain and gravity.

Soils capable of supporting sustainable AGRICULTURE have been an essential element in the development of stable human societies. The ability to produce sufficient food to support the population remains a critical factor in the well-being of a nation.

**soil acidity,** see SOIL PH.

**soil association,** *n.* **1.** a category in soil classification systems which consists of topographically related soils with similar soils developed within a region. The soil association may be likened to a plant ASSOCIATION.

**2.** a soil mapping unit in which two or more soil categories are combined as their restricted geographical occurrence cannot be accommodated by the scale of the map.

**soil atmosphere,** *n.* the atmosphere which occupies the PORE SPACES in a soil. The soil atmosphere is a continuation of the ATMOSPHERE above the ground although the levels of water vapour and carbon dioxide are higher and oxygen levels are lower due to the activity of plants and soil organisms. The process of diffusion with the above ground atmosphere usually prevents the build up of carbon dioxide in the soil to levels that are toxic to biological activity. In poorly drained soils, anaerobic conditions result in GLEYING which inhibits plant growth and soil organisms.

**soil buffer compounds,** *n.* the clay-humus carbonates and phosphates in a soil which enable it to resist substantial changes in SOIL PH. See CLAY-HUMUS COMPLEX.

**soil colour,** *n.* the colouration of a SOIL usually determined by the type and quantity of various iron compounds and organic material it contains. The colour of a profile and colour variations between horizons are important indicators when identifying soil types.

Under aerobic conditions, increasing levels of hydration in iron-rich soils are indicated by a gradation from red and brown to yellow soils. The alteration of iron compounds in anaerobic conditions causes olive-, blue- and grey- coloured soils. The upper horizons of a soil often change in colour from black to dark brown as the organic content decreases downwards. Pale-coloured soils often originate from the deposition of salts or calcium carbonate, the removal of iron compounds to leave a predominance of light-coloured minerals, or the lack of alteration of light-coloured underlying PARENT MATERIALS. SOIL PH and the level of activity of the CLAY-HUMUS COMPLEX also influences soil colouration.

The colour of the surface SOIL HORIZON can influence the temperature of the soil, dark soils often being several degrees

Celsius higher than light coloured soils. The warmer dark soils lose more moisture than light-coloured soils through evaporation.

**soil complex,** *n.* a restricted and intimate occurrence of soil categories within an area greater than that contained by a SOIL ASSOCIATION.

**soil conservation,** *n.* the protection of soil resources against the threat of SOIL EROSION. Various soil conservation techniques are utilized to protect agricultural soils including CONTOUR PLOUGHING, the construction of TERRACES, WINDBREAKS, CROP ROTATION, STUBBLE MULCHING, STRIP FARMING and STRIP CROPPING. In non-agricultural regions, soils can be protected through AFFORESTATION and the deliberate planting of species with extensive root systems which can bind the soil, for example, vines and deep-rooting grasses. Without such measures, severe erosion, often accelerated by overgrazing and use of environmentally unsuitable crops and management techniques, can lead to the development of BADLANDS as found in the southwestern USA. The increasing use of LAND USE CLASSIFICATION systems throughout the world is another conservation measure which identifies the restrictions on land use in an area.

Soil conservation must be an integral component of intensified agriculture. The choice of measures will depend on the form and extent of the processes of degradation which may include SALINIZATION, ALKALIZATION and compaction, as well as water- and wind erosion; they will also be dependant upon the economic feasibility of the proposed measures, and the responsiveness of farmers to the proposed control techniques. Soil conservation practices will not prevent erosion if the land use is fundamentally unsound. A combination of education and financial incentives to implement soil conservation and good husbandry will only succeed when individuals and authorities can be persuaded that current expenditure on conservation measures will accrue significant future benefits.

**soil creep,** see SOLIFLUCTION.

**soil degradation,** *n.* the damage or destruction of the soil structure resulting from the use of heavy machinery during cultivation, or by the application of unsuitable agricultural and/or land management techniques. The degradation of a soil inhibits root growth in plants and may encourage SOIL EROSION.

**soil erosion,** *n.* the accelerated removal of soil through various fluvial and aeolian processes, at a rate greater than it is formed through PEDOGENESIS. Soil erosion may occur in almost any part of the world although the most vulnerable areas are arid and semi-arid regions where natural processes have been greatly accelerated by human activity. In these environments, the loss of protective vegetation through DEFORESTATION, grazing and FIRE, and the effect of over-cultivation and COMPACTION causes the soil to lose its structure and coherence. As the soil dries it becomes increasingly susceptible to specific forms of erosion such as GULLY EROSION, SHEET EROSION and RAINDROP EROSION. In developed nations, vast amounts of money have been spent on various SOIL CONSERVATION techniques; in the USA, for example, $15 billion has been spent since the mid-1930s. In DEVELOPING COUNTRIES however, the drive for agricultural productivity and utilization of modern farming technology has often received more emphasis than the requirement for good soil conservation measures. The gradual DESERTIFICATION in many SAHEL countries has been closely linked to soil erosion.

Recent examination of soil erosion and the conservation measures designed to arrest land degradation have emphasized the political and economic context in which land users find themselves. In developing countries especially, soil erosion can be seen as a result of underdevelopment (that is, of poverty, inequality and exploitation), as a symptom of that underdevelopment and as a cause of underdevelopment through the reduced ability to produce, invest and increase productivity, leading in extreme cases to DROUGHT, desertification, FLOODS and FAMINE.

The full extent of soil erosion at the national and regional scale can only be estimated, due in part to the difficulty of accurately measuring its scale. An estimated 24 billion tonnes of sediment are deposited by the world's rivers into the oceans, although as much again may be trapped by dams or deposited on the land. The world's most sediment-laden river, the Huang or Yellow River in China accounts for the removal of some 1.6 billion tonnes of soil each year. Wind erosion similarly redistributes soil on a vast scale. Some 10% of Ethiopia's cropland area of 12 million ha is affected by erosion at a rate of soil loss of *circa* 42 t/ha/year, although on the Central Highland plateau almost half the land is badly affected by erosion. In countries such as Nepal, Peru and Turkey, the entire land area has been classed as affected by erosion, in the case of Nepal involving the annual loss of 1.66 billion tonnes per year (equivalent to 9.6 t/ha/year). Estimates by the FAO put the loss of productive land through erosion at 5-7 million ha/year.

**soil fertility,** *n.* the ability of a soil to supply the required type and amount of nutrients necessary for the optimum growth of a particular crop or vegetation system, when all other growing factors are favourable. SOIL PH is an important controlling factor in soil fertility. In undisturbed soil and vegetation systems the long term maintenance of soil fertility is assured by the existence of NUTRIENT CYCLES. In agricultural systems, the use of organic and inorganic FERTILIZERS can often sustain soil fertility levels and also bring inherently infertile soils into productive use, providing other factors such as temperature and water availability are also favourable.

**soil horizon,** *n.* a layer within a SOIL made distinctive by its colour, texture and mineral or HUMUS content and which results from the operation of pedogenic processes such as ILLUVIATION and ELUVIATION. Soil horizons may have sharply defined boundaries or may merge more gradually into each other (see Fig. 55). See H-HORIZON, A-HORIZON, B-HORIZON, C-HORIZON. See also SOIL COLOUR, PARENT MATERIAL, SOLUM.

**soil information system,** *n.* a method of storing, analysing and presenting information on soils through the use of computer technology. The information utilized in soil information systems is derived from traditional SOIL SURVEY and laboratory techniques.

Many countries, such as Canada and the Netherlands, have developed soil information systems to suit their specific requirements while the FAO is establishing an international soil data system which will incorporate information on regional climates, geology, vegetation patterns, land use and management. Soil information systems are particularly useful in LAND USE CLASSIFICATION for agriculture, engineering works, urban development, etc. See also GEOGRAPHICAL INFORMATION SYSTEMS.

**soil map,** *n.* a map which details the spatial variation of different soils in an area. Soil maps are constructed at various scales; for example, exploratory maps may be constructed at a scale of 1:1 000 000 whereas highly detailed soil maps of farms may be made at a scale of 1:10 000. The latter map type can be constructed only as a result of a detailed field survey whereas smaller scale maps can be prepared from aerial photographs or increasingly, from satellite-derived imagery such as that from the LANDSAT or SPOT systems. Soil maps provide an essential source of information for LAND USE CLASSIFICATION schemes.

**soil monolith,** *n.* a vertical section of a SOIL PROFILE which has been removed from a SOIL PIT and mounted for display and/or analysis.

**soil morphology,** *n.* the physical character of a SOIL PROFILE with particular regard to the type, arrangement and thickness of the SOIL HORIZONS and the structure, texture, consistency and porosity of each horizon.

**soil nutrient,** *n.* any of the various chemical elements found in soils which are essential for plant growth. Six main elements are utilized in plant growth and collectively are known as *macronutrients*: nitrogen, phosphorus, potassium, magnesium, sulphur and calcium. Other elements are only required in trace quantities but even so, they are essential for healthy plant growth; these *micronutrients* consist of iron, manganese, zinc, copper, cobalt, molybdenum, chlorine and boron.

To allow optimum plant growth, soil nutrients must be available in forms readily available by plants and in the required concentrations. The relative proportions of nutrients in a soil also influences the PRIMARY PRODUCTIVITY of a plant. The use of MANURES and FERTILIZERS can usually overcome a deficiency or depletion of soil nutrients.

**soil pH,** *n.* the degree of soil acidity or alkalinity. Most soils usually record a pH of between 3 and 9 with the variation largely determined by the type and concentration of chemical bases and the amount of organic material present in the soil. The substances derived from decomposing organic matter (mainly HUMIC ACIDS) cause the gradual acidification of a soil unless there is a considerable concentration of base nutrients.

Regions with high rainfall often have acidic soils due to the LEACHING of base nutrients and the domination of the CLAY-HUMUS COMPLEX by hydrogen and aluminium ions. Soils in arid areas are frequently alkaline in reaction as the limited effects of leaching allows the clay-humus complex to be dominated by soluble salts such as sodium and calcium. Soil pH is an important factor in determining the fertility of a soil, and in controlling plant growth and the type and number of soil organisms. In northern hemisphere, mid-latitude agricultural soils, the trend towards increasing acidity is counteracted by the application of lime.

**soil pit,** *n.* a hole dug vertically through the A- and B-horizons of a soil to expose a SOIL PROFILE. This enables the collection and visual observation of samples from different SOIL HORIZONS.

**soil productivity,** *n.* the capability of a soil to nurture a particular crop or vegetation type under a specific management system.

**soil profile,** *n.* a vertical section through a SOIL from which the various SOIL HORIZONS can be identified. The horizons within a soil profile are conventionally labelled O/H, A, B, and C, beneath which is the underlying PARENT MATERIAL (see Fig. 55). In practice, a soil profile may show great variation due to local conditions of drainage, topography, climate, land use and the age of the soil. See H-HORIZON, A-HORIZON, B-HORIZON, C-HORIZON.

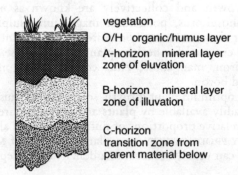

vegetation
O/H   organic/humus layer
A-horizon   mineral layer
zone of eluvation

B-horizon   mineral layer
zone of illuvation

C-horizon
transition zone from
parent material below

Fig. 55. **Soil profile.** The main features of a soil profile.

**soil structure,** *n.* the physical configuration of a soil, determined by the shape, size and arrangement of individual soil particles when bound together into larger units known as PEDS or AGGREGATES. Peds are formed through the FLOCCULATION of clays, the activity of soil organisms, and the cementing effect of the by-products of organic decomposition and inorganic substances such as iron oxides and calcium carbonate. The structure of a soil can be described according to the type, size and development of peds (see Fig. 56). Five main soil structures are recognized:

(a) *laminar* or *platy* structures characterized by horizontally disposed peds (the long axes of the soil particles are arranged horizontally). Such structures are usually found in compacted soils where the lack of vertical PORE SPACES can result in drainage problems.

(b) *blocky* structures consisting of cube-like peds common in LOAMS and in some GLEY SOILS.

(c) *prismatic* or *columnar* structures resulting in vertically disposed peds (the long axes of the soil particles are arranged vertically) and are associated with saline soils.

(d) *crumb* or *granular* structures consisting of rounded peds characteristic of cultivated soils with a high organic content.

(e) *structureless* soils, in which none of the previous features can be recognized.

COMPACTION due to overcultivation or the use of heavy machinery can cause irreparable damage to the structure of a soil. Restricted plant growth may result and waterlogging and erosion can occur. See POROSITY, SOIL TEXTURE.

| plate-like |  | platy | may occur in any part of the profile |
|---|---|---|---|
| prism-like | 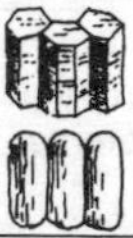 | prismatic (flat top) | common in subsoils of arid and semi-arid regions |
| | | columnar (round top) | |
| block-like |  | blocky (cube-like) | common in heavy subsoils of humid regions |
| | | blocky (subangular) | |
| spheroidal | | granular (porous) | characteristic of furrow slices |
| | | crumb (very porous) | |

Fig. 56. **Soil structure.** See main entry.

**soil survey,** *n.* the examination, description, analysis, classification and mapping of soils usually for specific use by farmers and foresters. Soil surveys take into account all the topographical, geological, climatological and vegetational data available for the area under examination. Field sampling to establish the boundaries of different soil types, and to obtain soil samples for laboratory analysis are important components of any soil survey. Such 'traditional' methods are now supplemented by data extracted from aerial photographs and more recently, from images produced by earth observation satellites such as LANDSAT, and SPOT.

**soil texture,** *n.* the size and comparative proportions of the constituent particles in a soil. The proportions of different sized particles can vary greatly between soil types and between individual SOIL HORIZONS, with the texture being largely determined by the PARENT MATERIAL and the type and rates of weathering to which it has been exposed. Mineral soils can usually be grouped into the three broad textural classes of CLAYS, SANDS and LOAMS. The

# SOIL TEXTURE

| Soil type | International (mm) | US Dept of Agriculture (mm) | British (mm) |
|---|---|---|---|
| Clay | <0.002 | <0.002 | <0.002 |
| Silt | 0.002–0.02 | 0.002–0.05 | 0.002–0.06 |
| Sand | | | |
| fine | 0.02–0.2 | 0.05–0.2 | 0.06–2.0 |
| coarse | 0.2–2.0 | 0.2–2.0 | |
| Gravel | >2.0 | >2.0 | >2.0 |

Fig. 57 **Soil texture.** Soil texture classes showing particle size in the most commonly used classification systems.

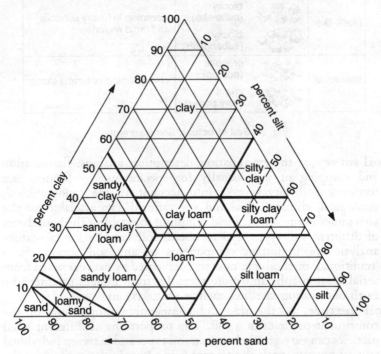

Fig. 58. **Soil texture.** The United States Department of Agriculture soil textural classes. The soil texture classification of a soil sample can be assessed according to the proportion of sand, silt and clay and which can be read off the three axes of the graph.

particle size ranges for each of these groupings vary considerably and depend upon the classification system used (see Fig. 57).

Through the feel of moist soil in the field and subsequent laboratory analysis, it is possible to allocate a soil to a textural class in accordance with its location within the textural triangle (see Fig. 58). The coarsest material, (gravel and cobbles) are excluded from the textural triangle as they rarely play an important role in pedogenic processes.

Soil texture is a vital component of the water-retention properties of a soil. Light textured, porous, easily cultivated sandy soils may be well-aerated but their water retention is limited and LEACHING may cause the loss of soluble nutrients required by agricultural crops. Heavy-textured clay soils have good water retention properties although high capillary POROSITY may sometimes cause waterlogging after heavy rainfall and may restrict cultivation.

**soil water,** *n.* the water that occupies the PORE SPACES of a soil and which originates from various sources including ground water, meltwater and PRECIPITATION. The amount of water within a soil is in a constant state of flux due to the loss of water via internal drainage and the addition of water from precipitation and/or OVERLAND FLOW. After a period of heavy rainfall or flooding the soil will be at saturation point but thereafter, drainage will reduce the amount of water, until after 6-24 hours the soil is said to attain FIELD CAPACITY. Providing no additional water is added to the soil, the drainage of excess water will continue at a reduced rate until the remaining *capillary water* is held within the pore spaces by increasing surface tension. This water is available for plant growth but ever-greater suctional forces must be exerted by the roots in order to remove the diminishing supply of water. Eventually, the plant roots are unable to remove further water from the pore spaces whereupon WILTING POINT is attained. See CAPILLIARITY.

**solanchak,** *n.* a grey-coloured soil characterized by very high salinity and alkalinity, low organic content, poorly differentiated horizons and a massive structure. Solanchak soils are typically associated with LOESS and ALLUVIUM parent materials. They may be found in a number of different climatic regimes while their topographic location shows an association with flat, low-lying areas or depressions in semi-arid or arid regions where saline ground water lies at or near the surface. Salinity is highest at the surface of the soil and in extreme circumstances, when the evaporation rate is high, the salts may form a visible surface crust.

There is little accumulation of salts in the zone of permanent saturation. Fine-textured solanchaks are usually more saline than coarse examples.

Due to their salinity, alkalinity and structural pattern, solanchaks have only limited agricultural potential. Flushing of the soils with fresh water may remove the toxic salts and addition of inorganic FERTILIZERS may be able to neutralize the presence of sodium (which is detrimental to the growth of most agricultural crops).

**solar constant,** *n.* the rate at which the sun's energy is received per unit area on a horizontal surface at the outer margins of the earth's atmosphere. The term 'constant' is misleading since its value fluctuates slightly; however, the average solar constant is 1388 $Wm^{-2}$. See also RADIATION, ALBEDO, INSOLATION.

**solar energy,** *n.* **1.** the radiant energy which originates from the sun. The sun can be considered a large thermonuclear device with a total energy output estimated to be 200 000 million, million times that of the largest nuclear reactor on Earth, and yet our planet intercepts only one fifty millionth of this energy. This amounts, on average, to $15.3 \times 10^8$ cal/m²/year (equivalent to 40 000 kW of electrical energy for every human on Earth). Due to latitudinal effects and cloud cover, the pattern of incoming solar radiation shows great variation. The UK, for example, receives on average only $2.5 \times 10^8$ cal/m²/year.

Of the solar energy which reaches the outer layer of the earth's atmosphere between 49% and 51% is lost through scattering or absorption within the atmosphere. Of the energy which reaches the surface, some 95-99% is absorbed by the oceans and is used to drive the HYDROLOGICAL CYCLE. The remaining 1-5% is used for PHOTOSYNTHESIS in green plants upon which all animal life is dependent.

**2.** the capture of the Sun's rays to provide a renewable source of energy for human use. Large-scale use of solar energy as a power source is still confined to experimental and research projects. Initial experiments in the USA and Israel which have shown promising results include those where concave mirrors positioned in hot desert areas are used to focus the sun's rays to raise the temperature of water sufficient to raise steam and generate electricity from steam turbines. Elsewhere and on a smaller scale, solar power has been successfully used to heat domestic water via solar panels, to charge batteries in locations remote from fixed power lines, to operate photo-electric cells, and to provide on-board power for satellite equipment. See ALTERNATIVE ENERGY.

**sol brun** or **sol brun acide,** *n.* the term extensively used by European soil classification systems for a BROWN EARTH soil.

**sol ferralatique,** *n.* the term used in the French soil classification system for a LATOSOL-type soil.

**solifluction** or **soil creep,** *n.* the slow downslope flow of saturated soil in PERMAFROST areas. Solifluction results when water released by the spring thaw in the ACTIVE LAYER cannot percolate downwards through the soil profile due to the impermeable nature of the permafrost.

**solonetz,** *n.* a grey-brown soil with high clay content and a saline B-HORIZON. The precise formation of solonetz soils is unclear but as they are often found close to SOLANCHAK soils it is possible that they represent a leached form of that soil. Crop production is difficult in these soils and their successful use depends much on local circumstances: in cool, semi-arid regions solonetz soils can be utilized for agriculture while in arid tropical areas, the lack of moisture makes IRRIGATION a necessity. Judicial application of water is necessary, however, in order to prevent the evaporation of salts from the B-horizon to surface layers. Heavy application of INORGANIC FERTILIZERS are also needed to counteract the inherent nutrient deficiencies of this soil type.

**solstice,** *n.* one of two occasions during the year when the sun is at its greatest distance from the equator. A solstice occurs on the 21 June (the *summer solstice*) and 21 December (the *winter solstice*). On the former date, the northern hemisphere receives its longest day while the southern hemisphere receives its shortest day. On 21 December, the situation is reversed. See also EQUINOX.

**solum,** *n.* the uppermost layers of the SOIL PROFILE affected by climate and vegetation. The term is most frequently applied to the A-HORIZON and the B-HORIZON.

**solution weathering,** see CHEMICAL WEATHERING.

**sourveld,** see VELD.

**Southerly Buster,** *n.* a strong, dry, cold, southerly WIND in southeastern Australia.

**southern lights,** see AURORA AUSTRALIS.

**species,** *n.* any of the taxonomic groups into which a GENUS is divided. A species is designated initalics by the genus name followed by its specific name, as in *Felix domesticus* (the domestic cat). See CLASSIFICATION HIERARCHY.

**spelaeology,** *n.* **1.** the scientific study of caves, with respect of their geological formation, flora, fauna, etc.

    **2.** the recreational pursuit of exploring caves.

**spit,** *n.* a bank comprising sand and/or shingle that projects along the coastline and eventually grows out across an ESTUARY or bay.

**spodosol,** *n.* the term used in the American soil classification system for a PODZOL-type soil.

**SPOT,** *n.* *acronym for* Système Pour l'Observation de la Terre, a French-based series of earth observation satellites, the first of which (SPOT-1) was launched in February, 1986. Like its American counterpart LANDSAT, SPOT-1 has a circular, near polar, sun-synchronous orbit, 832 km above the earth's surface with an inclination of 98.7°. The orbit repeat interval is 26 days, although due to the ability to direct the on-board optics 27° on either side of the central viewing line, imagery can be collected at approximately four-day intervals. The directable optics also allows SPOT-1 to scan away from cloud-covered areas and also to provide stereo-scopic imagery. SPOT-1 has a resolution imagery of 10 metres as opposed to the 30 metres of Landsat.

It is intended to provide long-term continuity of data through the anticipated launch of several additional satellites. SPOT-2 will duplicate SPOT-1 while SPOT-3 and SPOT-4 will have an enhanced ability to monitor changes in vegetation cover. See also EARTH RESOURCES TECHNOLOGY SATELLITE, COSMOS 1870.

**spring,** *n.* a natural flow of water occurring where the WATER TABLE intersects with the surface allowing the flow or seepage of GROUNDWATER onto the surface. The location of springs is dictated by surface topography and underlying geology, such as the boundary between layers of permeable and impermeable rock meeting the ground surface, or the impounding of groundwater movements by a DYKE. In wet climates, spring flow is usually continuous, whereas in drier climates flow may be intermittent.

**spring sapping,** see HEADWARD EROSION.

**spur,** see INTERLOCKING SPUR.

**SSSI,** see SITE OF SPECIAL SCIENTIFIC INTEREST.

**STABEX,** see LOMÉ AGREEMENTS.

**stability,** *n.* **1.** (Climatology) a condition in which a parcel of air which is forced to rise becomes cooler and heavier than the surrounding air and hence sinks to the lowest point within the air mass. Such a situation will be reached when the adiabatic LAPSE RATE of the uplifted air is greater than the ENVIRONMENTAL LAPSE RATE of the surrounding air.

**2.** (Ecology) the theoretical characteristic of ECOSYSTEMS and environments which for long periods of time show only minimal change (see CLIMAX COMMUNITY). In recent years the concept of stability in soil and vegetation systems has become discredited due to improved understanding of the changes which are constantly taking place within the environment. Stability in ecosystems is regarded as the exception rather than the rule.

**3.** (Animal ecology) a period of time when population numbers of a SPECIES become fixed around the CARRYING CAPACITY of the HABITAT. In the wild, extensive evidence of stability is rare. Indeed, stability is often associated with the management of animal populations by culling.

**stagnogleying,** see GLEYING.

**stand,** *n.* **1.** (Silviculture) any area of trees comprising one, or occasionally two species of similar or identical age. The 'stand' in forestry is synonymous to the 'field' in agriculture, representing an area in which uniform management techniques can be used.

**2.** any distinctive plant ASSOCIATION which may be identified at one or more sites. The identification of stands allows comparisons to be made between different vegetation communities.

**standardized birth rate,** see BIRTH RATE.

**standard man-day,** see MAN-DAY.

**staple,** *n.* a main element of the diet. Most communities have or have had at one time a single foodstuff that is consumed to a greater extent than any other and forms the basic dietary input. Wheat in the form of bread and pasta has been the staple in much of Europe and North America, while rice has dominated in Asia. Other cereal staples, generally eaten as a porrage meal, include maize in South America and parts of Africa, and sorghum and millet also in Africa. Root crops make up another important group of staples, for example, potatoes traditionally in the British Isles, especially Ireland, yams, cocoyams (taro) and increasingly cassava in West Africa, and cocoyams and sweet potatoes in the Pacific Islands. Only occasionally, as with plantains in parts of East Africa, do crops other than cereals or roots form the staple diet. Over-dependence on a staple crop can be nutritionally impoverishing, and in the event of crop failure, as with potatoes in Ireland in the 1840s, lead to widespread FAMINE.

**state farm,** *n.* a government-owned farm employing wage labour. In the USSR state farms or sovkhozi differ from COLLECTIVE FARMS in being larger and better resourced with the workers receiving a state-specified wage. The amalgamation and conversion of collective farms and the extension of the sown acreage in to the VIRGIN LANDS has seen their number grow from 4159 in 1940 to 22 313 by 1983, each farm averaging 16 600 ha. They have become a cornerstone of Soviet agro-industrial integration which ensures the direct delivery of produce to industrial plants for processing.

A number of THIRD WORLD governments have also experimented with state farms for the production of both food and industrial crops. In some cases, such as Tanzania and Mozambique, private estates and PLANTATIONS in foreign ownership have been taken over by the government and continue to be operated as large-scale production units with paid labour. In Ghana, the Ivory Coast and Ethiopia, state farms have been used to open up new agricultural areas. On the whole, the state farm in Sub-Saharan Africa has been marked more by ideological rhetoric than economic success.

**static rejuvenation,** see REJUVENATION.

**steam coal,** see BITUMINOUS COAL.

**steam fog,** *n.* a type of FOG formed by the movement of cold air over a body of warmer water. Evaporation from this body is rapidly condensed in the cold overlying air, and gives the appearance that the water is steaming. In very cold air, the water vapour may freeze to form *ice fog*. Compare ADVECTION FOG, RADIATION FOG.

**stenotypic species,** see ENVIRONMENTAL GRADIENT.

**steppe,** *n.* the traditional wild grasslands of Eurasia now removed or much altered by cultivation or grazing. The steppe lands of Russia extend over some 260 million ha (equivalent to 12% of the land area of the USSR) and are now the main locations for cereal crop production. The steppes are the Eurasian equivalents of the North American PRAIRIES. They form broad, rolling landscapes, which are treeless apart from SHELTER BELTS and ribbons of woodland which follow the river valleys.

In their original form the steppe lands were covered by drought-resistant grasses with many brightly coloured flowering herbs. Distinct regional variation existed in the original steppe flora as moisture and precipitation levels decreased from west to east. Similarly, while the characteristic soil type of the steppe land was the black CHERNOZEM, as precipitation became less this was replaced by chestnut-brown soil and ultimately, in the driest areas, SOLONETZ soils occured.

**stochastic process,** see SIMULATION STUDY.

**stock farming,** see ANIMAL HUSBANDRY.

**stocking density,** see FEEDLOT.

**stocking rate,** *n.* the number of animals that can be supported in a given area over a given time period. See CARRYING CAPACITY.

**stoma,** see TRANSPIRATION.

**stone stripes,** see PATTERNED GROUND.

**stop-and-go determinism,** see ENVIRONMENTAL DETERMINISM.

**storm,** *n.* **1.** any violent meteorological disturbance characterized by high winds and precipitation. There are several types of storm including thunderstorms, snowstorms, and hailstorms.
**2.** a force 11 wind on the BEAUFORT SCALE.

**straight fertilizer,** see FERTILIZER.

**strategic opportunist,** *n.* any plant or animal species which shows great aptitude in exploiting an opportunity for growth and/or increase in number. Species with a rapid breeding cycle, such as insects, are often well-adapted to respond quickly to short-lasting, favourable periods of growth, while humans, helped by their technological achievements, are best suited to altering environmental circumstances to ensure maximum survival of their species.

**stratigraphical column,** see EARTH HISTORY.

**stratification,** *n.* **1.** (Geology) the arrangement of horizontal layers of sediment that comprise SEDIMENTARY ROCK.
**2.** (Oceanography) the arrangement of distinct temperature zones in a body of water. See THERMOCLINE.
**3.** See LAYERING.

**stratification,** see LAYERING.

**stratocumulus,** *n.* a grey or white low-altitude CLOUD mass of globular visual appearance usually of extensive form and associated with the passage of a WARM FRONT. This cloud type is often associated with continuous drizzle.

**stratopause,** *n.* the transitional zone of maximum temperature between the STRATOSPHERE and the MESOSPHERE, found at an altitude of approximately 50 km (see Fig. 9). Compare TROPO-PAUSE, MESOPAUSE

**stratosphere,** *n.* the layer of the ATMOSPHERE that lies between the TROPOSPHERE and the MESOSPHERE. The base altitude of the stratosphere varies with latitude, beginning at 9 km at the Poles and increasing to 16 km in equatorial regions. The stratosphere extends for around 50 km and ends at the STRATOPAUSE. Air temperature rises with altitude in this layer and clouds are a rare occurrence. The stratosphere contains most of the atmosphere's OZONE which concentrates around an altitude of 22 km.

**stratovolcano** see VOLCANO.

**stratum,** *n.* any of the distinct layers into which SEDIMENTARY ROCK is divided.

**stratus,** *n.* a whitish-grey continuous layer of low-altitude CLOUD associated with the mixing of warm and cold masses of air by the wind. A light drizzle is often associated with stratus cloud forms.

**stream capacity,** see FLUVIAL TRANSPORTATION.

**stream competence,** see FLUVIAL TRANSPORTATION.

**strip cropping,** see STRIP FARMING.

**strip farming,** *n.* **1** the division of a large field into elongated strips, each of which is farmed by separate owners or tenants. Strip farming was widespread in the FIELD SYSTEMS of medieval Europe, and led to much farm FRAGMENTATION. It was largely replaced by ENCLOSURE as the main land-holding pattern in Europe.

**2.** a SOIL CONSERVATION technique in which alternate rows or strips of different crops are sown across a slope in furrows following the contours of the land to counteract SOIL EROSION. Contour strip farming or cropping may also alternate cropped and FALLOW strips, with the pattern reversed each year to distribute the benefits of fallowing more evenly. The cover provided by the fallow strips arrests run-off from adjoining cultivated strips. See CONTOUR PLOUGHING.

**strip grazing,** see GRAZING MANAGEMENT.

**strip mining,** *n.* a method of surface mining mineral deposits. On level or gently undulating land, area strip mining may be carried out in which OVERBURDEN, the overlying layers of soil or rock, is stripped away before a trench is dug to remove the mineral deposit. A parallel trench is then dug and its overburden is placed in the first trench, and so on, leaving a series of spoilbanks of waste material. Contour strip mining is practised in areas of more pronounced relief, and involves the cutting of benches or terraces on the hillslope. Overburden from each new terrace is deposited on the terrace below. See OPENCAST MINING, DEEP MINING.

**strontium-90,** *n.* an ISOTOPE used in nuclear power sources and produced from atmospheric nuclear explosions as a fission product. It has a HALF-LIFE of 28 years and resembles the element calcium which, in certain circumstances, it can replace. Strontium 90 can accumulate in FOOD CHAINS; for example, a typical pathway for the isotope would be:

$$\text{atmosphere} \rightarrow \text{precipitation} \rightarrow \text{soil} \rightarrow \text{grass} \rightarrow$$
$$\text{cow} \rightarrow \text{milk} \rightarrow \text{human bone tissue}$$

Strontium-90 can cause leukaemia in humans and fears have been expressed that residents in close proximity to nuclear power stations are subjected to bombardment by strontium released from the power stations. However, government-sponsored research has suggested that strontium-90 levels around nuclear power stations are not hazardous. The main source of strontium 90 is undoubtedly that of atmospheric testing of nuclear devices.

**structural reform,** *n.* any change which affects the essential components of an agricultural system, specifically the nature of LAND TENURE, FARM SIZE, farm layout, or the composition of the farming population. Poor agrarian structures reflect an imbalance between the factors of production, and prevent realization of the full potential of the agricultural resource base. Thus, structural reform measures are directed towards increasing the efficiency and economic viability of farming. The concept of viability is not easily defined, but frequently implicit in reform policies is the aim of creating holdings that will provide sufficient income for the farmer and his family (see FAMILY FARM). This may be achieved by the reform of tenure, including a redistributive LAND REFORM in areas with gross inequalities in the pattern of landholding. Elsewhere, and especially in Europe, structural reform includes FARM ENLARGE-MENT where farms are too small to be viable and where many holdings can support only PART-TIME FARMING, particularly in Italy and West Germany; land CONSOLIDATION where farms consist of scattered parcels of land; and measures aimed at altering the human resources of the farm sector such as encouraging retirement and retraining in order to reduce the number of farmers or to improve management skills.

**structure plan,** *n.* a statutory UK planning document required under the terms of the TOWN AND COUNTRY PLANNING ACT (1971). The structure plan is the policy statement of the region (in Scotland) or county (in England and Wales) in respect of the spatial allocation of land for various uses within its territory in forthcoming years. The written strategy is accompanied by a map or *key diagram* portraying the spatial implications of future land use. The structure plan provides the framework within which more detailed decisions about particular sites are made by the lower tier district authorities in their statutory LOCAL PLANS. The pattern of land use in the structure plan is the design for the INFRASTRUCTURE required to house the projected future population, to provide social services, and to provide opportunities for economic development. Unitary DEVELOPMENT PLANS will eventually replace structure plans within metropolitan districts in England and Wales. Outside metropolitan areas, it is likely that in future the district will be the major plan-making level and that here too structure plans will be phased out.

**stubble mulching,** *n.* the retention of crop stubble and other organic residues between harvests to form a surface cover that restricts SOIL EROSION.

**subglacial transportation,** see GLACIAL TRANSPORTATION.

**sublimation,** *n.* the process whereby a solid changes state to become a gas without first melting. Sublimation causes the recession of patches of snow without any meltwater being produced.

**sublittoral zone,** see BENTHIC ZONE.

**submarine canyon,** *n.* a steep-sided, V-shaped trench that dissects the CONTINENTAL SHELF and may continue down the CONTINENTAL SLOPE to terminate at the edge of the DEEP-SEA PLAIN. Trenches can be up to 300 m in depth and the Hudson River Canyon, for example, extends for 320 km. Submarine canyons have many origins and can be result from faulting, rising sea levels drowning a former river valley and the action of TURBIDITY CURRENTS.

**subsistence farming,** *n.* the production of food and other necessities to satisfy the needs of the farm household or extended family. Surplus production may be exchanged but this can be distinguished from partly commercialized farming where there is a prior intent to produce for sale while still retaining the greater part of gross farm output for domestic consumption. In the strictest sense, subsistence farming is a self-sufficient and self-contained farming system. However, it is becoming increasingly rare with the more widespread use of external inputs, and market participation. Now largely confined to THIRD WORLD countries, it is generally associated with farming methods such as SHIFTING CULTIVATION and ROTATIONAL BUSH FALLOW.

Typical characteristics of both subsistence and partly commercialized farming in developing countries include: the maximization of output rather than profits; the maximization of YIELDS by weight or calorific value per hectare rather than output per head; low levels of capital inputs; the sole use of family labour; crop diversity to minimize risk and ensure variety of diet (see MULTIPLE CROPPING); a dependence on crop production rather than livestock rearing; and small farm size, often with considerable FRAGMENTATION.

**subsoil,** see B-HORIZON.

**substratum,** *n.* **1.** (Ecology) any object or material upon or within which a plant or animal organism lives. Not all organisms possess a substratum. Aquatic life forms which float or swim (for example, PLANKTON) and which never come to rest on a solid surface do not posses a substratum. The most common substrata comprise weathered rock debris, soil, wood, plant and animal surfaces, the meniscus of liquids, (particularly water) and a host of manufactured materials such as paper, leather, oils, foodstuffs and clothing. Parasitic organisms have adapted to yet other highly specialized substrata such as the internal membranes of host animals or plants.

**2.** (Geology) the solid rock underlying soils, gravels etc. See BEDROCK.

**subtropical jet stream,** see JET STREAM.

**subtropical vegetation,** *n.* vegetation found between the latitudes 15°-35°N and S of the equator. The characteristic vegetation consists of extensive grassland or SAVANNA. Trees are infrequent although never entirely absent. Wherever burning can be prevented tree regeneration can be rapid. The climatic regime shows distinct seasonal variation based on the migration of the sun between the tropics. In summer, when the sun is overhead, average monthly temperatures exceed 21°C and are usually accompanied by abundant precipitation. In winter, when the sun is in the opposite hemisphere, the dry, 'cool' season prevails with an average monthly temperature of no lower than 17°C. Annual rainfall varies from as little as 500 mm to a maximum of 2500 mm. Compare TEMPERATE VEGETATION.

**suburb,** *n.* a residential district situated on the outskirts of a city or town. Suburban development has been made possible by improved transportation links, in paticular by the suburban railway in the period 1890-1914 and thereafter by road transport. Until 1973, suburban 'sprawl' was a feature around many large cities, but traffic congestion and the vastly increased cost of petrol has resulted in the redevelopment of the older suburban areas near the city centre. See COMMUTER, NEW TOWN, PERI-URBAN FRINGE, RURAL-URBAN CONTINUUM.

**successional cropping,** see MULTIPLE CROPPING.

**suitcase farming,** *n.* a farming system found in the USA in which the farmer, although owning more than one holding, does not reside on any one farm but moves, with the farm machinery, from one to another at critical times of the farming year, such as ploughing and harvesting. Suitcase farmers tend to live closer to their holdings than do sidewalk farmers (see SIDEWALK FARMING). Suitcase farming is mostly associated with cereal cultivation, as GRAIN CROPS generally require minimal attention between sowing and harvesting

**supercooling,** *n.* the retention of a substance in its liquid state at a temperature below its FREEZING POINT. Water can be supercooled providing it remains undisturbed. Supercooled water is common in

## SUPERIMPOSED DRAINAGE

clouds which lack the HYGROSCOPIC NUCLEI to allow it to form as ice; this represents a potential hazard to aircraft since in certain circumstances, the supercooled water can form ice deposits on the fuselage thus altering the weight distribution of the aircraft. The movement of supercooled water droplets by a slight wind may also result in the development of a RIME on exposed trees, hedges, fences, cars etc.

**superimposed drainage,** *n.* a river drainage system originally developed on a land surface since removed by EROSION and subsequently imposed on older underlying rocks. Superimposed drainage often bears little relation to the nature and structure of the present land surface. Examples of superimposed drainage are found in the English Lake District and in the Appalachians in the USA. See also ANTECEDENT DRAINAGE, DRAINAGE PATTERN.

| Age X | Males | | Females | |
|---|---|---|---|---|
| | lx | ex | lx | ex |
| 0 | 10000 | 68.2 | 10000 | 74.5 |
| 5 | 9742 | 65.1 | 9798 | 71.0 |
| 10 | 9719 | 60.2 | 9783 | 66.1 |
| 15 | 9699 | 55.4 | 9770 | 61.2 |
| 20 | 9651 | 50.6 | 9751 | 56.3 |
| 25 | 9599 | 45.9 | 9728 | 51.4 |
| 30 | 9551 | 41.1 | 9699 | 46.6 |
| 35 | 9494 | 36.3 | 9659 | 41.8 |
| 40 | 9405 | 31.6 | 9595 | 37.0 |
| 45 | 9258 | 27.1 | 9489 | 32.4 |
| 50 | 9013 | 22.8 | 9324 | 27.9 |
| 55 | 8595 | 18.8 | 9076 | 23.6 |
| 60 | 7915 | 15.2 | 8713 | 19.5 |
| 65 | 6819 | 12.1 | 8157 | 15.7 |
| 70 | 5513 | 9.4 | 7306 | 12.2 |
| 75 | 3935 | 7.2 | 6049 | 9.2 |
| 80 | 2354 | 5.4 | 4385 | 6.8 |
| 85 | 1051 | 4.0 | 2504 | 5.0 |

Fig. 59. **Survivorship curve.** A life table for the UK for the years 1964-66. The lx column shows the numbers surviving at the beginning of each successive 5-year age interval, starting with 10 000 new-born babies. Column ex shows the average expectation of life at those ages.

418

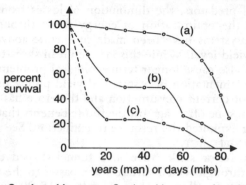

Fig. 60. **Survivorship curve.** Survivorship curves for (a) human females, and of a predatory mite (*Cheyletus eruditus*) reared in the laboratory at (b) 20°C, 65% relative humidity and (c) at 25°C, 85% relative humidity.

**supersaturation,** *n.* the condition of a body of air of which the RELATIVE HUMIDITY exceeds 100%. Supersaturation is a relatively common in the LOWER ATMOSPHERE and develops as a result of the lack of particulate matter in the atmosphere, such as pollen and dust, around which water droplets can condense. See SATURATION.

**supraglacial transportation,** see GLACIAL TRANSPORTATION.

**surface water gleying,** see GLEYING.

**survival zone,** see TOLERANCE.

**survivorship curve,** *n.* a curve plotted on a graph showing the percentage survival of a species based upon a known starting population size and extending over the entire life history of that species. Examples of survivorship curves are given in Fig. 60. The number of individuals which die with the passage of time may be expressed in various ways, for example, as a percentage of the whole population per year, or as the percentage of age groups within the total population. Age-specific survival rates are commonly recorded as *life tables* (see Fig. 59).

**suspended load,** see FLUVIAL TRANSPORTATION.

**sustained yield,** *n.* the amount of harvestable material which can be removed from an ECOSYSTEM over a long period of time with no apparent deleterious effects on the system. The concept of sustained yield is fundamental to the successful operation of an ECODEVELOPMENT policy. In practice, few ecosystems are allowed to operate at a sustained yield level and as a result of management policies which involve the addition of inorganic FERTILIZERS, the removal of predators, the diminution of losses from disease and decay and the optimization of energy flows through the FOOD CHAIN, ecosystems have been made to operate above that of the sustained yield level. While this may result in short-term financial gain it can also cause longer term ecological problems.

The implementation of a sustained yield policy may involve a reduction of current consumption and therefore has an economic cost that must be offset against the future benefits that derive from not having exploited the resource to extinction. See MULTIPLE USE SUSTAINED YIELD ACT.

**swallow hole** or **sink hole** *n.* a funnel-shaped depression in limestone through which surface water passes to the WATER TABLE (see KARST). Swallow holes are often 15-20 m deep with precipitous sides and are formed by the gradual CHEMICAL WEATHERING of the calcareous bed rock.

**swamp,** *n.* a permanently waterlogged area of overgrown vegetation, such as the Everglades in Florida, USA. If drained, swamps are often highly fertile.

**swarm,** see DRUMLIN.

**swash,** *n.* the forward movement of water up a beach resulting from a breaking wave.

**swidden,** see SHIFTING CULTIVATION.

**symbiosis,** *n.* a relationship between members of two different SPECIES which results in mutual benefit. A symbiotic relationship often permits the survival of a species in a hostile environment. For example, conifer trees can grow on soils with a very low nutrient content due to the presence of symbiotic mycorrhiza algae which make the scarce nutrients available to the tree rootlets; in return, the rootlets provide a SUBSTRATUM for the algae. Compare COMMENSALISM, PARASITISM.

**syncline,** *n.* a trough-like downfold in rock strata, often the result of long-term addition of eroded materials transported by rivers from the land to areas of marine deposition. Compare ANTICLINE.

**synecology,** *n.* the study of the structure, development, function and distribution of whole plant COMMUNITIES. Synecological studies often provide the means of identifying the most important problem species within a plant community. Compare AUTECOLOGY.

**synergism,** *n.* the combination of two harmless substances to produce an impact greater than the sum of the separate effects. For example, ACID DEPOSITION is the result of a synergistic process.

Industrial chimneys may emit harmless elemental sulphur which quickly joins with oxygen to form sulphur trioxide ($SO_3$). This substance is still of low hazard value but when it joins with water molecules in the atmosphere ($H_2O$), a dilute solution of sulphuric acid is formed ($H_2SO_4$) and causes considerable environmental damage when precipitated as acid rain.

**synoptic chart,** see WEATHER CHART.

**system building techniques,** see HIGH-RISE DEVELOPMENT.

# T

**taiga,** see BOREAL FOREST.

**tailings,** see DEEP MINING.

**talus,** see SCREE.

**tarn,** *n.* a mountain lake occupying the floor of a CIRQUE and dammed by a glacial MORAINE.

**tar sand,** see OIL.

**taxon,** see FAMILY.

**technological optimum** or **technological fix,** *n.* the reliance on technological processes to solve human problems. Due to complexity of these problems, such solutions may be only short-term or cosmetic, and may result in other problems. The term is often used pejoratively by ecologists and conservationists of industrialists committed to the use of high-tech engineering processes to exploit more of the biosphere RESOURCES with little regard for the environmental impact of these processes. The greatest reliance on the technological fix is that shown by the advocates of nuclear power for which there is currently no prospect of a safe means of disposal for NUCLEAR WASTE. Compare DOOMSDAY SYNDROME.

**tectonic estuary,** see ESTUARY.

**tectonism** or **diatrophism,** *n.* the deformation and movement of rock in the EARTH'S CRUST. Most tectonic movements occur slowly over long time periods but some, such as EARTHQUAKES, may be sudden and violent in nature. Other tectonic movements include folding and faulting which may result in the uplift or subsidence of land. Compression or tension caused by tectonic forces may also generate lateral movements. OROGENESIS and EPEIROGENESIS are two types of tectonism. See FOLD, FAULT.

**temperate vegetation,** *n.* vegetation characteristic of the temperate latitudes (typically 35°-50° N and S, although less extensive in the southern hemisphere). The climatic regime shows distinct seasonal variation with winter conditions which may include one month with an average temperature of 0°C. More commonly, winters are not as severe, with an average January temperature of 2-4°C. Summers are long (6 months) and may be hot, with a July average temperature of 18°C. Precipitation usually attains a winter maximum although convectional storms can produce substantial summer rainfall.

The dominant plant life form is the HARDWOOD tree which may be either evergreen or deciduous. The original forests were immensely valuable in terms of timber and animal resources and were so over-exploited that by the end of the 19th century that very little of this vegetation type remained. The forests were replaced by grasslands and arable agriculture.

**tenancy,** see LAND TENURE.

**tenement,** *n.* a residential building containing individual flats or apartments which are accessed by way of a common stairway and entrance door. The term is usually applied to buildings of four- or five-storey height which were built before the mid-20th century. In the UK such buildings are common only in Scotland where this style of housing has a long tradition. The term often has connotations of a slum, although in Scotland most social groups can be found living in tenement homes displaying a wide range of sizes and styles.

**Tennessee Valley Authority (TVA),** *n.* an agency of the US Government, established in 1933 as part of President F.D. Roosevelt's 'New Deal' programme. Located in Tennessee and neighbouring states in the southern Appalachians, the TVA has been responsible for coordinating a comprehensive programme of power development, flood control, improved navigation, AFFORESTATION, agricultural rationalization, SOIL CONSERVATION, and industrial rehabilitation and diversification within the Tennessee and Cumberland River basins.

Although subsequently reined in by waning political enthusiasm, the TVA has achieved notable progress in regional agricultural and industrial development, and in INFRASTRUCTURE provision, especially of power supplies. The basic TVA strategy of water control and resource development has been much copied in postwar river basin planning, although not in the USA itself, often with technical assistance from TVA civil engineers.

**terminal moraine,** see MORAINE.

**terrace,** *n.* a platform or field of level ground constructed on a hillslope to permit crop cultivation. The retaining walls, running parallel to the contour, vary in height with the angle of incidence of the slope. Terraces are usually associated with IRRIGATION systems both in dry areas, such as Yemen and Peru, and the more humid areas of WET-RICE CULTIVATION. The low walls that ring and separate each terrace control the flow of water and permit a more thorough soaking of the soil. They also trap down-slope movement of soil, so reducing the risk of SOIL EROSION and creating MICRO-ENVIRONMENTS suitable for the cultivation of crops such as olives and grapes on otherwise rock-strewn, shallow-soiled Mediterranean hillsides. They also facilitate the harvesting of PERMANENT CROPS such as tea.

Terrace cultivation is widely distributed geographically and culturally, and has been practised for several millennia. Today, terraces are most common among the rice PADDIES of southern China, Taiwan, Japan, Indonesia, and the Philippines. However, in many areas, especially in developed countries, the cost of maintaining terraces while continuing with largely mechanized farming practices has become prohibitive, leading to widespread abandonment, as witnessed in German vineyards and Mediterranean olive groves. In parts of Africa, such as northern Cameroon and central Nigeria, the decay of terrace cultivation has followed the abandonment of highland refuges by tribes no longer threatened by hostile neighbours.

**terracette** or **sheeptrack,** *n.* a small terrace on a slope formed by SOLIFLUCTION or SLUMPS. Terracettes may be up to 30 cm in height and width and usually occur in parallel groups following the contours of a slope. Cattle and sheep often use terracettes as paths across slopes and this may lead to the growth of the feature.

**terra rosa,** *n.* a red CLAY-rich soil which develops on PARENT MATERIAL with a high calcium carbonate content. Terra rosa soils comprise the insoluble residue products of the chemical weathering of limestone and as such they are usually very old. They are most common in semi-arid regions which border the Mediterranean Basin and in parts of Australia. The red colouration arises from the oxidized (ferric) state of the iron compounds in the parent material. They form fertile agricultural soils.

**territorial waters,** *n.* any area of sea over which an adjacent country claims jurisdiction. Under international law, territorial waters extend for 5 km from the high water mark, although in 1977 the United Nations Law of the Sea Convention designated *exclusive economic zones* of 370 km within which countries have the sole right to exploit the mineral and fish resources off their coastline. See LAW OF THE SEA.

**territory,** *n.* any area which provides a living space for plants and animals. The concept of territory is best displayed by animals, many of which have elaborate behavioural patterns related to the delimitation of territory, for example, patrolling and scent marking by large mammals or the use of song posts by birds. Any territory so marked will be rigorously defended by its occupant against intruders.

The territory of a plant is the area occupied by the furthest extremities of the root system below ground and the area occupied by the stems, leaves and branches of the above ground portion of the plant.

**tertiary consumer,** see FOOD CHAIN.

**Tertiary period,** *n.* the geological PERIOD that followed the CRETACEOUS PERIOD around 65 million years ago and that ended about 2 million years ago when the QUATERNARY PERIOD began. The Tertiary is the first period of the CENOZOIC ERA and is split into five well-defined EPOCHS. There was a major phase of mountain building and associated volcanic and igneous activity in Europe, North America, Africa and Asia. A variety of invertebrate marine fossils, many of which can be found in the contemporary fauna, are used to date and correlate Tertiary aged rocks. At the end of the Tertiary, the climate underwent a pronounced cooling which culminated in the ICE AGES of the PLEISTOCENE EPOCH. See EARTH HISTORY.

**thalweg,** see RIVER PROFILE.

**thermal depression,** *n.* an area of low pressure that results from the intense heating of continental areas during the day leading to a convectional rising of air currents and a drop in surface atmospheric pressure. Thermal depressions usually occur only during the summer. Small examples of thermal DEPRESSIONS may be found over islands and peninsulas whilst larger examples are located over the Iberian peninsula and Arizona in the USA.

**thermal erosion,** *n.* the environmental degradation of the PERMAFROST resulting from human activity. The removal of surface vegetation cover, which has an insulating effect for the underlying permafrost, prior to the construction of buildings, roads and bridges as well as the laying of underground services can result in the downward extension of the ACTIVE LAYER. During the summer, this disturbance can cause the active layer to become a quagmire with the thawed soil being very susceptible to erosion by meltwater. Thermal erosion can be prevented by constructing roads and railways on insulating gravel beds which protect the underlying permafrost. Houses and other buildings along with services can be erected on piles which allows the free movement of air beneath them and prevents damage to the permafrost.

**thermal expansion and contraction,** see PHYSICAL WEATHER-
ING.

**thermal pollution,** *n.* the release of substances into the environment which results in an undesired rise in ambient temperature
above that possible by an input of natural solar radiation. The main
sources of thermal pollution include waste heat from domestic
space heating and from most industrial processes, and from waste
hot water discharges from electrical power stations. The heat may
be released into the atmosphere, oceans or rivers whereupon it
accumulates to such an extent that it causes unwanted changes to
the natural flow of heat energy. For example, the amount of
oxygen dissolved in river or lake water is inversely proportional to
the water temperature. If hot water from electricity generating
stations is allowed to enter rivers or lakes it can cause the de-
oxygenation of the water and lead to the suffocation of aquatic
animals. Large cities release so much thermal pollution that they
modify the local climate by producing a HEAT ISLAND.

**thermal spring,** see HOT SPRING.

**thermocline,** *n.* a permanent layer of water within lakes and
oceans that displays a rapid decline in temperature and which
separates an upper warmer layer (the *epilimnion*) and a lower colder
layer (the *hypolimnion*). Compare INVERSION.

**thermosphere,** *n.* the layer of the ATMOSPHERE that lies above the
MESOSPHERE and the MAGNETOSPHERE, extending from around 80
to 500 km above the earth's surface. Solar radiation causes
ionization of atmospheric gases which leads to the phenomenon of
the AURORA AUSTRALIS and the AURORA BOREALIS.

**therophyte,** see RAUNKIAER'S LIFEFORM CLASSIFICATION.

**therophyte climate,** see BIOCHORE.

**Third World, the,** *n.* those DEVELOPING COUNTRIES located
mainly in Africa, Asia and Latin America that are neither industrial
market economies (the *First World*) nor centrally planned economies (the *Second World*). The term is thus a convenient shorthand for
a grouping of supposedly non-aligned countries with apparently
similar low levels of material development. However, the term is
so fraught with inconsistencies and incompatible groupings of
countries that if it is to be used at all, it is best defined in a general
sense rather than with wholly misplaced precision.

It can be shown that few Third World countries are genuinely
non-aligned, and that former colonial ties and dependent economic
relations with DEVELOPED COUNTRIES ensure that the Third World
cannot be regarded as an entirely separate entity. The countries of

the Third World, equally loosely termed as developing economies, less developed countries and the South, reveal considerable diversity in their geography, economic problems and prosperity, levels of industrialization, political ideology, and form of government. The United Nations now identifies a group of 31 least developed countries, 21 of them in Africa, which some regard as constituting a *Fourth World*, while the WORLD BANK makes a threefold division of developing countries on basic income criteria.

**Thirty Percent Club,** see ACID DEPOSITION.

**throughflow,** *n.* the lateral and downslope subsurface movement of water through the soil. The erosive effect of throughflow is usually limited due to its low velocity. Throughflow often becomes concentrated in natural 'pipes' in the soil to form *percolines* which may flow into rivers or cause springs. Compare OVERLAND FLOW, RUNOFF.

**thunder,** *n.* the deep rumbling sound during a THUNDERSTORM caused by the rapid heating and expansion of the air surrounding a flash of LIGHTNING. Thunder is usually audible up to 16 km away from its point of origin but can sometimes be heard up to 65 km away.

**thunderhead,** see CUMULONIMBUS.

**thunderstorm,** *n.* a STORM consisting of THUNDER, LIGHTNING, heavy precipitation and the massive vertical development of CUMULONIMBUS cloud, which results from the formation of strong CONVECTION cells. These convection currents are usually triggered by the intense heating of the earth's surface, orographic processes or from the passage of a COLD FRONT.

In a developing thunderstorm, the initial updraughts are supplemented by the heat released from the condensation of atmospheric water vapour. These updraughts may occasionally exceed speeds of 30 m/s and can rise to heights of 12 km. Initially, the strength of the updraughts prevents the fall of water droplets formed by condensation although as the droplets grow in size gravity overcomes this effect. The sudden fall of heavy rain or hail with the accompanying occurrence of thunder and lightning indicates the mature stage of a thunderstorm. Most convection cells have a lifespan of approximately 1 hour and gradually the friction of the falling rain establishes down draughts which, along with the depletion of water vapour in the cell, causes the gradual dissipation of the storm.

Most thunderstorms consist of several convection cells which self-propagate when the downdraughts from two degenerating cells converge, causing the warm air trapped in between to rise and initiate a new cell. Multicell thunderstorms can be up to 8 km in width and 100 km in length.

Thunderstorms occur in most latitudes apart from polar regions and are particularly common in tropical latitudes. Mid-latitude thunderstorms are largely confined to the summer and early autumn.

**tidal range,** see TIDE.

**tidal wave,** see TSUNAMI.

**tide,** *n.* the cyclic and daily changes in the elevation of the OCEAN surface caused by the gravitational attraction exerted by the sun and moon. The difference between high and low tide is known as the *tidal range* and may vary from less than 1 m on the open ocean to 15 m nearer the coast. Tidal movement is currently under investigation in several countries as a source of ALTERNATIVE ENERGY.

**tillage,** *n.* the process by which soil is prepared to form a seedbed with conditions favourable for CROP growth.

*Primary tillage* is the breaking up and loosening of the soil to a depth of between 15 and 90 cm, usually by ploughing which may also overturn and cover crop residues. Deep ploughing may also improve rooting and drainage conditions. *Secondary tillage* is the finer preparation of the seedbed. Weeds and crop residues are cut up by discs and harrows operating at shallower depths than primary tillage implements. The soil may be pulverized by rollers to break up the aggregates and eliminate air pockets, thus ensuring the contact between seed and moist soil that is necessary for germination.

With increased MECHANIZATION of agriculture, there has been a tendency for soil structure to break down under extensive tillage, especially with compaction in areas of heavy and poorly drained soils. In drier zones the soil may be left more susceptible to crusting with problems of impeded water intake, increased run-off and reduced water storage for crop use. To counter these problems, methods of *minimal tillage* have been introduced in which the number and intensity of tillage operations is reduced. Narrow rows may be tilled but the spaces in between are left unploughed. Wheel-track ploughing occurs when seed is planted in soil firmed by the tractor's weight. In an extreme form of reduced cultivation, *direct drilling* or *zero-tillage* may be employed in which a narrow slit is cut in the soil and seed and fertilizers are sown directly without further seedbed preperation. Crop residues, usually from LEY FARMING, either decompose in situ (known as MULCH tillage), or are controlled by herbicides. Although the covering vegetation is killed, a dense mat of plant debris helps to conserve moisture and reduce run-off likely to cause SOIL EROSION. Minimal tillage methods have further advantages in reducing labour and fuel costs, but are only appropriate in soils already in good structural condition.

Minimal tillage has spread rapidly in the USA, especially in the drier areas of the south, expanding from 10% of the harvested CROPLAND in 1972 to an estimated 33% by the mid-1980s.

**timber line,** see TREE LINE.

**time lag,** *n.* the interval of time that exists between the beginning of an event and the response which occurs as a result of the operation of that event. All events, whether natural or as a result of human activity, make an impact either upon the physical components of an ECOSYSTEM or upon species within an ecosystem. The complex inter-relationships which exist between the changing environmental circumstances and ecosystems ensure that the time lag between the occurence of an event and the reaction to that event may take many years to become apparent. For example, lung cancer in a human due to cigarette smoking can take 10-20 years to manifest itself and because of interdependency between environmental variables the strength of the relationship may be statistically uncertain. It is now thought likely that the most stable ecosystems have the most well-developed time lag network.

**tolerance,** *n.* the ability of a plant or animal to survive under changing environmental circumstances. The concept of tolerance

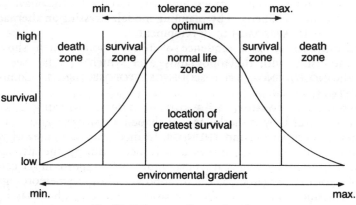

Fig. 61. **Tolerance.** See main entry.

is illustrated in Fig. 61. Under normal conditions organisms will exist within the *normal life zone*, within which optimal living conditions exist; survivorship levels are highest in this zone. On either side of the normal life zone, conditions for survival rapidly decline when environmental factors such as light, temperature or water availability increase or decrease to a point where the probability of death is at least as high as the probability of survival; the organism is now in the *survival zone*. If environmental conditions continue to deteriorate then survival rates fall until the organism enters the DEATH ZONE or *lethal zone*.

**toll,** *n.* a charge levied on vehicles using bridges or special roads, such as the TURNPIKES in the USA, or certain MOTORWAYS in European countries.

**top-down approach,** *n.* a strategy of development based on the hypothesis that development is driven by external demand and innovation impulses, and that the benefits of development will spread from a few dynamic sectors and locations to the rest of the system.

Top-down approaches, or development 'from above', emphasize the maximization of economic growth rates, and strong centralized direction and control of investment. Projects tend to be capital-intensive, dominated by high technology, and favour the 'large project' approach. At the heart of such strategies is the notion of polarized growth and the belief that implanted urban-industrial growth centres will stimulate regional development towards greater spatial equilibrium. Top-down approaches have dominated regional and national development strategies in most DEVELOPING COUNTRIES in recent decades. Since the mid-1970s, they have been increasingly challenged by planning models stressing an alternative BOTTOM-UP APPROACH to development.

**toposequence,** *n.* a sequence of related soils which show a relationship with the topography over which they lie. See also CHRONOSEQUENCE, LITHOSEQUENCE, TOPOSEQUENCE. Compare CATENA.

**tornado** or **twister,** *n.* a narrow, violent whirling STORM, characterized by a dark, funnel-shaped CUMULONIMBUS cloud, heavy precipitation and THUNDER, formed around an area of very low air pressure. Tornadoes usually occur during the afternoon when surface heating is at its maximum. Wind speeds in the centre of a tornado, regarded as the most powerful of the meteorological disturbances, have never been measured accurately, but may be as high as 550 km/h. Tornadoes may travel across land at up to 60 km/h but are of limited duration, rarely lasting more than a few hours.

The precise reasons for the formation of tornadoes is not known but they seem to form when a large mass of moist warm air meets a mass of cooler air. The resulting INSTABILITY is supplemented by the strong heating of the earth's surface which occurs during the spring and summer.

Such conditions are common in such places as the Mississippi Basin in the USA where warm moist air from the Gulf of Mexico meets cool air from the interior of the continent. Tornadoes are also found in other parts of the world including Australia, the Indian subcontinent, and more rarely in Canada and Europe.

Despite being very narrow at their base, which may be as narrow as 100 m in diameter, tornadoes can cause considerable loss of life and damage to property, such as the tornado which struck Edmonton, Canada in July 1987. The most damaging tornado of recent times was that which moved across Missouri, Illinois and Indiana in March 1925 killing 689 people and injuring more than 2000. In some areas where tornadoes are frequent early warning systems have been established and buildings are constructed to withstand the severe structural stresses associated with tornadoes. Protective storm cellars are a common feature in these areas. Compare HURRICANE, WHIRLWIND.

**Town and Country Planning Acts,** *n.* a series of British legislation dating from 1910 which provide the basis of the present planning system. The main aims of the acts were:

(a) to define by means of DEVELOPMENT PLANS the policies for the planning of each area;

(b) to bring DEVELOPMENT generally under the control of a local planning authority or central government department;

(c) to empower public authorities to aquire and develop land to meet planning objectives and to extend the scope and scales of financial aid for these purposes;

(d) to provide for special AMENITY issues, for example, buildings of architectural merit or HISTORIC BUILDINGS, the creation of tree preservation orders and ADVERTISMENT CONTROL.

**townscape,** *n.* the visible composition of an urban area that conditions its overall character. Elements which comprise a townscape include the scale and massing of the buildings, their architectural style, the nature and colour of the building materials used, the location of open spaces and greenery, the textures of the footpaths, the street furniture and the means of handling vehicular traffic. Perception of the character of the townscape in various parts of a city forms an important frame of reference against which planners develop standards of appropriateness for the renovation or replacement of buildings and thus for the overall control of DEVELOPMENT. See also BUILT ENVIRONMENT.

**toxic waste,** *n.* any discarded material, commonly from industrial or commercial processes, capable of causing injury or deaths to living organisms. The term is sometimes confused with *hazardous wastes* which refers to all substances, including toxic wastes, that present an immediate or long-term human health risk or which pose a health risk to the environment.

Nearly all wastes are potentially hazardous or toxic if present in sufficient quantity and establishing the level at which this occurs requires considerable research. It is not surprising, therefore, that the identification of minimum toxicity levels frequently lags behind the commercial and industrial use of a potentially harmful substance and that few internationally agreed toxicity levels exist; for example, in the Netherlands, a cyanide level of 50 mg/kg is considered toxic whereas in Belgium, a concentration of 250 mg/kg must be reached before toxicity level is reached. In practice, most countries draw up lists of specific substances only for which there is specific evidence linking them to adverse human health and environmental effects. It is probable that many potentially toxic wastes still await discovery.

As industrial processes use increasingly sophisticated materials then so the complexity of wastes increases. Safe dumping grounds for wastes becomes ever more vital in order that leakage of wastes is not to occur with the consequent contamination of water courses, groundwater and food chains. See CONTROLLED TIPPING.

**trace element,** *n.* any of various chemical elements that occur in very small amounts in organisms and are essential for many physiological and biochemical processes.

**trace fossil,** see FOSSIL.

**traction load,** see BEDLOAD.

**Trade Winds,** *n.* the winds which blow from the subtropical high pressure HORSE LATITUDES towards the equatorial zone of low pressure (see Fig. 70). The location of the Trade Wind belts alter according to the season: they blow from the north-east in the northern hemisphere and the south-east in the southern hemisphere, and are most persistant over the oceans. Variations in air pressure patterns may cause alterations in the direction of the Trades Winds over the continental landmasses.

**traffic capacity,** see TRAFFIC FLOW.

**traffic flow,** *n.* the flow of vehicles or people through a given TRANSPORT NETWORK. Traffic flows are measured in terms of the quantity of vehicles or pedestrians passing a given point on one or several links on a network within a given time; this provides a *traffic capacity* value for the traffic flow. The variety of vehicles using roads is sometimes generalized into 'passenger car units'. Measures of traffic flow entering or leaving a particular area, such as a city centre, are made by a 'cordon count' where a ring of survey points is established along a circle which intersects all routes giving access to the area. Where volume approaches the designed capacity of a section of the network, congestion occurs. See also TRAFFIC INTENSITY INDEX.

**traffic intensity index,** *n.* a measure of the average waiting time of a vehicle within a TRAFFIC FLOW.

**traffic lane,** *n.* a track marked on a highway to accommodate one vehicle width. MOTORWAYS in the UK for example, usually have two or three traffic lanes on each carriageway separated from each other by linear road markings. Lanes are used to impose discipline on a stream of vehicles and thus position groups of vehicles to their best advantage for making turns or leaving the highway safely. Lanes also economize in the use of road space, permitting maximum packing of vehicles into the given road width.

**traffic management,** *n.* the control and organization of TRAFFIC FLOWS along a transport network to ensure an efficient movement of vehicles. Traffic flows may be managed either through physical means (that is, by adaptations to the road structure) or by systemic means (that is, by control procedures). Physical adaptations include: blocked off streets; automatic traffic lights; grade separation of vehicles at road junctions; or spatial separation of different modes of travel which might otherwise conflict (for example, cars and pedestrians in a shopping area). Systemic means include central computer control of lights and direction signs over wide areas of a network; signalling systems on a rail network, or AIR TRAFFIC CONTROL SYSTEMS.

**transect sampling,** *n.* a sampling technique in which a base line is drawn across at least two observable areas of zoning and along which data relating to the changing conditions is recorded. Transects showing soil and vegetation variations can be used in areas displaying altitudinal variation, in coastal environments where the effects of sea salt rapidly diminish inland, or on industrial waste land where toxic substances in the soil or air show rapid change. Many different types of transect have been developed, for example, a *line transect* along which regularly spaced QUADRATS can be established, or a *belt transect* comprising two parallel transect lines between which data can be collected either continuously or at intervals. See also PLOTLESS SAMPLE.

**transhumance,** *n.* the seasonal or periodic migration of pastoral farmers and their livestock in search of grazing, usually between two areas having distinctly different climatic and ecological conditions. Both areas generally have permanent living accommodation, a characteristic that may be used to distinguish transhumance from the semi-nomadism of tropical pastoralists whose seasonal wanderings from a permanent base rarely involve a stay in any other location long enough to warrant more than temporary shelters.

Transhumance is characteristic of upland regions. In tropical zones, the movement, mainly of cattle, is generally downhill in the rainy season, except in areas prone to flooding. In temperate zones, the movement is uphill in the summer months, returning to the valleys for the winter. In Switzerland and Norway, for example, advantage is taken of the short GROWING SEASON on the high pastures or ALPS to graze cattle and sheep while allowing valley pastures to recover for winter grazing or to be cropped for winter fodder. Although no longer on the scale of previous centuries, transhumance is still practised around the Mediterranean and between the Danube Plain and the Carpathian mountains. See also NOMADISM.

**transition zone,** see BIOSPHERE RESERVE.

**translocation,** *n.* the movement of materials, often in solution, through a SOIL PROFILE. Clay particles, plant nutrients and dissolved iron compounds are common examples of substances that can be translocated between different SOIL HORIZONS. In climates in which PRECIPITATION exceeds the rate of EVAPOTRANSPIRATION, translocation will result in the washing down of materials and may result in the formation of an ironpan or claypan (see HARDPAN). When evapotranspiration exceeds precipiation the dominant movement through the horizons will be upwards and salts may accumulate at the surface in a salt pan. Translocation incorporates a variety of processes including LEACHING, ELUVIATION, ILLUVIATION, CALCIFICATION, SALINIZATION, ALKALIZATION.

**transpiration,** *n.* the loss of water vapour from the internal spaces of a plant through pores (*stomata*) located mainly on the undersurfaces of leaves. Transpiration lowers leaf temperatures and is a vital factor in the distribution of minerals throughout the plant. The transpiration rate is dependent upon external environmental factors such as air temperature, wind speed and humidity, and upon internal factors such as growth rate and the life history stage of the plant. If the transpiration rate exceeds the capacity of the plant root system to absorb water from the soil then the plant reaches WILTING POINT, a condition which quickly leads to the death of the plant. In agriculture, addition of water by IRRIGATION can offset the attainment of wilting point. Transpiration rate is greatest on warm, breezy summer afternoons when, for example, a large oak tree can transpire up to 450 litres of water and a cabbage, 1 litre. However, because of difficulties in separating transpiration rates from EVAPORATION rates the two values are often combined as the EVAPOTRANSPIRATION rate.

**transportation,** *n.* the movement of soil and rock fragments by the action of rivers, glaciers, sea and wind. See SLUMP, CREEP, SOLIFLUCTION, MUDFLOW, EARTHFLOW, FLUVIAL TRANSPORTATION, GLACIAL TRANSPORTATION.

**transport network** or **communication network,** *n.* an interconnected system of fixed lines of movement between ACTIVITY NODES. The main means of carrying traffic are roads, footpaths, railways, and inland WATERWAYS. Sea-travel routes and AIR CORRIDORS also form parts of transport networks but are neither fixed in position nor visible. Each section of fixed line forms a link in a national network connecting junction points at which travellers change mode of travel or terminate a journey.

**trawling,** *n.* a method of fishing in which a large bag-shaped net is dragged or trawled at deep levels behind a specially adapted boat. The older form of trawl net had a 10 m wooden beam across its mouth to hold it open. The modern net has a mouth of approximately 25 m and and has large flat plates (*otter boards*) attached to either side of it; as the net is dragged along, the water pressure forces the otter boards apart and so keeps the net open. Trawling is the single most important method of FISHING in terms of catch. Compare SEINE NETTING, LINE FISHING.

**tree crop,** see PERMANENT CROP.

**tree line** or **timber line,** *n.* the altitudinal limit of tree growth. Below the tree line, trees are erect but often as the tree line is approached the trees become stunted and distorted by the wind and may eventually assume a horizontal, prostrate, mat-like form known as *krummholz*. Above the treeline, exposure, frost damage and snow cover all prevent tree colonization.

# TRELLIS DRAINAGE

*Inverted tree lines* can sometimes occur in large inter-MONTANE valleys. Here, the accumulation of very cold air in the valley bottoms can prevent trees from growing and, in effect, a dual treeline is formed, a lower line marking the top edge of cold air accumulation and an upper line marking the point of excessive exposure.

**trellis drainage,** see DRAINAGE PATTERN.

**Triassic period,** *n.* the geological PERIOD that followed the PERMIAN PERIOD around 225 million years ago and that ended 195 million years ago when the JURASSIC PERIOD began. The Triassic was the first period of the MESOZOIC ERA. A continuous mountain building phase extended from Permian to Triassic times throughout much of Europe and North America. The fossil record is incomplete, with marine invertebrates being the most commonly found forms. Dinosaurs continued to evolve throughout the period. Economic resources found in Triassic rock include LIMESTONES and evaporites such as ROCK SALT, ANHYDRITE and GYPSUM. See EARTH HISTORY.

**trickle filter** or **percolating filter,** *n.* a bed of inert material, usually rock fragments, used to filter and purify polluted water. The water is trickled through the rock bed which is covered by aerobic BACTERIA which oxidize and digest the dissolved organic pollutants, so purifying the water. See SEWAGE, ACTIVATED SLUDGE PROCESS.

**trigger factor** or **critical factor,** *n.* any environmental factor which has increased or decreased to the point where it precipitates a hitherto latent response or process. The onset of trigger factors may be swift, as when a wind above a critical speed causes vegetation to be blown over. Alternatively, the trigger factor may accumulate slowly in the environment, causing change to occur but which goes unnoticed until a threshold is reached, at which damage symptoms become evident; for example, AIR POLLUTION levels may gradually increase resulting in internal damage to vegetation until a critical value is passed and symptoms of visible damage are seen.

**trophic level,** *n.* **1.** any of a series of distinct feeding (or nourishment) levels in a FOOD CHAIN. Most food chains comprise three separate trophic levels: primary AUTOTROPH producers, primary consumers (HERBIVORES), and secondary consumers (CARNIVORES). Until comparatively recently, the relationships between the trophic levels were poorly understood. In 1942, the American biomathematician, Lindemann, constructed a theoretical energy flow model in which the flow of energy between trophic levels was predicted (see Fig. 62). Few researchers dispute this general model, although much additional detail has since been added, notably the role of the DECOMPOSER organisms.

   2. the nutrient content of a body of water, in particular the quantity of nitrogen and phosphate present.

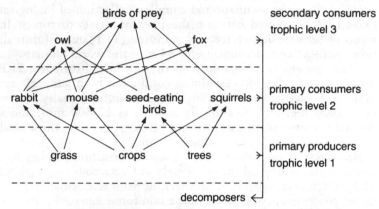

Fig. 62. **Trophic level.** The arrangement of species in a food chain into feeding levels. The arrows show main directions of movement of energy and materials

**tropical-alpine,** see PARAMO

**tropical cyclone,** see HURRICANE

**tropical desert vegetation,** *n.* a type of vegetation characteristic of the true hot DESERTS. Its distribution is confined to relatively small areas of the Atacama and parts of western Australia and to the Sahara and Arabian deserts where it is best developed. Two main groups of plants exist:

   (a) the ANNUALS, which rely on the chance occurence of rain to allow germination of dormant seeds and the completion of the life cycle often in the space of four or five weeks;

   (b) the PERENNIALS, which survive either by storing water in time of short-lasting plenty (*succulence*) or by extreme adaptation to limit water loss (*xeromorphism*).

   Much tropical desert vegetation relies for its survival on ground water which reaches the surface at oases or wadis. Many of the common oases species have very deep roots, for example, the mesquite bush (*Prosopsis juliflora*) whose roots can extend to 50 m. Oases have formed important supply points for desert tribes and permanent settlement has sometimes been possible based on the cultivation of the date palm, fig and wheat.

**tropical grassland,** see SAVANNA.

**tropical rain forest** or **selva,** *n.* the dense forest found in tropical areas of high rainfall. It is the most extensive natural vegetation type, extending over some 8% or 17 million km² of the planet and contains the most luxuriant and complex collection of biological COMMUNITIES located between the lines of the two tropics. In terms of BIOMASS output, tropical rainforest (TRF) outperforms all other FORMATIONS, accounting for 34% of the world biomass.

There are specific climatological criteria which distinguish TRF areas from all other BIOMES: almost continuous high temperatures (annual average above 21°C); high annual rainfall (usually greater than 2000 mm and occasionally as high as 12 500 mm) and a diurnal variation in climate which is often greater than the annual variation. See Fig. 63. for a climatic summary.

Rainforest is one of the most ancient formations, being first noted in the geological record as early as Cretaceous times (about 120 million years ago). It is, however, difficult to calculate the age of the contemporary members of the rain forest due to the absence of annual rings because of the constancy of the growing season.

The vertical structure of TRF shows a distinctive LAYERING (see Fig. 64 on p. 440). Stratification is believed to be a response to the intense competition for favourable light conditions. Up to three tree strata can exist above a variable shrub layer. Often the distinctive layers are blurred by the profusion of climbers and EPIPHYTES which interconnect the layers. The foliage of the upper stratum often displays xerophytic characteristics in response to the very high leaf temperatures (50°C) which may be recorded on the upper side of the canopy.

Undisturbed TRF characteristically has between 40-100 different tree species per hectare. There are considerable species differences between the different geographical areas of rain forest. About 70% of all species are trees and up to 90% of biomass comprises woody tissue. TRF has an average net PRIMARY PRODUCTION figure of 2000 g/m² per year and a maximum value of 5000 g/m² per year.

The rain forests also support a vast animal population. Probably as few as one in six of all species have yet been named.

About 12 million hectares of TRF are being lost each year (equivalent to the size of England) and if this rate of depletion continues, rain forest will have disappeared by 2050 AD. In

Amazonia alone, 2.5 million ha of rain forest is removed each year to make way for cattle ranching; much of this area is burned with no attempt to extract the valuable hardwood timber. Elsewhere, however, the forest is removed by commercial logging companies anxious to obtain the mahogany, teak, ebony and meranti for sale to northern industrial countries.

| TRF subtype | Rainfall | Mean temperature | Diurnal variation |
|---|---|---|---|
| Evergreen TRF | Minimum of 100 mm in all months. >2500 mm per annum. Short dry phase up to 14 days | 24–25°C (monthly) | 2–12° |
| Cloud forest | Great monthly variation. Very high annual fall, up to 4500 mm. | >21° (monthly) | 8–15° |
| Tropical summer rain zone | Pronounced seasonality. 400–700 mm per annum. | | |
| Deciduous TRF | Pronounced dry season of up to 3 months. 900–1400 mm per annum. | >21° (annual) | |

Fig. 63. **Tropical rain forest.** Summary table of tropical rain forest climatic subtypes.

Loss of the tropical rainforest represents an ecological disaster of major proportions for the forest serves to stabilize the ALBEDO, temperature, humidity and carbon dioxide levels of the planet. Furthermore, the rain forest species contain the genetic blueprints of our evolution, both past and future and provides a renewable resource base of immense financial significance. See also MEDICINE (FROM PLANTS).

**tropical vegetation,** *n.* vegetation located between the lines of the two tropics, that is between latitudes 23.5°N and S. The term has a broad non-specialist currency which takes no account of the fact that tropical-type vegetation can occur beyond the tropics while large areas within these boundaries have non-tropical-type vegeta-

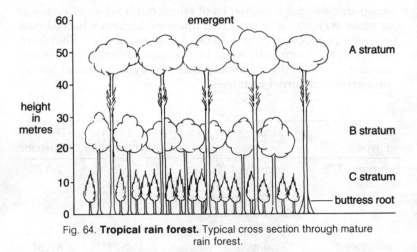

Fig. 64. **Tropical rain forest.** Typical cross section through mature rain forest.

tion. See TROPICAL RAIN FOREST, TROPICAL DESERT VEGETATION, SAVANNA.

**tropopause,** *n.* the transitional zone between the TROPOSPHERE and the STRATOSPHERE. It varies in altitude from approximately 18 km above the equator to 6 km above the Poles, and is characterized by a temporary stabilization of the LAPSE RATE (see Fig. 9). Compare STRATOPAUSE, MESOPAUSE.

**tropophyte,** *n.* any plant (usually PERENNIAL) which displays a distinct variation in its annual life cycle based upon the availabilty of water. Two phases can usually be distinguished: an active growth phase when water is available and a dormant phase in time of water shortage. Mid-latitude DECIDUOUS FOREST species are tropophytes, displaying an active summer growth phase which contrasts with a period of winter dormancy brought about by a fall in temperature which makes the soil water temporarily unavailable.

**troposphere,** *n.* the lowest layer of the ATMOSPHERE, extending between a height of 9-16 km above the earth's surface. The MESOSPHERE continues beyond the TROPOPAUSE, the upper boundary to this layer. Almost all weather phenomena occur in the troposphere, together with most atmospheric water vapour, dust and contaminants.

440

**trough of low pressure,** *n.* the narrow elongated extension of a DEPRESSION, which is contained between two areas of higher air pressure (see Fig. 65).

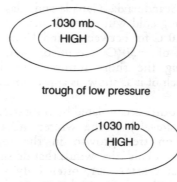

trough of low pressure

Fig. 65. **Trough of low pressure.**

**truck farming,** see MARKET GARDENING.

**true desert,** see DESERT.

**truncated spur,** *n.* a spur whose lower end has projected into a preglacial river valley and has been partially removed by the downvalley progress of a GLACIER. Truncated spurs are commonly associated with U-SHAPED VALLEYS (see Fig. 66). See also INTER-LOCKING SPUR

**tsunami,** *n.* a rapidly moving and often large seawave generated by submarine earthquakes, landslides or volcanic activity. Tsunamis may be up to 200 km in length and in deep water can travel up to 90 km per hour. On the oceans, a tsunami may result in an almost imperceptible swell of 1.5m but as the wave approaches a coastline its height may increase to up to 30 m. Large tsunamis can cause serious flooding in low-lying coastal areas and may result in considerable loss of life and damage to property. In 1755, 60 000 lives were lost when a tsunami devastated the city of Lisbon in Portugal. More recently, an earthquake in Chile was responsible for a tsunami that caused the deaths of 1000 people across the Pacific. To lessen the effects of tsunamis, the *Seismic Sea Wave Warning System* has been established in the Pacific. The system relies on seismographic stations around the Pacific monitoring the location and magnitude of earthquakes, establishing the direction of any resulting tsunami, estimating the travelling time of the wave to any vulnerable coastline and alerting the relevant authorities. Tsunamis are often mistakenly referred to as *tidal waves*.

**tundra,** *n.* a treeless region confined mostly to latitudes beyond 60°N of the equator and south of the polar ice, that is, the northern regions of Siberia, Scandinavia, Canada and Alaska. The tundra is characterized by long cold winters with mean monthly temperatures falling to $-10°C$ for between 6 and 10 months of the year, with temperatures of $-30°C$ not uncommon. Mean monthly temperatures during the short summer remain below 10°C. PRECIPITATION, much of it as snow, is approximately 250-300 mm per annum.

The intense DROUGHT accentuated by the extreme low temperatures (*physiological drought*) of the winter months is the major LIMITING FACTOR on tree growth in the region. The near continuous blowing of cold, dry winds that dessicate plant tissues, and the presence of PERMAFROST, often only a few centimetres below the surface also restrict plant growth. Vegetation is characterized by mosses, lichens, sedges and rushes, interspersed with grasses and flowering herbs. Towards the southern margin of the tundra low woody plants and DECIDUOUS dwarf shrubs, such as alder, birch and willow can be found.

With the summer thaw of the top soil the badly drained terrain becomes highly water-logged and unstable. The long daylight hours in summer provide a brief period during which PRIMARY PRODUCTIVITY is good, and is sufficient to support large populations of emerging insects, migrant birds and herds of herbivores, such as caribou, moose, elk, and reindeer.

Tundra BIOMES are extremely fragile. In some areas the delicate ecological balance has been disrupted by the overhunting of large herbivores which has resulted in a reduction in natural carnivorous predators and the excessive growth of ungrazed vegetation; however, subsequent conservation efforts have in turn led to a rapid increase in herbivores and severe problems of overgrazing. Furthermore, the threat to the environment of one of the world's last remaining WILDERNESS AREAS from activities such as oil exploration is considerable.

**turbidity current,** *n.* the movement of water containing SEDIMENT in turbulent suspension down the CONTINENTAL SLOPE, along a SUBMARINE CANYON or the sloping bottom of a LAKE. Turbidity currents form after EARTHQUAKES or when a river in flood empties its unusually heavy BEDLOAD into a lake or the sea. The flow of these currents can break submarine cables and destroy jetties.

**turnpike,** *n.* a road in the USA, the use of which necessitates payment of a TOLL which varies according to the distance travelled, the load or number of passengers per vehicle or the weight or engine capacity of the vehicle. See also MOTORWAY.

**TVA,** see TENNESSEE VALLEY AUTHORITY.

**twister,** see TORNADO.

**2,4-D,** *n. Trademark.* 2,4-dichloro phenoxyacetic acid, a hormonal HERBICIDE comprising synthetic organic compounds with properties akin to natural plant growth regulators. Developed in America in 1942, its use as a herbicide was not recognized for some years.

Application of 2,4-D can cause rapid tumour-like overgrowth of annual, broadleaved weeds of cultivated land. This rapid growth results in grotesque twisting and contortion of stems and leaves and to the production of masses of minute rootlets within a week of application. This is followed by a collapse of the plant's natural growth regulating system. 2,4-D is absorbed by the roots and shoots and moves rapidly through the plant. Despite much research, there is still no complete understanding of how 2,4-D works at the biochemical level. Upwards of 250 varieties of 2,4-D exist, some being of 'universal' application while others are species-specific.

2,4-D is cheap to produce and easy to apply; it has a wide safety margin, is BIODEGRADABLE and is virtually non-poisonous to humans and animal. Its active life is usually no more than 28 days.

**2,4,5-T,** *n. Trademark.* trichlorophenoxy acetic acid, a HERBICIDE developed in 1942 and used for the control of woody plants and brush wood and for selective weed control in conifer plantations. Easily leached from the surface layers of the soil it is useful for controlling deep-rooted weeds and shrubs. This herbicide appears to perform similar functions to the natural plant growth hormone, indole acetic acid. It has no known effects on animals and does not persist in soils, disappearing after a few weeks. However, since 1969 the US government has restricted the use of 2,4,5-T on food crops after evidence linking it to fetus deformities in mice and rats.

**typhoon,** *n.* a HURRICANE occurring along the western margins of the Pacific Ocean.

# U

**ultrabasic rock,** *n.* any IGNEOUS ROCK usually of plutonic or hypabyssal origin that contains very low levels of silica compared to iron and magnesium minerals. See also BASIC ROCK. Compare ACID ROCK.

**UNCLOS** III, see LAW OF THE SEA.

**underground,** *n.* a railway network housed in tunnels below ground level, accessed through stations at ground level that are connected to the train level by lifts, stairs or escalators. The tracks sometimes emerge at ground level at the outer edges of the network, where, at the time of building, the land was free of buildings and conventional surface tracks could be laid. Resort to the expensive construction of underground railways has been necessary in order to reduce the heavy traffic congestion which now exists on the streets of most large cities. Major cities like London, New York (*subway*), Paris (*metro*) and Berlin (*U-bahn*) have long-established underground rail systems. Others, such as Prague and San Francisco, have constructed underground systems in recent years.

**underground drainage,** *n.* an underground river system usually found beneath KARST or chalk geology. Surface drainage is often confined to ephemeral streams which quickly disappear via SWALLOW HOLES.

**underpass** *n.* a pedestrian or traffic route passing beneath another routeway which passes over it at ground level. Underpasses are sometimes constructed where a new road has divided a farm, to permit the safe passage of animals to and from the farm buildings. They are also widely used in NEW TOWNS to segregate pedestrian movement from vehicular movement.

**underpopulation,** *n.* a shortage of people in an area such that the available resources cannot be exploited to the full or to the best advantage, resulting in lower per capita real income than might be achieved under OPTIMUM POPULATION conditions. Several THIRD WORLD countries claim to be underpopulated in that their small domestic market and even smaller aggregate purchasing power inhibits industrialization. Compare OVERPOPULATION.

**underfit river,** see MISFIT RIVER.

**undiscovered resources,** *n.* presently unidentified mineral deposits of unknown quality and quantity but which are thought to exist given current geological data.

**UNESCO,** *n. acronym for* United Nations Educational, Scientific and Cultural Organisation, founded in 1945 to reduce international tension by encouraging the interchange of scientific and cultural ideas, and by improving education. In the late 1960s, UNESCO launched its MAN AND THE BIOSPHERE PROGRAMME which differed from most other multinational conservation projects in that it concentrated on long-term scientific research and the monitoring of the environment to detect small-scale, cumulative changes.

**uniformitarianism,** see ZONAL SOIL.

**unitary development plan,** see DEVELOPMENT PLAN.

**unleaded petrol,** *n.* petrol (or gasoline) containing a reduced amount of the ANTIKNOCK ADDITITIVE tetraethyl lead. See also AUTOMOBILE EMISSIONS.

**unloading,** see PHYSICAL WEATHERING.

**upland farming,** *n.* **1.** a form of DRY FARMING carried out on unterraced hillslopes in the tropics, chiefly for the cultivation of rice.

**2.** a type of farming on higher ground in temperate regions where ANIMAL HUSBANDRY predominates. Upland farms, in contrast to lowland farms, have poorer PASTURE production, less frequent reseeding, less fertilizer application, and poorer drainage. However, the general environment is less severe than in areas where HILL FARMING predominates.

**upper atmosphere,** *n.* the section of the ATMOSPHERE which lies above the STRATOPAUSE and incorporates the MESOSPHERE, THERMOSPHERE and EXOSPHERE (see Fig. 9). Compare LOWER ATMOSPHERE.

**urban aid,** *n.* any policy measure designed to tackle the decline of an industrial city and the problems of DEPRIVATION and dereliction often found at the heart of major conurbations in DEVELOPED COUNTRIES. Measures that assist in arresting the erosion of the urban economic base include regional industrial policies that impinge on urban areas without being specifically urban initiatives, and more focused programmes directed at the INNER CITY. They may be concerned with industrial development or with schemes designed to revive inner areas through the provision of funds to community groups and towards the rehabilitation of housing stock and blighted environments.

Urban policy in the USA since the 1960s has ranged from urban renewal and the model cities programme of 1966 aimed at coordinating and concentrating the provision of federal resources and involving local communities in improving their physical environments and living standards, to the more direct cooperation between the federal government and private investors that emerged in the 1970s. After 1977 local authorities in areas suffering a high level of deprivation could obtain Urban Development Action Grants to encourage joint schemes between the public and private sectors aimed at revitalizing local neighbourhoods economically, socially and environmentally.

In the UK, the Urban Programme, introduced in 1968 in response to growing concern over race relations and urban deprivation, provided central government grants for approved projects in areas of special social need, that is, those areas with a high proportion of poor housing, the unemployed, ethnic minorities, large families, and children in care. Projects aimed at improving housing and employment opportunities, however, did not qualify for assistance. Overall, the impact of such aid on inner cities was negligible. It was not until the mid-1970s that the extent of the economic decline of the inner city was fully recognized. The amount of urban aid was increased with the 1978 Inner Urban Areas Act which defined a limited number of local authorities, designated *partnership authorities*, as having severe urban problems. There has been a shift in resources towards job creation, training, small-scale industry and economic regeneration. Recent initiatives include the setting up of Urban Development Corporations in Merseyside and London's dockland, the provision of Urban Development Grants to encourage private sector investment, the designation of ENTERPRISE ZONES, and the creation of ENTERPRISE TRUSTS.

**urban climate,** *n.* the MICROCLIMATE developed over urbanized areas and often characterized by a HEAT ISLAND effect, AIR POLLUTION and alterations in local wind, humidity, and precipitation patterns due to the topography of the urban landscape.

**urban ecosystem,** *n.* a concept which proposes that a CITY is analogous to a living organism. The urban environment provides the means of life support for all the people living therein. This comprises housing, employment, commercial opportunity, recreational and leisure facilities, health care and transport. The analogy, first developed by the Chicago school of social ecology, also proposes that the life of a city shows parallel development to an organism: a city grows by importing materials and energy, it produces wastes which must be disposed of, and if poorly governed, can become unhealthy and ultimately die.

**Use Classes Order,** *n.* an order in UK planning law setting out the classes of land use within which change can take place without constituting DEVELOPMENT but which may still, in some cases, require planning permission. The local authority is empowered to impose conditions on proposed changes within categories if it so desires. Governments can loosen planning control through alteration to the Use Classes Order broadening its categorization so that each class encompasses a wider variety of potential users. Compare GENERAL DEVELOPMENT ORDER. See DEVELOPMENT CONTROL.

**use right,** see LAND TENURE.

**U-shaped valley,** *n.* a valley with steep sides and a flat floor formed by a VALLEY GLACIER widening, deepening and straightening a former river valley (see Fig. 66). U-shaped valleys are found in glaciated regions throughout the world including the FIORDS of

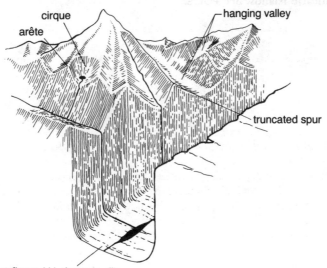

Fig. 66. **U-shaped valley.**

Norway, the glens of Scotland and the valleys of South Island, New Zealand.

The broad flat floors of U-shaped valleys allows easy access into mountainous areas although in some instances the valley gradient is so flat as to cause poor drainage conditions. Besides lines of communication, valley floors are usually utilized for settlement and agriculture. The development of commercial forestry may allow the productive use of steep valley sides. Fast flowing streams and waterfalls on the valley sides are often used in the generation of hydro-electric power. The spectacular scenery of U-shaped valleys and other associated glacial features has lead to the development of tourist resorts in many glaciated areas.

**usufruct,** see LAND TENURE.

**utilized agricultural area,** see FARM SIZE.

**uvala,** *n.* a steep-sided depression in KARST areas formed by the coalescence of several DOLINES or the collapse of underground caverns. Uvalas may be more than a kilometre in diameter. The bottom and sides of uvalas may be covered with varying thicknesses of clays and silts that can support productive agriculture in an otherwise totally barren limestone area; a good example is to be found near the monastery at Lluch on the island of Majorca in the Ballearic Islands. See POLJE.

# V

**vadose zone,** see WATER TABLE.

**valley glacier,** see GLACIER.

**valley train,** see OUTWASH PLAIN.

**valley wind,** see ANABATIC WIND.

**vandalism,** *n.* the deliberate disfiguring or destruction of parts of the visible environment. Common forms of vandalism include rubbish dumping, GRAFFITI, cutting down trees or bushes, setting fire to buildings and putting telephones out of action. The causes of vandalism are not fully understood but are often attributed to the social problems associated with DEPRIVATION in the INNER CITIES; most perpetrators of acts of vandalism are young people residing in cities. Repair of vandalized property can amount to as much as 2% of the gross annual expenditure of a city or regional budget.

**Van't Hoff's 'law',** *n.* a broad rule extrapolated from a chemical equation (*Van't Hoff's isochore*) which states that the rate of increase in a chemical reaction generally doubles for every 10°C rise in ambient temperature. Application of this rule explains why the role of CHEMICAL WEATHERING of rocks and soils and the rapidity of decomposition of organic matter in equatorial regions is so much greater than that in mid- and high-latitudes. Named after the Dutch chemist Jacobus Hendricus Van't Hoff (1852-1911).

**vascular tissue,** see CONDUCTING TISSUE.

**vector,** *n.* **1.** any organism that conveys a disease-producing organism from one host to another, either within or on the surface of its body, such as the malaria-carrying mosquito.

**2.** any physical quantity which possesses not only a magnitude but also a direction, for example, an ocean current has both a speed of movement and a prevailing direction of travel. Techniques for displaying information on physical vector data are called *vector analysis*.

**vegetation,** *n.* the mass of plant forms in an area. In practice, vegetation can be differentiated into variably sized units or plant COMMUNITIES.

**vegetation classification,** *n.* a complex group of techniques such as nearest neighbour analysis that allows the detection of pattern amongst the VEGETATION of an area. Vegetation classification, as distinct from LINNEAN CLASSIFICATION, is frought with ambiguity and misinterpretation, brought about in part by linguistic misunderstanding between researchers of different nationalities and also by different interpretations placed upon the vegetation units of ASSOCIATION and FORMATION. Some plant ecologists claim that pattern does not exist in natural vegetation and that variation occurs either randomly or continuously along a gradient extending between two extreme HABITATS, (the concept of the *continuum*).

**vein** or **lode,** *n.* a fissure within COUNTRY ROCK into which minerals have been deposited by solution. Veins are usually associated with igneous activity but may occasionally have sedimentary origins. The ores of many metals commonly occur in veins, such as gold, silver and tin.

**veld,** *n.* a vegetation type dominated by grasses which occupies extensive areas of the high plateau of South Africa. The elevation of the region (1500-2000 m above sea level) and the distance from sea (1000 km) result in a climate characterized by summer rainfall and winter drought with severe night frosts. Burning of the vegetation has also been a persistant feature of the region as has grazing by wild and domesticated herbivores.

Veld vegetation comprises grass species mainly of Themada varieties. In the most degraded areas, brought about by overgrazing, *sourveld* occurs in which Aristida and Eragrostis varieties dominate. Veld has a low PRIMARY PRODUCTIVITY and owes much of its existence to the malpractices of farmers of European origin. See KAROO. Compare PRAIRIE, PAMPAS, STEPPE.

**vermin,** *n.* any animal or insect species that is troublesome to humans, domesticated animals or agricultural crops. Typical vermin include rodents and fleas.

**vertebrate,** *n.* any member of the subphylum Vertebrata which possesses an internal bony or cartilaginous skeleton, particularly a back bone. All animals including fish, amphibians, reptiles, birds and mammals (including humans) belong to this group. Compare INVERTEBRATE.

**vertical termperature gradient,** see LAPSE RATE.

**virgin land,** *n.* any land that has never been cultivated. The opening up of virgin lands is associated historically with agricultural LAND COLONIZATION and voluntary migration. More recently, governments and other agencies in many parts of the world have encouraged the development of virgin land in order to increase agricultural output or to meet the demand for land from groups in over-populated regions elsewhere in the country.

The most striking example of an attempt to increase the area of land under cultivation has been the Virgin Lands Campaign of 1954-56 in the USSR in which more than 40 million ha of MARGINAL LAND in Kazakhstan and western Siberia were ploughed up, primarily for cereal production. Despite the provision of IRRIGATION involving major engineering schemes, drought and soil deficiencies have caused the subsequent abandonment of some areas in favour of improving yields in less marginal climatic zones.

On an even greater scale globally has been the loss of TROPICAL RAIN FOREST to agricultural activities through spontaneous and government-directed settlement. In Brazil, for example, poor and landless families from the semi-arid and over-crowded north-east of the country have been offered free plots of virgin forest in the Amazonian states of Rondonia and Acre, causing unprecedented environmental destruction in the search for short-term gains from unsustainable agricultural practices.

**visitor pressure,** *n.* the impact of large numbers of visitors on conserved areas such as NATIONAL PARKS and AREAS OF OUTSTANDING NATURAL BEAUTY. Visitor pressure results in litter, footpath erosion, increase in noise levels and the disappearance of plants and animals from their traditional habitats. Such adverse effects can be mitigated by careful management policies involving the creation of visitor centres, interpretation centres, craft centres, way-marked trails and walks, all of which can channel visitors away from areas most vulnerable to damage from over-use.

**viticulture,** *n.* a form of HORTICULTURE concerned with the cultivation of the grapevine. Viticulture had its origins in western Asia, was known to the ancient Egyptians 6000 years ago, and spread out across the Mediterranean Basin as Greek settlements spread westwards. The vineyards of Europe outside the Mediterranean region are the oldest areas of specialized horticulture, the Romans having been influential in the northward spread of viticulture. Europe, together with Turkey accounted for 61% of the world output of 64.4 million tonnes of grapes in 1984. The spread of European settlement has seen the growth of viticulture in California, Argentina, Chile, South Africa, Australia and New Zealand.

**volatile,** *n.* a gaseous element or compound dissolved in MAGMA as a result of the high pressures within the earth's crust; examples include water, carbon dioxide and chlorine. Volatiles return to the gaseous state during volcanic eruptions. Some of these volcanic gases are highly toxic and it is thought many of the deaths from the Mount Pelee disaster in 1902 resulted from asphyxiation. In 1985, over 1700 lives were lost at Lake Nyos in Cameroon due to the emission of poisonous volcanic gases.

**volcanic neck** or **volcanic plug,** *n.* a mass of solidified LAVA in the vent of a dormant or extinct VOLCANO. The erosion of the surrounding volcanic cone results in the resistant neck forming an upstanding steep-sided landform. Historically, fortifications have been constructed on volcanic necks, such as Dumbarton Rock and Edinburgh Rock in Scotland.

**volcanic rock,** see EXTRUSIVE ROCK.

**volcano,** *n.* a VENT or FISSURE in the earth's surface from which LAVA and volatiles are extruded. Fissure eruptions result in the formation of *lava platforms,* such as the Deccan Plateau in India, whilst vent eruptions form a conical hill or mountain. Volcanoes can be classified according to topographical form:

(a) *shield volcanoes* are dome-shaped with angles of slope ranging from 2° to 10° and are formed entirely by fluid lava flows. Some shield volcanoes can be very large such as Mauna Lua in Hawaii, USA which is around 96 km long, 48 km wide and rises 9000 m from the ocean floor.

(b) *cinder cones* consist only of PYROCLASTIC MATERIAL with angles of slope around 35°. Cinder cones rarely exceed 500 m in height.

(c) *composite cones* or *stratovolcanoes* are steep-sided and result from the alternating eruption of lava and pyroclastic material, as with, for example, Mount Saint Helens in the USA and Mount Etna in Italy.

The nature of a volcanic eruption is dependant upon the composition and gas content of the magma and may range from the gentle eruption of rapidly moving, fluid lava with little explosive activity to the highly explosive eruption of viscous lava and the formation of NUÉE ARDENTES. It is thought that volcanic eruptions are fed by localized reservoirs of MAGMA within the earth's crust. As the reservoir discharges, eruptions subside and other volcanic activity begins, such as HOT SPRINGS and FUMAROLES. When not active, the pipe leading to the vent of a volcano is usually plugged with solidified lava forming a VOLCANIC NECK.

The distribution of volcanic activity is associated with OROGENE-SIS and LITHOSPHERIC PLATE boundaries. The major volcanic areas are the Caribbean, the Eastern Mediterranean, the African Rift Valley, Iceland and the circum-Pacific zone, also known as the

Ring of Fire due to the widespread volcanic activity which almost encompasses the Pacific.

Volcanoes which have not erupted in historical times are considered to be *extinct*. This can sometimes be a dangerous assumption as magma reservoirs can sometimes take thousands of years to recharge between eruptions, with the result that apparently extinct volcanoes can erupt without warning, such as the Heimaey eruption in Iceland in 1973. A volcano which has been known to erupt but has no obvious external activity is said to be *dormant*.

**V-shaped valley,** *n.* a river valley in which FLUVIAL EROSION has resulted in a characteristic V-shaped cross section. The angle of the V is dependent upon the relationship between the river's vertical erosion and the WEATHERING and EROSION of adjacent hillslopes. V-shaped valleys tend to occur in the upper reaches of a river since lateral erosion frequently opens out the valley cross section in its lower reaches. Compare U-SHAPED VALLEY.

**vulnerable organism,** see RED DATA BOOK.

# W

**wadi,** *n.* a steep-sided dry valley with only intermittent stream flow, found in semi-arid and desert areas. Wadis were probably formed by FLUVIAL EROSION under previously more humid climatic conditions. Under present climatic conditions wadis convey water only during the occasional but heavy rainfall which causes flash floods. On these occasions wadis carry large quantities of SEDIMENT onto adjacent plains to form ALLUVIAL FANS. The threat of flash floods prevents the use of wadis for settlement or routeways.

**Wallace line,** *n.* a zoogeographical division separating the distinct placental mammal fauna of the Oriental faunal region from the more ancient marsupial mammals of the Australian zoofaunal region. The line also has a significance for floral regions(see Fig. 67). Named after the English naturalist Alfred Russel Wallace (1823-1913). See also ANIMAL REALM.

**warm front,** *n.* the boundary between an advancing warm air mass which is overriding a cold air mass (see Fig. 68 on p. 456). This usually occurs at the front of the WARM SECTOR in a DEPRESSION. The gradient of a warm front is less steep than that of a COLD FRONT. The gradual upward movement of moist warm air along the warm front results in the condensation of water vapour and the formation of CLOUD. Continuous rain for up to several hours precedes the passage of a warm front and is usually accompanied by a rise in temperature, a drop in air pressure and a change in wind direction. See FRONT. See also OCCLUDED FRONT.

**warm occlusion,** see OCCLUDED FRONT.

**warm sector,** *n.* the wedge-shaped region of warm air in a DEPRESSION which lies between the WARM FRONT and COLD FRONT. In the northern hemisphere warm sectors taper to the north while in the southern hemisphere they taper to the south. Mild, equable and sunny weather is usually associated with the warm sector. Over time, the warm sector reduces in size as the more rapidly moving cold front overtakes the warm front to form an OCCLUDED FRONT.

**waste factor,** see WASTE TIP.

**waste load,** see WASTE TIP.

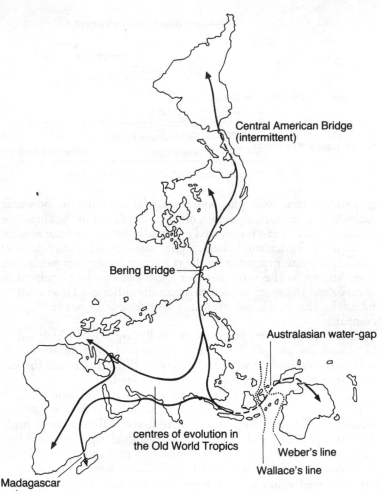

Fig. 67. **Wallace line.** The centres of evolution and dispersal routes for land invertebrates since the Mesozoic (arrows main dispersal routes).

**waste tip,** *n.* a dumping place for unwanted rubbish. The dumped material may originate from a variety of sources. It may be material removed as OVERBURDEN in open cast quarrying operations, or rock surrounding mineral seams underground which has to be removed

455

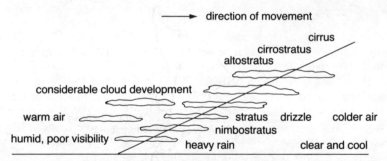

direction of movement

cirrus
cirrostratus
altostratus

considerable cloud development

warm air · stratus · drizzle · colder air

nimbostratus

humid, poor visibility · heavy rain · clear and cool

Fig. 68. **Warm front.** Cross-section through warm front.

to permit extraction. The waste material may also be domestic rubbish; in such cases, tipping is often into hollows in the landscape so that the waste can be covered eventually by soil and the area reinstated as agricultural land. The amount of domestic and municipal waste generated by the consumer society per person per year, known as the *waste factor* or *waste load* can be calculated in advance and the space needed to dump the anticipated load must be found by the municipal cleansing department or waste disposal company.

Chemical industries may produce polluting waste which has to be stored in one place for safety reasons. Chemical waste can pollute the ground for years and it may be difficult to re-use tip sites even after the visible rubbish has been cleared.

The impact on the local landscape varies with the form of tip. If it is built of coarse-grained material, such as colliery waste, the material will have a high angle of rest, and will produce a high, conical tip. Chemical reactions within coal tips may produce noxious gases and spontaneous combustion may cause smouldering fires within the dumped material which burn under cover for years.

**water-cooled reactor,** *n.* a type of NUCLEAR REACTOR in which water is used as a core coolant and also, in most cases, as a MODERATOR. Water-cooled reactors were pioneered in North America, first in military and later in commercial programmes. Water-cooled reactors are of two types:

(a) *pressurized water reactor (PWR)*. Heat generated in the nuclear core is removed by water which circulates at high pressure through a primary circuit. Heat is transferred from the primary to a secondary circuit via a heat exchanger, thereby generating steam in the secondary circuit which can be used to power turbines. PWRs are relatively inefficient due to the technical difficulties in working at pressures greater than $1.5 \times 10^7$ Pa (2000 psi). About 60% of the heat energy generated escapes as the coolant is 'dumped' usually into the sea, lake or river alongside which PWRs are inevitably built (see THERMAL POLLUTION).

(b) *boiling water reactor (BWR)*. These differ from PWRs in that the pressurized water acts not only as a coolant and moderator but also is allowed to boil within the reactor itself. The BWR works at $7.5 \times 10^6$ Pa (1000 psi) and the steam, having passed through a turbine, is condensed and the water re-used, thus avoiding some of the problems of 'dumping' large volumes of very hot water into the environment. See ADVANCED GAS-COOLED REACTOR, MAGNOX REACTOR, NUCLEAR REACTOR.

**water devil,** see WHIRLWIND.

**waterfall,** *n.* the sudden vertical descent of a river. Waterfalls may develop at the junction of a HANGING VALLEY or at the edges of plateaus, or on resistant outcrops of rock, fault escarpments and sea cliffs. The presence of waterfalls usually impedes river navigation unless a series of LOCKS are constructed to bypass the feature.

**water meadow,** see MEADOW.

**waterparting,** see WATERSHED.

**watershed, water parting** or **divide,** *n.* **1.** an imaginary line that separates the source streams (*headwaters*) that flow into different river systems. A watershed often follows an irregular course and may be sharply defined by the crest of a ridge. Conversely, it may be indeterminate in an area of undulating relief. Through HEADWARD EROSION and RIVER CAPTURE it is possible for rivers to encroach on other CATCHMENT AREAS, and any watershed should therefore be regarded as only temporary. In the USA, the term *divide* is synonomous with a watershed. A *continental divide* is a line which separates rivers that drain towards opposite sides of a continent.

**2.** a line used to demarcate a single river's drainage basin.

**watershed management** or **drainage basin management,** *n.* the coordinated planning of development within a drainage basin or CATCHMENT AREA. Watershed management may involve the allocation of water resources for a single use, such as IRRIGATION, or the design of an integrated basin development scheme that incorporates the use of water for industry, agriculture, domestic purposes and river navigation. Watershed management takes into account the prevailing geology, topography and hydrology of drainage basins. Examples of large-scale watershed management schemes include the TENNESSEE VALLEY AUTHORITY scheme in the USA and the Indus/Tarbela irrigation project in Pakistan.

**water table** or **phreatic zone,** *n.* the upper surface of the zone of permanent GROUNDWATER saturation. The depth of the water table varies according to the season and climatic factors, the overlying topography and the nature of the bedrock. Above the water table, in the *zone of aeration* where pore spaces are partly filled with air, water drains down under gravity; below the water table in the *zone of saturation* where pores are completely filled with water, water moves up under pressure. Between these, there exists a zone of intermittent saturation, the *vadose zone* (see Fig. 69).

The water table is largely recharged by the downward PERCOLATION of water through the overlying vadose zone. If percolation is impeded by a localized layer of impermeable rock then a subsidiary area of saturation, known as a *perched water table*, is established. The water table tends to follow overlying relief and when it intersects with the ground surface springs, marshes and lakes may develop.

The use of wells for groundwater abstraction may result in a localized lowering of the water table if the rate of extraction exceeds percolation. Major engineering works, such as dams and canals, can cause significant changes in the level of the water table. When a dam or canal is constructed, water percolates into the surrounding rock causing the water table within the vicinity to rise. Adjacent low-lying land may then be permanently saturated with damaging effects on settlements and agriculture. In arid areas, such a rise in the water table may be beneficial, allowing farmers to irrigate formerly dry land. See also AQUIFER, ARTESIAN WELL.

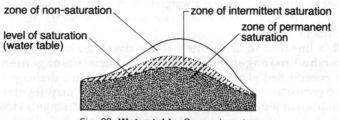

Fig. 69. **Water table.** See main entry.

**water vapour,** *n.* a colourless, ordourless gaseous form of water which mixes perfectly with the other gases of the air. Water vapour enters the atmosphere following the evaporation of fresh and salt water, condensing into clouds and fog before being released from the atmosphere as PRECIPITATION, (see HYDROLOGI- CAL CYCLE). Water vapour effectively absorbs the sun's radiant energy and as such the presence of water vapour in the TROPOS- PHERE stabilizes the temperature of the earth. HUMIDITY is the concentration of water vapour in the air and is of prime importance as an environmental factor. The water vapour content of the atmosphere varies greatly from as low as 0.02% in hot deserts to 1.8% in humid equatorial regions.

**waterway,** *n.* a natural or artifical network of water channels used for travel or transport. These are mostly river or lake systems but include the canals which were built to interconnect towns for freight transport purposes in the 19th century. The Great Lakes and the St Lawrence Seaway of North America represents one of the most important waterways in the world.

Since the 1960s inland water-borne recreation has developed significantly, and stretches of abandoned canal have been rehabili- tated to create long distance routes for recreational cruising. Disused canals have become important havens for many plant and animal species which have been ousted from their traditional habitats by intensification of agriculture and the expansion of urban land use. See also TRANSPORT NETWORK.

**wave,** *n.* an undulating ridge of water moving across the surface of a body of water, such as an ocean, sea or lake. Generally caused by wind action, the size of a wave is determined by its FETCH and by the wind's speed and duration. As waves enter shallow water, the lower section of the wave is retarded by friction from the ocean floor whilst the upper portion of the wave continues forward causing the wave to break. After breaking, the water which moves up a BEACH is known as the SWASH while the return flow which occurs under gravity is called the BACKWASH.

*Constructive waves* cause the deposition of material carried in the sea and result when the swash is stronger than the backwash. Erosion through *destructive waves* is caused by a stronger backwash. See MARINE EROSION, MARINE DEPOSITION.

**wave power,** *n.* the generation of electrical power from the stored kinetic energy contained within ocean waves. Energy output varies greatly as it fluctuates with the state of the sea. Most schemes are still at the experimental stage because sites with sufficient wave height are limited, construction and operation costs are high, and equipment is easily damaged by storms and saltwater corrosion. Research on ocean wave power in the UK was abandoned in 1986 as results up to that time indicated that it would make little contribution to world electricity production. See ALTERNATIVE ENERGY.

**WCS,** see WORLD CONSERVATION STRATEGY.

**weather,** *n.* the state of the atmosphere resulting from the combined effect of clouds, PRECIPITATION, temperature, HUMIDITY, visibility, wind velocity, and barometric pressure. Weather is treated on two bases: *geographic* (global, regional, or local) and *chronological* (long- or short-term).

Much of the equipment of the meteorologist consists of the products of 17th- and 18th-century science. The 20th century has contributed radar, dating techniques, very sensitive photography, and computer modelling. It is now possible to track storms by radar and to use satellites to photograph developing weather patterns. See WEATHER FORECASTING, WEATHER STATION. See also CLIMATE.

**weather chart, weather map** or **synoptic chart,** *n.* a map which indicates a variety of meteorological information for a given area at a particular time. Weather charts usually contain details on air temperature, wind speed and direction, pressure and features such as FRONTS, CYCLONES and DEPRESSIONS. The information on weather charts is gathered by WEATHER STATIONS, satellites, meteorological rockets and ballons, and radar.

**weather forecasting,** *n.* the prediction of future weather conditions, usually through the application of sophisticated computer models to present weather conditions. Short-range forecasts are the most accurate and cover the following 24 hours with an indication of the outlook for up to 72 hours. Medium-range forecasts cover up to four days ahead whilst long-range forecasts are usually made for a month. General short-range forecasts are issued regularly each day in most countries on radio and television and in newspapers. Specialist forecasts are also prepared for shipping, airlines, motorists, farmers and for sports such as skiing and climbing.

**weathering,** *n.* the disintegration and decomposition of rock and REGOLITH at or near the earth's surface. Three types of weathering are recognized: PHYSICAL WEATHERING, CHEMICAL WEATHERING and BIOLOGICAL WEATHERING. The rate of weathering is dictated by the chemical and physical nature of rock, climate and topography.

**weather map,** see WEATHER CHART.

**weather station,** *n.* any one of a network of meteorological observation posts where weather data are recorded for use in forecasting. A variety of instruments are used to record atmospheric parameters such as air pressure, temperature, humidity, wind, cloudiness, precipitation, sunshine and visibility. Weather stations can be classified according to the number of parameters observed and the frequency and sophistication of measurements. The global network of weather stations consists of over 7000 land stations and around 4000 passenger, merchant and specialist weather ships which make observations at sea. The distribution of weather stations on land and at sea is uneven, with few stations in remote unpopulated regions.

**weed,** *n.* any plant growing without human encouragement among cultivated plants on, for example, agricultural land, parks or domestic gardens. Dandelion and nettles are commonly occurring examples of weeds.

**Westerlies,** *n.* the winds which blow from the subtropical high pressure HORSE LATITUDES towards the lower pressures of the mid-latitudes between 35° and 65°N and S. The location of the Westerlies wind belts alter according to the season. They often blow from the south-west in the northern hemisphere and are responsible for the eastward tracking of DEPRESSIONS and ANTI-CYCLONES although their strength, persistance and direction is complicated by the topography of the landmasses and the associated variations in patterns of air pressure. The flow of the Westerlies is disrupted over Asia by the occurence of the MONSOON. In the southern hemisphere, the Westerlies, which blow from the north-west, are largely unaffected by landmasses and are stronger and more persistant, particularly in the ROARING FORTIES.

**wetland,** *n.* any area of low-lying land where the WATER TABLE is at or near the surface for most of the time, resulting in open-water habitats and waterlogged land areas. Wetlands are typically found in estuaries, along rivers with little vertical descent or in uplands where natural drainage is impeded due to extensive boulder clay deposits or by drainage channels disrupted by GLACIAL WATERSHED BREACHING. Wetlands may also form in areas of very high rainfall and low evaporation. Depending on their location wetlands can be freshwater, brackish or saltwater habitats. Those located alongside estuaries can be extremely fertile due to the replenishment of nutrients washed in by the constant flow of water. Net PRIMARY PRODUCTIVITY of lowland wetlands may be as high as 4000 g/m²/yr and equal that of TROPICAL RAIN FOREST. Upland wetlands, conversely, can be OLIGOTROPHIC due to the anaerobic conditions and the isolation of the mineral soil by a layer of acidic MOR peat. Productivity of these sites may be as low as 400 g/m²/yr.

**wet-rice cultivation,** *n.* a type of agriculture involving the cultivation of rice in PADDIES. In contrast to upland or dry rice, which is grown like any other GRAIN CROP, wet-rice must be covered by water for much of the growing season. The need for level land, as in deltas and river flood plains or on artificially constructed TERRACES restricts the geographical distribution of wet-rice cultivation, although its economic importance in Asian agriculture is unmatched by other crops.

Most areas of wet-rice cultivation support a high density of PEASANT farmers intensively operating small fragmented paddyfields within a largely SUBSISTENCE FARMING economy. Rice is often the only crop grown, and soil fertility is maintained by green manuring (see MANURE), by the application of NIGHT SOIL, animal droppings and crop residues, and by the paddy's unique MICROENVIRONMENT. In rain-fed areas without supplementary IRRIGATION, a crop other than rice may be taken from the drained field during the dry season. Labour inputs are invariably high, with high marginal returns to labour, that is, the system continues to reward ever-greater inputs of labour long after other types of farming would have collapsed from over-intensification (see INTENSIVE FARMING. Since the mid-1960s, areas of wet-rice cultivation have benefited from the introduction of GREEN REVOLUTION technology.

**WFS,** see WORLD FOOD SURVEY.

**whaling,** *n.* the hunting and killing of any of the species of very large marine mammals of the order Cetacea to provide whalebone, a variety of waxy substances for use in textile finishes, candles, cosmetics and perfume, blubber, and blubber oil.

The most heavily exploited species in the early 1980s was the minke whale with an annual catch of up to 12 000, 85% of which were caught by Japan, the USSR and Norway. Other species include the longfin pilot whale (3062 of which were caught in 1982, 87% of these by the Faeroe Islanders), the Beluga whale (1811), Bryde's whale (802, 60% by Japan), and the sperm whale (621, 71% by Japan). In 1970, 22 904 sperm whale were registered caught, but action by the International Whaling Commission (IWC) to conserve stocks has markedly reduced the annual catch.

In 1982, the IWC imposed a moratorium on commercial whaling but with few regulatory powers, it has been unable to prevent some whaling nations, notably the Japanese, from continuing the harvest, ostensibly for scientific purposes. Opponents of the whaling industry claim that although scientific expeditions to assess whale stocks are within the letter of the IWC convention, it is merely a way of maintaining an industry until the commercial ban is reassessed in 1990. Pressure on whaling nations has been strongest from IWC members such as the UK and the USA whose own commercial industries ceased in 1963 and 1972 respectively. The concern is that action to force the major whaling nations of Japan, the USSR, Norway, South Korea and Iceland to comply with the convention may push them out of the IWC altogether, leaving the Commission with little effective control over whaling, and no more than a conservational voice.

**whirlwind, land devil** or **water devil,** *n.* a narrow column of whirling air formed around a centre of very low air pressure. Whirlwinds form due to the localized atmospheric INSTABILITY caused by the intense warming of the earth's surface during the summer. Water devils usually form over land and travel over water where they quickly collapse, although sometimes they can form over water when a cool breeze flows over a warm water surface. Short-lived land devils are common in desert and semi-arid lands but may occasionally occur in temperate countries such as the UK. Whirlwinds are smaller in scale and are of less intensity than TORNADOES but can still cause localized damage to property. Compare HURRICANE.

**wilderness area,** *n.* any extensive area of underutilized land which, having escaped conversion to agricultural land use, has now assumed great ecological and conservation value. In the USA, the Wilderness Act of 1964 inaugurated the *National Wilderness Preservation System* comprising 54 areas within which no formal land use is permitted; economic activity is also prohibited unless sanctioned by presidential decree in times of national emergency.

The main drawback to the establishment of wilderness areas is that few countries can afford the luxury of setting aside areas in which no formal land use takes place. Only those nations of great land area (the USA, USSR, China, Australia, and Brazil) have sufficient unused land for designation as wilderness areas.

**wildlife park,** see WILDLIFE RESERVE.

**wildlife reserve** or **wildlife park,** *n.* any specific area of land set aside for the prime purpose of providing a refuge for species of wild animals. In DEVELOPING COUNTRIES these areas are usually NATIONAL PARKS whereas in the developed world the wildlife reserve can be a small area of land in private ownership within which breeding programmes for endangered species are undertaken.

# WILLY-WILLY

Ideally, wildlife reserves should be provided for all the major ECOSYSTEMS of the world; however of the 200 or so biogeographical provinces, 12% are not represented by any reserve or park. Almost half the wildlife reserves (totalling more than 100 million ha) are located in the northen boreal and tundra regions of North America while the least number of reserves are tropical moist forest, grasslands and Mediterranean ecosystems.

**willy-willy,** *n.* **1.** a HURRICANE occurring over north-west Australia.

**2.** a WHIRLWIND in Australian deserts.

**wilting point,** *n.* the point at which a plant becomes unable to extract sufficient water from the soil via its roots to compensate for the loss of moisture through its leaves by TRANSPIRATION. Wilting point is reached at variable times following rainfall depending on the SOIL TEXTURE which determines a soil's water retention capacity. Water still remains in the soil even though the wilting point has been reached and is called *hygroscopic water*; it is unavailable for plant growth because it is retained in the PORE SPACES by a considerable surface tension force. See also SOIL WATER, FIELD CAPACITY, INFILTRATION.

**wind,** *n.* the usually horizontal movement of air from an area of higher air pressure to an area of lower pressure. The rate of this equalizing flow is determined by the PRESSURE GRADIENT, with wind strength being measured on the BEAUFORT SCALE. The occurrence of winds may range in scale from temporary and localized LAND BREEZES and SEA BREEZES to the persistant major surface wind belts such as the TRADE WINDS, the WESTERLIES and the POLAR EASTERLIES.

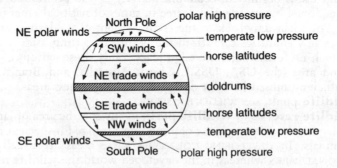

Fig. 70. **Wind belt.** Diagram showing the earth's wind belts and the directions in which they blow (arrowed).

**wind belt,** *n.* any of the major latitudinal wind systems such as the Westerlies or Trade Winds. See Fig. 70.

**windbreak,** *n.* an artificial screen or natural obstacle such as a line of trees or hedge, which limits the effects of winds on houses, crops or animals. Windbreaks can also reduce SOIL EROSION from dry, sandy soils. See also SHELTERBELT

**wind deposition,** *n.* the dropping of soil and rock debris transported by the wind. Deposition occurs when the wind is slowed by obstacles such as rocks, trees and fences or when velocity decreases naturally. Wind deposition provides an important source of material for the formation of DUNES, and under optimum conditions leads to the formation of extensive LOESS deposits.

**wind erosion,** *n.* the progressive removal of soil and rock debris by the wind. Wind erosion is most prevalent in arid and coastal areas but can occur in any climate if there is only sparse vegetation cover and a lack of moisture to bind surface materials. Two types of soil erosion can be distinguished:

(a) *deflation*, which involves the removal of fine, dry unconsolidated material by the wind. The effects of deflation may range from minor features such as sand-dune hollows to large scale DUST BOWLS. In the past, many agricultural areas were badly affected by deflation, although modern SOIL CONSERVATION techniques have greatly reduced the problem;

(b) *abrasion*, which entails the scouring and undercutting action of material transported by wind. On artificial structures the effects of abrasion are largely restricted to within a few metres of the ground. Wind abrasion may result in the scouring of paintwork on cars and the 'frosting' of windows. Protective metal casings are often used around the bases of electricity and telephone poles to prevent such abrasion. See WIND TRANSPORTATION.

**windgap,** *n.* **1.** a dry section of a valley which has resulted from RIVER CAPTURE of the headwaters of a river (see Fig. 54).

**2.** a COL or gap in a line of hills which may have been formed as in (1) or which may be the result of glacial breaching during the PLEISTOCENE EPOCH. Gap sites are often used by lines of communication and if large enough may even be the location of villages and towns. A major windgap is the Hudson-Mohawk Gap which links the New York CONURBATION with its HINTERLAND.

**wind power,** *n.* power produced from windmills (to drive machinery) and from wind turbines (to generate electricity). Examples of traditional wind-powered machines are windmills and wind pumps, the latter still being extensively used in remote areas of DEVELOPING COUNTRIES. In the developed world, much attention has been given to the improvement of design of the wind-driven generator. Large experimental machines with blade diameters of 100 m and capable of producing up to 3-4 MW of electrical power have been built. Ideal locations for such wind-powered generators are on islands such as the Shetland Islands in Scotland or in coastal areas, for example in Holland, where a steady prevailing wind makes generation of electricity a practicality. However, it is unlikely that wind generation of electrical power will be able to contribute more than 10% of the total demand for electrical power of a region. See ALTERNATIVE ENERGY.

**wind transportation,** *n.* the movement of soil and rock debris by the wind. Three types of transportation can be distinguished:

(a) *traction* which involves the rolling and sliding of material along the ground by the wind;

(b) *saltation* which is the movement of material in a hopping motion;

(c) *suspension* which results from turbulent wind flow, with most material being carried within a few metres of the ground.

In desert areas, the transportation of material in DUST STORMS and SANDSTORMS may greatly reduce visibility.

**windward,** *n.* the side or direction facing into the wind. Windward slopes are usually exposed sites and may receive substantially higher rainfall totals and lower average temperatures than LEEWARD sites. For example, the windward slopes of the Southern Alps in the South Island of New Zealand receive in excess of 3200 mm rainfall per annum on 175 - 200 RAIN DAYS per year. On the leeward side, occupied by the Canterbury Plains 130 km to the east, the annual rainfall is less than 800 mm and the number of rain days between 75 and 100. See RAINSHADOW.

**worker-peasant farming,** see PART-TIME FARMING.

**World Bank,** *n.* the popular name for the *International Bank for Reconstruction and Development* (IBRD), and its affiliate, the *International Development Association* (IDA). The IBRD was established by the United Nations in 1945 to assist the economic development of member nations through the provision of investment funds on favourable terms. The Bank has become the most important multilateral AID institution, in terms of the size of its lending and in its influence on the lending decisions of governments, and on the domestic policies of recipient countries.

Membership of the World Bank is restricted to member countries of the International Monetary Fund, and currently numbers 151 countries. Financial contributions to the capital of the Bank are made in proportion to each member's relative economic strength, on which basis voting rights are allocated. Thus, US foreign policy interests have a powerful influence on the destination of World Bank loans both geographically and sectorally. In 1985-86, one-third of disbursed funds were committed to Brazil, India and Indonesia. Lending up to June 1986 exceeded $126 billion, some 40% of which had been directed towards energy and transport INFRASTRUCTURE provision, as against only 22% on AGRICULTURE and RURAL DEVELOPMENT.

The IDA, with 134 members, became operational in 1960 specifically to provide assistance to the poorer DEVELOPING COUNTRIES in the form of long term loans at little or no interest. By June 1986 some $40 billion had been lent, 38% on agriculture and rural development. In 1985-86, 45% of funds were invested in South East Asia and 36% in Africa.

**World Conservation Strategy (WCS),** *n.* a policy document published in 1980 which presented a single integrated approach to global problems. WCS was sponsored by the major international conservation organizations such as INTERNATIONAL UNION FOR THE CONSERVATION OF NATURE, WORLD WIDE FUND FOR NATURE, FAO, and UNESCO.

The strategy is based on three concepts:

(a) that species and populations, whether plant or animal, must be helped to retain their capacity for self-renewal;

(b) that the basic life support systems of the planet, including climate, the water cycle and soils must be conserved intact if life is to continue;

(c) that genetic diversity is a major key to the future wellbeing of the earth and must also be maintained.

The document stressed the need to appreciate that mineral and FOSSIL FUEL resources were finite and that the exploitive nature of industrial development with its associated unchecked release of pollutants into the biosphere was certain to lead to a serious deterioration in its quality.

More than 30 countries have already become involved in preparing development plans which incorporate the recommendations of the WCS. Funding agencies such as the WORLD BANK and the United Nations are being urged to make money available to developing countries only when they present development plans which are based on the sustainable policies outlined in the WCS document.

**world cultural and natural heritage site,** see WORLD HERITAGE SITE.

**World Food Programme,** *n.* an organization, established in 1961 and run jointly by the FAO and the Economic and Social Council of the United Nations to administer the movement and distribution of UN FOOD AID and emergency relief. Between 1963 and 1986 more than 1300 projects in 120 countries, mostly of the 'food-for-work' variety or for feeding especially vulnerable groups, had been approved at a cost of $7.6 billion. A further $1.8 billion had been distributed in almost 900 emergency relief operations in 96 countries. The Programme estimates a long-term annual global food aid need of around 15 million tonnes, even without adverse environmental conditions such as DROUGHT to contend with.

**World Food Survey (WFS),** *n.* an analysis, conducted by the FAO, of trends in food production and supply against the background of increasing world population; the WFS also reviews the most recent evidence regarding the incidence of undernourishment in the world's population.

The Fifth WFS, published in 1986 following earlier reports in 1946, 1952, 1963 and 1977, identified a widening gap between the haves and have-nots despite an increase in global FOOD SUPPLY during the 1970s. It revealed that by 1980 the number of undernourished people in the world was at least 335 million, and perhaps even as many as 494 million, depending on the criteria used in determining the critical cut-off point below which undernourishment can be said to exist (see MALNUTRITION). On this upper estimate, almost one-quarter of the population of developing countries were suffering from the disabling effects of insufficient food: hunger or starvation, high incidence of disease, deteriorating ability to work, and the rise of psychological and mental impairment. See also FOOD BALANCE SHEET.

**world heritage site,** *n.* a natural area defined by the INTERNATIONAL UNION FOR CONSERVATION OF NATURE AND NATURAL RESOURCES as being of outstanding universal importance. Sites are nominated by countries which belong to the World Heritage Convention. Evaluation of possible sites is based upon the presence of the following criteria:

(a) a major characteristic of the earth's evolutionary history;

(b) a significant on-going geological process;

(c) a unique or superlative natural phenomenon, formation or feature;

(d) a habitat for endangered or rare species of plants and animals required for the survival of the species.

World heritage sites should not be confused with *world cultural and natural heritage sites* which aim to protect natural and man-made features such as the Galapagos Islands, the Pyramids, the Taj Mahal, and the Great Wall of China.

**World Resources Institute (WRI),** *n.* a policy research centre established in Washington DC in 1982 to help governments, international organizations and the private sector to meet basic human needs and foster economic growth without undermining the natural resources and environmental diversity of the biosphere. The institute research programme includes work on TROPICAL RAIN FORESTS, sustainable agriculture, climatic change, health care and resource and environmental information systems. WRI is funded from private donations, the United Nations, government agencies and industrial sources.

**World Wide Fund for Nature (WWF),** *n.* an international NON-GOVERNMENT ORGANIZATION established in 1961 to raise funds for conservation projects through public appeal. The World Wide Fund for Nature (until May, 1988, called the *World Wildlife Fund*) works in conjunction with the INTERNATIONAL UNION FOR THE CONSERVATION OF NATURE AND NATURAL RESOURCES and has been responsible for the implementation of key international laws and agreements on conservation such as the CONVENTION ON INTERNATIONAL TRADE IN ENDANGERED SPECIES, for the creation of NATIONAL PARKS, and for saving nearly 30 endangered animal species such as the giant panda.

**World Wildlife Fund,** see WORLD WIDE FUND FOR NATURE.

**WRI,** see WORLD RESOURCES INSTITUTE.

# XYZ

**xerophyte,** *n.* any plant with special characteristics such that it can survive in climates with a pronounced dry phase. Vegetation which grows in very cold climates may also display xeromorphic features.

Xerophytes can display some but rarely all of the following features: reduction of leaf size; thickening of the cuticular covering of leaves and stems; thick corky bark; leaves which are protected by hairs; stomata that are sunken and protected by enlarged guard cells; stomatal openings which may be filled with a wax-like substance; leaves that can be rotated to remain 'edge-on' to the sun (*heliotrophic*). Compare HYDROPHYTE. See also XEROSERE.

**xerosere,** *n.* any vegetation SUCCESSION developed on dry soil as opposed to rock, as for example the succession of plants on sand dunes.

**xylem,** see CONDUCTING TISSUE.

**yellowcake,** see FUEL ROD.

**yield,** *n.* the weight or volume of production of an agricultural commodity per unit of some limiting factor or resource. In practice, crop production is generally given in weight of output per unit of land area, while animal output is measured per head of livestock. Yield, when expressed per unit of labour or other resource input, is more commonly referred to as PRODUCTIVITY.

Agricultural developments in the present century have shown marked improvements in yields in response to SELECTIVE BREEDING, increasing application of FERTILIZERS, HERBICIDES and PESTICIDES, and more efficient farm management. In the USA total crop production rose by 97% between 1950 and 1981 with only a 3% increase in CROPLAND. In the EC wheat yields rose from 2330 kg per hectare in 1955 to 5600 kg/ha in 1984 (including Greece) while potato yields rose from 17.3 tonnes/ha to 32 tonnes/ha over the same period. In the UK average milk yields per dairy cow rose from 3331 litres in 1963 to 4965 litres in 1983. Not all areas of the world have shared the experience of increasing yields. Grain output per hectare in Nigeria and Sudan fell by 6% and 38% respectively between 1950-52 and 1983-85 to 714 kg and 479 kg, although in part this reflects the extension of agriculture on to highly MARGINAL LAND.

Yields show marked regional, national and continental variations according to environmental conditions, cultivation practices, and levels of inputs. Average grain yields across Europe and North America exceeded 3800 kg/ha in 1980-82 as against 1450 kg/ha in South Asia and 1050 kg/ha in Africa.

Conventional measures of yield are poor indicators of food output as crops vary in their provision of calories per unit weight. For example, 100 g of cassava (manioc) contains only 109 calories, whereas the same weight of millet provides 345 calories. A more realistic measure of yield or crop productivity is the calorific output per hectare, which may be expressed as a percentage of the highest valued crop of the region. In West Africa, for example, yams and maize produce weight yields per hectare of 75% and 10% that of cassava, but calorific yields per hectare of 65% and 33% respectively.

**yield class (YC),** *n.* (Silviculture) the maximum MEAN ANNUAL INCREMENT, measured in cubic metres and irrespective of age, which a STAND of trees is capable of attaining during its lifetime. It is usual to refer to YC values by quoting a number, thus YC16, which implies that a stand, irrespective of species type has the potential to grow at a maximum mean annual increment of about 16 cubic metres per annum.

YC values are compiled from data collected from sample forests and form the basis of *yield class tables* which provide the standard values of tree growth rate against which other growth rates can be compared. Management of a forest, including thinning routines and final clearance dates are based upon the YC data. YC values are of relevance mainly in even-aged, single species forests typical of the managed coniferous forests which are now found in most industrialized countries. See CURRENT ANNUAL INCREMENT, FOREST MANAGEMENT.

**zero-grazing,** see GRAZING MANAGEMENT.

**zero population growth (ZPG),** *n.* a situation in which the birth rate is constant and equal to the death rate, the *age structure* (the proportion of the population at each age group) is constant, and the growth rate is zero.

**zero-tillage,** see TILLAGE.

**zircaloy,** see FUEL ROD.

**zonal soil,** *n.* a type of soil believed by 19th century pedologists to be the product of the interaction of macro-climate upon geology and distributed over a vast continental area. Such a concept was taught by the Russian soil scientists Dokuchaev and Sibirtzev and followed the prevalent ideas of *uniformitarianism* (the belief that the physical world was controlled by a small number of variables) and which dominated the thinking of geologists and biologists at the end of the 19th century. Apart from being an over-simplification of reality, the zonal system had the unfortunate effect of restricting, in the mind of the student, the distribution of each zonal soil within the limits of its climatic zone. Compare INTRAZONAL SOILS and AZONAL SOILS.

**zone of aeration,** see WATER TABLE.

**zone of intolerance,** see DEATH ZONE.

**zone of saturation,** see WATER TABLE.

**zone of subduction** or **subduction zone** *n.* the zone where the theory of PLATE TECTONICS suggests that converging LITHOSPHERIC PLATES collide, resulting in one or both plates being forced downwards into the MANTLE and melted to form MAGMA. When two oceanic plates collide a DEEP-SEA TRENCH indicates the presence of a zone of subduction. As one plate is overridden and descends into the mantle an ISLAND ARC forms on the other plate. The zone of subduction is also marked by a deep-sea trench when oceanic and continental plates collide. The oceanic plate is forced under the more buoyant continental plate whose edge is subjected to intense deformation and METAMORPHISM. A highly folded mountain range results on the edge of the continental plate; a good example is the Andes in South America. If two continental plates collide then a highly folded mountain range may develop with associated volcanic activity as the magma formed by the melting plates in the subduction zone below rises to the surface. The Himalayas were formed by the collision of the Eurasian and African-Arabian-Indian plates. Earthquakes and volcanic activity are commonly found in most subduction zones. See FOLDED MOUNTAIN.

**zoning,** *n.* the deliberate regulation of LAND USE by allocating specific land use types to particular areas, or zones. Zoning is most frequently undertaken to enhance the quality of life for inhabitants of an area to ensure for example, that industrial development does not occur adjacent to a residential area. However, zoning can be deliberately used to prevent one sector of society from making contact with another, such as the segregation of the black, coloured and white population of South Africa.

**zoogeography,** *n.* the study of the geographical distribution of animals. It includes an analysis of the present distribution of animals and their historical evolution. Zoogeography is a less well developed branch of geography than its companion, BIOGEOGRA- PHY. See ANIMAL REALM.

**zooplankton,** *n.* the animal constituent of PLANKTON, consisting mainly of small CRUSTACEANS and fish larvae. The zooplankton is intermixed with the PHYTOPLANKTON upon which it feeds. One of the most common forms of zooplankton are the small crustaceans called copeopods. They are a major source of food for the commercially important fish species (such as herring, sprat, and pilchard). The euphausiid group of zooplankton occur in vast numbers in the Antarctic Ocean, forming the KRILL which are the main source of food for whales.

**ZPG,** see ZERO POPULATION GROWTH.